Semiconductor and Metal Binary Systems

Phase Equilibria and Chemical Thermodynamics

Semiconductor and Metal Binary Systems

Phase Equilibria and Chemical Thermodynamics

V. M. Glazov and L. M. Pavlova

Moscow Institute of Electronic Engineering
Moscow, USSR

Technical Editor

E. A. D. White

Late of Imperial College of Science and Technology
University of London
London, United Kingdom

CONSULTANTS BUREAU
NEW YORK AND LONDON

Library of Congress Cataloging in Publication Data

Glazov, Vasiliĭ Mikhaĭlovich.
    Semiconductor and metal binary systems: phase equilibria and chemical
thermodynamics / V. M. Glazov and L. M. Pavlova; technical editor E. A. D.
White.
        p.   cm.
    Translated from the Russian.
    Bibliography: p.
    Includes index.
    ISBN-13:978-1-4684-1682-4   e-ISBN-13:978-1-4684-1680-0
    DOI: 10.1007/978-1-4684-1680-0

    1. Phase diagrams. 2. Thermodynamics. 3. Binary systems (Metallurgy) 4.
Semiconductors. I. Pavlova, L. M. II. White, E. A. D. III. Title.
QD503.G575   1989   89-7261
621.381´52—dc20   CIP

This translation is published under an agreement with the Copyright
Agency of the USSR (VAAP)

© 1989 Consultants Bureau, New York
Softcover reprint of the hardcover 1st edition 1989
A Division of Plenum Publishing Corporation
233 Spring Street, New York, N.Y. 10013

PREFACE

This book is devoted to the fundamentals of the theoretical analysis of phase equilibrium diagrams. Phase diagrams are known to play an important role in metallurgy and materials science, chemical engineering, petroleum refining, etc. A study of phase diagrams can help in choosing the optimal composition of mixtures and alloys and in determining the appropriate conditions for their thermal treatment, as well as in determining the efficiency of such processes as distillation, rectification, zone refining, and controlled crystallization for the separation and purification of materials. In spite of this, the extensive thermodynamic information which can be extracted from phase diagrams has scarcely been utilized until recently, due to the comparatively poorly developed foundations of the analysis of phase equilibria.

We have attempted to present a general picture of the thermodynamic analysis of phase diagrams, and to demonstrate the broad possibilities of this approach in elucidating the nature of the interaction of the components and the structure of the phases. This book summarizes research carried out at the Moscow Institute of Electronic Engineering over the past decade. Extensive summaries of published data are also included.

In the course of our work we have made extensive use of modern computing methods, which allowed solutions to be obtained to many problems.

It is our hope that our work will play a decisive role in the development of the subjects mentioned, and that it will serve as a basis for the extensive application of the thermodynamic technique to the analysis of equilibrium in metallic, semiconductor, salt, organic, and other systems.

We express our sincere thanks to Ya. I. Gerasimov, Corresponding Member of the Academy of Sciences of the USSR, and V. B. Lazarev, with whom we have repeatedly discussed the problems dealt with in this monograph, and to V. B. Ufimtsev for reviewing it. In addition, we appreciate the helpful discussions we have had with our colleagues E. B. Sokolov and A. S. Pashinkin, and the help of L. I. Étazhov and L. A. Sergeev in the layout of the manuscript.

FOREWORD TO THE
ENGLISH EDITION

A new upsurge has recently occurred in the development of the Gibbsian thermodynamics of heterogeneous equilibria, associated with the broad vistas opened up by the application of computer techniques to solving problems hitherto considered insoluble. The direction of this development is most clearly indicated by the subtitle on the jacket of the international journal *CALPHAD*: *Computer Coupling of Phase Diagrams and Thermochemistry*. This includes both calculations of phase diagrams from known thermodynamic phase functions, as well as the solution of the inverse problem, namely, calculations of thermodynamic properties and phase structure characteristics from phase-equilibrium data.

To facilitate the solution of these problems, the traditional apparatus of chemical thermodynamics had to be reconsidered, and a whole series of new relations have been obtained. The most useful of these turned out to be the well-known van der Waals differential equation generalized to multicomponent and multiphase systems in Storonkin's monograph [13]. In writing this book, published by Metallurgiya Press in the USSR in 1981, we have attempted to unify the theoretical and experimental work on this problem.

The publication of the English edition, which has been corrected for errors and inconsistencies, is a source of great satisfaction to us, as it will, we hope, promote further development in the field of chemical thermodynamics. We would like to express our thanks to Plenum Publishing Corporation for undertaking the work of translating and publishing our monograph.

It is our hope that this book will be received with interest by the scientific community.

V. M. Glazov
L. M. Pavlova

CONTENTS

Chapter 1 SOME GENERAL RELATIONS OF CHEMICAL
THERMODYNAMICS 1

1. Systems, Their Properties and Classification. Thermodynamic
Parameters and the Equation of State. Thermodynamic Processes....... 1
2. Partial Molar Quantities and Their Properties............................. 4
3. The Zeroth, First, and Second Laws of Thermodynamics................. 6
4. Entropy Relations and Some Expressions Involving Derivatives
of Thermodynamic Functions 11
5. Characteristic Thermodynamic Functions and the
Gibbs–Helmholtz Equation 15

Chapter 2 THEORY OF CHEMICAL POTENTIALS AND THE
GENERAL THEORY OF THERMODYNAMIC
EQUILIBRIUM 23

1. Fundamental Theorems and Equations. The Gibbs–Duhem
Equation.. 23
2. The Equilibrium Principle of Gibbs.................................... 32
3. Conditions for Phase Equilibrium. Real and Virtual Components 34
4. Conditions for Chemical Equilibrium. The Law of Mass Action......... 39
5. Criteria of Phase Stability in Heterogeneous Systems.................... 43
6. The Principle of Equilibrium Displacement
(the Gibbs–Le Chatelier Principle).. 53
7. Displacement along Equilibrium Lines. Generalized van der Waals
Differential Equations 57
8. The Adequacy of Various Deductions of the Principle of
Displacements along Equilibrium Lines.................................... 61

Chapter 3   PHASE EQUILIBRIA IN TWO-COMPONENT SYSTEMS   69

1.   General Considerations on State Diagrams   69
2.   The Isobaric–Isothermal Potential as a Function of Temperature,
      Pressure, and Concentration...   72
3.   Thermodynamic Derivation of Basic Types of State Diagrams
      in Two-Component Systems   79

Chapter 4   FUNDAMENTALS OF THE THERMODYNAMIC THEORY
             OF SOLUTIONS OF CONDENSED SYSTEMS...................   105

1.   Generalized Concept of Solution. Thermodynamic Classification
      of Solutions...   105
2.   Thermodynamic Functions of Mixing. Basic Properties and
      the Law of Ideal Solutions   107
3.   The Law of Nonideal Solutions. Excess Thermodynamic Functions.
      Classification of Deviations from Ideality...................................   114
4.   The Concept of a Regular Solution. Lattice Model of Solutions.
      Strictly Regular Solution and Its Modifications   123
5.   Formal Description of the Thermodynamic Properties
      of Real Systems..   131
6.   Associated Solution Model..   142

Chapter 5   THERMODYNAMIC ANALYSIS OF PHASE DIAGRAMS
             WITHIN VARIOUS APPROXIMATE THEORIES
             OF SOLUTIONS..   147

1.   Analysis of Phase Equilibria Curves in Two-Component
      Systems Based on the Theory of Ideal Solutions   147
2.   Interrelation of Various Types of Phase Equilibrium Diagrams
      and the Nature of Intermolecular Interactions..............................   157
3.   Analysis of Liquidus and Solidus Curves in Systems with
      Restricted and Unrestricted Solubilities   175
4.   Thermodynamic Justification of the Retrograde Solidus. Calculation
      of the Retrograde Solidus in Various Approximations   184
5.   Analysis of the Liquidus Curve of a Congruently Melting Compound
      A_mB_n in Various Approximations of Ideal Solutions......................   195
6.   Equation of the Liquidus Curve and Thermodynamic Analysis
      of Systems with Incongruently Melting Compounds......................   210
7.   Various Methods of Estimating the Exchange Energy
      from Experimental Data   214
8.   Calculation of Thermodynamic Properties of Alloys
      from Data on Phase Equilibria...   222
9.   Choice of Model Solution Based on Matching
      with Experimental Phase Equilibrium Data   235

Chapter 6   THERMODYNAMIC ANALYSIS OF THE DISSOCIATION
             OF CHEMICAL COMPOUNDS 257

1.   Shape and Position of Maxima in the Liquidus
      and Solidus Curves 257
2.   Thermodynamic Basis of Various Estimation Procedures for the
      Dissociation Parameters of Congruently Melting Compounds 261
3.   Estimates of the Degree of Dissociation of Chemical Compounds
      in Various Approximations of the Thermodynamics of Solutions 273
4.   Dissociation of Binary Compounds in Quasibinary Solutions
      of Three-Component Systems 285

CONCLUSION 289

REFERENCES 293

Chapter 1

SOME GENERAL RELATIONS
OF CHEMICAL THERMODYNAMICS

1. SYSTEMS, THEIR PROPERTIES AND CLASSIFICATION. THERMODYNAMIC PARAMETERS AND THE EQUATION OF STATE. THERMODYNAMIC PROCESSES

As is well known, thermodynamics is a universal method capable of dealing with a large variety of processes, allowing one to generalize observed phenomena in order to establish their interrelations. Extensive applications of this method to the analysis of processes involving changes in the chemical composition of material systems have led to the emergence of a distinct discipline, called chemical thermodynamics. J. W. Gibbs may be said to be the father of chemical thermodynamics [1].

The profound changes accompanying the redistribution of masses in reacting materials have, in their turn, had a strong influence on the development of the thermodynamic method.

The region of applicability of this method is thus extraordinarily wide. Accordingly, it is desirable to formulate some general ideas which may be abstracted from the specific behavior of various substances, and to consider material bodies and their interactions as the sphere of application of the thermodynamic method as a whole. Before giving an account of the fundamentals of chemical thermodynamics, it will thus be necessary to formulate a series of basic concepts and definitions, whose precise form will be dictated by the stringent interrelations arising from the concrete application of the thermodynamic method to various material objects.

Any material (or its components), or an aggregate of such objects, hypothetically separated from the surroundings, will be called a system. A thermodynamic system is one in which material exchange (i.e., mass or heat exchange) can take place between its constituent parts. Thermodynamics can be applied only to macroscopic systems, comprising large numbers of molecules.

1

A system has precise spatial boundaries separating it from its surroundings. These boundaries may be either real physical surfaces or fictitious mathematical surfaces. Systems may be either homogeneous or inhomogeneous. In the former case, one may have either macroscopic or microscopic (local) homogeneity, corresponding to the system possessing either separate homogeneous parts, or a continuously varying degree of uniformity (as for a concentration gradient).

In a physicochemical description it is expedient to express the composition of systems in terms of the ratio of the number of moles of each substance to the total number of moles of all the substances composing the system. Let the system contain n_1 moles of substance A_1, n_2 moles of A_2, etc. Then the mole fraction of, for example A_1, is defined by

$$x_i = n_i / \sum_i n_i , \tag{1.1}$$

where, clearly,

$$\sum_i x_i = 1. \tag{1.2}$$

It is convenient to introduce the idea of "component" instead of the term "substance," defined as substance comprising only one sort of atoms or molecules. This definition is provisional and does not allow for the possible occurrence of chemical reactions. The concept of "material object" is thus conveniently replaced by that of "phase," meaning any homogeneous body distinguished from its immediate surroundings by sharply changing properties on its boundary. This definition is also preliminary and its rigorous thermodynamic basis will be supplied after the exposition of the thermodynamic method.

Depending on the degree of isolation of systems from their surroundings, one may have isolated, closed, and open systems. The isolated system is one that has absolutely no interaction with the surrounding medium. A system that does not exchange its constituent particles (atoms, molecules, ions) with the surroundings, but which interacts with the latter in other ways (e.g., heat exchange, mechanical work, radiation), is called a closed system. Open systems can exchange with their surroundings all or some of their constituent material parts.

The state of a system may be defined by the sum total of its properties. Quantities characterizing any macroscopic property of a system are called thermodynamic parameters. These can be either external or internal, depending on whether they are determined by the interactions between the system and an external body, or by the interactions and states of particles internal to the system. Although initially one may choose a set of independent thermodynamic parameters in many ways, one may not change sets arbitrarily in the course of solving a problem. Any transformation from one set to another must be effected by means of well-defined mathematical rules.

The independent thermodynamic parameters are subject to direct variation (for example, temperature, pressure, molar or specific volume, concentration, etc.).

One may distinguish intensive and extensive parameters. Intensive quantities, such as all molar and specific properties, temperature, pressure, etc., are independent of the mass. They may have one value throughout the system, or may vary from point to point. Their magnitudes are nonadditive. The magnitudes of extensive properties are proportional to the mass. Examples are the volume of the system, its mass or the number of moles, etc. These are all additive. Since intensive properties characterize the system in a given state, they usually serve as independent thermodynamic parameters.

If we limit our study of thermodynamic properties to simple systems, namely if we do not consider strong external fields (gravitational, electrostatic, or magnetic) and if, in addition, the system is at rest, then its state is completely specified by the triad: volume V, pressure p, and absolute temperature T. Consequently, the volume, pressure, and temperature must be interrelated. The most general form of this relation may be expressed as

$$f(p, V, T) = 0. \tag{1.3}$$

Equations relating state parameters to one another are called equations of state. Relation (1.3) is the thermal equation of state.

When a system transits from one state to another, the change in its properties does not depend on the nature of the transition, but only on the initial and final state of the system, i.e., on the thermodynamic parameters in these two states. Thus, a change in the state of a system may be described by considering only the initial and final states.

The assertion that a change of property is independent of the path is mathematically equivalent to the statement that an infinitesimal change in the property is a perfect differential. One knows from analysis that the integral of a perfect differential does not depend on the path of integration. The corollary of this is that if the change in any quantity is independent of the way it is effected, then this quantity is a property of the system. An ideal gas may serve as an example of a simple system. Its behavior can be described by just two of the three state parameters, the third being fixed by the thermal equation of state.

Combining the laws of Boyle–Mariotte and of Gay-Lussac–Charles, we obtain the gas equation

$$pV = nRT, \tag{1.4}$$

where R is the universal gas constant and n the number of moles.

A mixture of ideal gases also satisfies an equation of the form (1.4). In this case n becomes the total number of moles in the mixture, whereas p is, by Dalton's law, the sum of partial pressures of the gases forming the mixture, i.e.,

$$p = p_A + p_B + p_C + \dots, \tag{1.5}$$

where p_A, p_B, and p_C are the partial pressures of the components.

We now introduce tentatively one of the most important ideas of thermodynamics, namely "thermodynamic equilibrium." A system is in thermodynamic equilibrium if its state does not change in the course of time, i.e., if its parameters retain constant values. It should be emphasized that in thermodynamic equilibrium only the parameters characteristic of the system in question remain constant. The idea of "absolute" equilibrium or of "equilibrium under all conceivable changes" is physically meaningless.

Equilibrium processes are characterized by the fact that the properties of the system undergoing equilibrium changes of state are throughout related by equations of state such as (1.3). In contrast, for nonequilibrium processes the very concepts of pressure, temperature, and other state parameters have scarcely any meaning, since at the same moment they may assume totally different numerical values at different points.

In addition, thermodynamic processes may be reversible or irreversible. The concept of reversible and irreversible changes was introduced by Planck in 1879 [2].

An equilibrium process is reversible if, when carried out in the opposite direction, the system reverts to its initial state without in any way affecting the surroundings. The process is irreversible if this condition is not fulfilled. To the latter category belong all nonequilibrium processes, as well as equilibrium processes which can be reversed, but not without affecting the surrounding medium.

The concept of a "reversible process" is more general than that of an "equilibrium process." Indeed, an equilibrium process occurring in a particular system may be accompanied by nonequilibrium processes in the systems coupled to it.

If a thermodynamic system undergoes a process first in one and then in the opposite direction, and if, as a result of the latter, it returns to the initial state, then the system is said to have undergone a cyclic process.

In thermodynamics one also distinguishes between isothermal, isobaric, isochoric, and adiabatic processes, depending on the conditions prevailing during the process.

2. PARTIAL MOLAR QUANTITIES AND THEIR PROPERTIES

Let us consider partial molar quantities, which play an exceedingly important role in the thermodynamics of homogeneous systems comprising two or more components.

Let Y be any extensive property of a homogeneous system containing n_i moles of materials A_i. Its dependence on temperature, pressure, and the number of constituent particles may be written as

$$Y = f(T, p, n_1, n_2, ..., n_k). \tag{1.6}$$

The parameters defined by

$$\bar{y}_i = (\partial Y / \partial n_i)_{T, p, n_{j \neq i}} \tag{1.7}$$

are termed the partial molar quantities of species A_i. In contrast to molar quantities, the partial molar quantities may be either positive or negative. The reason is that, whereas molar quantities characterize the properties of a system, partial molar quantities describe their changes.

We now introduce some equations relating partial molar quantities to one another. Since any extensive property is a homogeneous function of the first degree in the independent variables $n_1, n_2, ..., n_k$, Euler's theorem allows one to write

$$\sum_i \frac{\partial f(T, p, n_1, n_2 ..., n_k)}{\partial n_i} n_i \equiv f(T, p, n_1, n_2, ... n_k). \tag{1.8}$$

This identity may be written, on account of (1.7), as

$$Y = \sum_i n_i \bar{y}_i. \tag{1.9}$$

Equation (1.9) relates an extensive property to the partial molar quantities pertaining to the material constituents of the system. It shows that the value of an extensive quantity is determined if one knows the partial molar properties and the amounts of the separate substances.

Differentiating (1.9) with respect to n_j ($j = 1, 2, \ldots, k$) at constant p and T and using (1.7) for each j, we obtain

$$\sum_i n_i \, (\partial \bar{y}_i / \partial n_j)_{p, T} = 0 \tag{1.10}$$

or

$$\sum_i n_i \, (\partial \bar{y}_j / \partial n_i)_{p, T} = 0. \tag{1.11}$$

This shows that the values of the partial molar quantities are not independent. For a two-component system Eq. (1.10) becomes

$$n_1 \, (\partial \bar{y}_1 / \partial n_1)_{p, T} + n_2 \, (\partial \bar{y}_2 / \partial n_1)_{p, T} = 0. \tag{1.12}$$

Equations (1.10) and (1.12) describe the relation between an increment in any quantity $\bar{y}_i$ and other quantities of this type when an infinitesimal change occurs in the composition of the system at constant pressure and temperature.

Equations (1.9) and (1.10) express the so-called partial molar conditions in chemical thermodynamics. By Euler's theorem Eq. (1.11) is the condition that the partial molar parameters are homogeneous functions of zero degree, so that for each i, we have

$$\bar{y}_i \, (T, p, ln_1, ln_2, \ldots, ln_k) = \bar{y}_i \, (T, p, n_1, n_2, \ldots, n_k). \tag{1.13}$$

This means that increasing the scale of the system by a factor l leaves the partial molar quantities invariant, which in turn shows that they are intensive parameters. This is because they do not depend on the aggregate number of particles but only on the composition, i.e., on the relative number of constituent particles. Hence, if to a homogeneous system one simultaneously adds such quantities of the different substances as to preserve their relative compositions, then the partial molar quantities retain their former values. Like other intensive properties, partial molar quantities are determined by the pressure, temperature, and relative number of constituent particles or composition of the system.

For one mole of the system Eqs. (1.9) and (1.10) may be transcribed into the following form:

$$Y = \sum_i x_i \bar{y}_i, \tag{1.14}$$

$$\sum_i x_i \, (\partial \bar{y}_i / \partial x_j) = 0. \tag{1.15}$$

Applying (1.15) to a binary system, we obtain, taking x_2 as the independent variable,

$$x_1 (\partial \bar{y}_1/\partial x_2)_{p,T} + x_2 (\partial \bar{y}_2/\partial x_2)_{p,T} = 0 \qquad (1.16)$$

or

$$(\partial \bar{y}_1/\partial x_2)_{p,T}/(\partial \bar{y}_2/\partial x_2)_{p,T} = -x_2/x_1. \qquad (1.17)$$

It follows from (1.17) that the derivatives $(\partial \bar{y}_1/\partial x_2)$ and $(\partial \bar{y}_2/\partial x_2)$ have always opposite signs; for $x_1 = x_2 = 0.5$ the slope of the curves $\bar{y}_1 = f(x_2)$, $\bar{y}_2 = f(x_2)$ are equal and differ only in sign; if one of the curves has a maximum, then the other has a minimum at the same composition.

The behavior of partial molar quantities for an infinitesimal dilution may also be considered by means of Eq. (1.17). For example, let us assume that $x_2 \to 0$. Then two situations may arise: $(\partial \bar{y}_1/\partial x_2)_{p,T}$ may either approach zero, or it may stay finite. In the latter case the derivative $(\partial \bar{y}_2/\partial x_2)_{p,T}$ should tend to infinity, since its derivative remains finite as the ratio x_2/x_1 tends to zero. In other words, if $x_2 = 0$, then the curve of $\bar{y}_1$ is horizontal while that of $\bar{y}_2$ is vertical.

We shall make extensive use of the relations derived in this section in the solution of concrete problems involving phase equilibria in two-component systems.

3. THE ZEROTH, FIRST, AND SECOND LAWS OF THERMODYNAMICS

It is commonly assumed that systems are in thermal equilibrium if no heat exchange takes place in the absence of thermal insulation between them. It is quite obvious that differently heated bodies in contact will reach identical temperatures with the passage of time. This conclusion, arrived at with the development of thermometry, is one of the four postulates of thermodynamics and is termed the law of thermal equilibrium (zeroth principle or zeroth law).

In modern thermodynamics this law has been formulated by Fowler in the following way: if system 1 is separately in thermal equilibrium with systems 2 and 3, then 2 and 3 are in thermal equilibrium with one another.

Let us consider three simple systems, 1, 2, and 3, capable of exchanging heat with one another. We write the condition of thermal equilibrium of systems 1 and 2 as

$$f^{12} (p^1, V^1; \; p^2, V^2) = 0. \qquad (1.18)$$

Similarly, between 1 and 3 one has

$$f^{13} (p^1, V^1; \; p^3, V^3) = 0. \qquad (1.19)$$

Furthermore, by the zeroth law, 2 and 3 should also be in equilibrium:

$$f^{23} (p^2, V^2; \; p^3, V^3) = 0. \qquad (1.20)$$

Since all the parameters are interdependent, relation (1.20) should follow from (1.18) and (1.19). Mathematically this is possible if the variables considered will occur in combinations as certain functions. For example, p^1 and V^1 should occur as a function $x(p^1, V^1)$, p^2 and V^2 as $y(p^2, V^2)$, and p^3 and V^3 as $z(p^3, V^3)$. Then Eqs. (1.18)–(1.20) may be transformed into the form:

$$f^{12}(x, y) = 0, \tag{1.21}$$

$$f^{13}(y, z) = 0, \tag{1.22}$$

$$f^{23}(x, z) = 0. \tag{1.23}$$

It is now possible, in principle, to eliminate one of the three variables in any two of these equations. For example, (1.21) and (1.22) may each be solved for y and the results equated.

If one uses the other pairs of equations in a similar way, one finally arrives at the equality of the three functions

$$x(p^1, V^1) = y(p^2, V^2) = z(p^3, V^3). \tag{1.24}$$

This is the basis for the assertion that there exists a system property $t(p, V)$ such that conditions $t^1 = t^2$ and $t^1 = t^3$ imply $t^2 = t^3$. This property is the empirical temperature. The thermal equilibrium principle has thus led to the analytic criterion for thermal equilibrium defined by the equality of temperatures. This rule enables one to compare the temperature of two bodies and supplies a criterion to determine whether and in what direction heat flow occurs.

The works of Rumford, Darwin, Gay-Lussac, Mayer, and Joule–Thomson [3] have furnished irrefutable evidence for the equivalence of heat and work. Especially valuable was the work of Mayer, which was in essence the first formulation of the first law of thermodynamics and provided an entirely correct explanation of Gay-Lussac's famous experiments. Mayer also calculated the mechanical equivalent of heat for cyclic processes.

Following Mayer's discoveries, systematic and precise experiments by Joule proved conclusively the concept of the mechanical equivalent of heat for systems undergoing cyclic processes. On this basis one may draw the following conclusion: In a closed system the sum total of all energies is a constant and in the process of their conversion into one another, energy is neither lost nor gained. One has thus arrived at the law of conservation of energy, one of the most important universal laws of nature. Clausius called this law the first law of thermodynamics.

From the above principle of equivalence follows the impossibility of building a perpetuum mobile that would perform continuously without any expenditure of energy. The principle of equivalence may be written in the general form

$$\oint \delta Q = \oint \delta A, \tag{1.25}$$

where δQ and δA are infinitesimal quantities of heat and work. The sign $\oint$ denotes integration over a closed contour.

We now consider the most important consequences of the equivalence of work and heat. Let us imagine an isolated system for which the internal energy U is a constant, so that $dU = 0$. We now increase the temperature of the system by passing some heat into it. If the volume is maintained constant, no expansion work will be performed and all the heat will be converted into changing the internal energy. If heating causes a volume change, then the heat supplied will be expended into changing the internal energy as well as into performing external work, i.e., work of expansion. Consequently,

$$\delta Q = dU + \delta A \tag{1.26}$$

or, in integral form,

$$Q = \Delta U + A. \tag{1.27}$$

Let us agree to consider heat passing into the system positive, and out of the system, negative. Then (1.27) may be written as

$$Q = (U_{II} - U_I) + A, \tag{1.28}$$

i.e., the heat passed goes into increasing the store of internal energy from U_I to U_{II}, and into the performance of external work. Equation (1.28) is the mathematical expression of the first law of thermodynamics.

If the system performs only expansion work, Eq. (1.28) has the form

$$Q = U_{II} - U_I + \int_{V_I}^{V_{II}} p\,dV. \tag{1.29}$$

For constant p this gives

$$Q_p = U_{II} - U_I + p\,(V_{II} - V_I) \tag{1.30}$$

or

$$Q_p = H_{II} - H_I = \Delta H, \tag{1.31}$$

$$dQ_p = dH. \tag{1.32}$$

Therefore, the heat absorbed at constant pressure is equal to the change in enthalpy of the system if the only form of work is that of expansion.

According to the first law of thermodynamics, in the interconversion of energy from one form to another the total energy of an isolated system is conserved. However, although this principle applies universally to all natural phenomena, it has nothing to say on which of these may or may not take place. The answer to this question is given by the second law of thermodynamics. In the fundamental work of the French engineer Carnot [4] in 1824, entitled "Reflexions sur la puissance motrice du feu et sur les machines propres a developper cette puissance" ("Reflections on the motive power of fire," etc.), a first, incomplete, attempt was made to formulate what is today called the second law of thermodynamics. Carnot tackled three basic problems: 1) finding the conditions necessary for converting heat into work; 2) finding the condition under which this conversion will be most efficient; 3) deciding whether the efficiency of a heat engine giving the maximum effect depends on the working material. Carnot noted that all reversible heat engines have the same efficiency independently of the working substance, but depending only on the temperatures between which the heat engine operates. Carnot's idea was formulated mathematically by Clapeyron (1834), who was also the first to describe graphically the Carnot cycle, which consists of two isotherms and two adiabatics of an ideal gas.

Subsequent formulations of the second law were obtained by Clausius (1850) and by Thomson (1851) [4].

The Clausius formulation is the following: heat cannot be transferred spontaneously from a cold to a hot body. Subsequently, Clausius replaced "spontaneously" by the expression "without compensation," i.e., without some associated change in the thermodynamic state of the working substance or of other bodies participating in the process. Such a formulation of the second law is known as the Clausius postulate. The veracity of the Clausius postulate in its first form is self-evident and is corroborated by a host of experimental data connected with our observations. The Clausius postulate should be understood in its broader aspects. As repeatedly pointed out by Clausius, this statement should not simply mean that heat cannot pass directly from a colder to a hotter body. This is quite natural and already follows from the definition of temperature. The modern, broader meaning of the Clausius postulate is that, in any process whatsoever, heat cannot be passed from a colder to a warmer body by whatever means, unless other, "compensating" changes take place. As noted by Planck [2], only by using this wider interpretation may one arrive at conclusions concerning any of the natural processes. Planck formulated the second law thus: It is impossible to construct a periodically operating machine which would cause no other changes except the lifting of a weight and the cooling of a heat reservoir.

According to Planck the irreversibility of natural processes as a whole may be expressed as follows: Processes which cannot be completely reversed by any method are called irreversible. All other processes are reversible. For example, the process of heat generated by friction cannot be reversed by any method whatsoever. That is to say, if a given weight joined to a suitable paddle-wheel mechanism converts work into heat by causing friction in water or mercury, then it is absolutely impossible to carry out the process in the reverse direction such that the weight and the associated mechanism return to their former position, since part of the energy in the form of heat is irreversibly dissipated into the surrounding medium.

Reversible processes occur infinitely slowly. Processes occurring at measurable speeds cannot be reversible, since on account of their rate the internal distribution of energy cannot manage to follow exactly the changes in the external conditions. This implies a certain departure from thermal equilibrium which is inevitably connected with causes of irreversibility. Since a reversible process is conceived as a series of equilibrium states which does not affect the surroundings, it can proceed both directly and in the opposite direction.

Thus, reversible processes, though realizable only under ideal conditions which cannot be met in practice, nevertheless have a certain reality in the sense that they do not contradict the laws of nature. Although they can never be completely attained, they can be reproduced approximately with an arbitrary degree of accuracy. Clearly all real processes are to some degree irreversible. A measure of the degree of irreversibility of any process in thermodynamics is provided by a new function of state called entropy, introduced by Clausius.

According to Clausius, the degree of irreversibility of a process should be defined not by means of the amount of heat transferred Q, but rather in terms of the reduced heat Q/T, where T is the absolute thermodynamic temperature [5]. The value of Q/T in an irreversible process represents the growth of the entropy of the system. If the initial and final entropies in the process are S_1 and S_2, then

$$S_2 - S_1 = Q/T. \tag{1.33}$$

The physical meaning of entropy is quite complicated. By expressing the change of entropy in an irreversible process as the difference in entropies at the end and beginning of the process, we have already implied that entropy is a property of the system, i.e., that it is a function of the state parameters. Hence, the change in entropy is independent of the path the process takes and is determined solely by its initial and final parameters.

Considering entropy as a function of state and including the composition among the state parameters, we have

$$dS = (\partial S/\partial T)_{p,n}\, dT + (\partial S/\partial p)_{T,n}\, dp + \sum_i (\partial S/\partial n_i)_{T,p,n_j}\, dn_i. \tag{1.34}$$

Entropy is an extensive property obeying additivity. The mathematical proof of the existence of entropy may be followed clearly in the method of Clausius.

From a step-by-step consideration of a Carnot cycle [4], it follows that the efficiency η of a reversible engine may be expressed as

$$\eta = (Q_1 - Q_2)/Q_1 = (T_1 - T_2)/T_1, \tag{1.35}$$

where Q_1 is the heat absorbed by the working substance from the heater and Q_2 the heat rejected by it to the cooler, while T_1 and T_2 are the maximum and minimum temperatures of the working substance exchanging heat with the corresponding heat sources.

On the basis of Eq. (1.35) we have, for any reversible cycle,

$$\oint (\delta Q_{\text{rev}}/T) = 0.$$

Since the integral of $\delta Q/T$ over a closed contour vanishes, so that the change of this quantity is independent of the path, $\delta Q/T$ is a perfect differential of some function S of the state parameters, called entropy by Clausius, as we have seen.

Accordingly, for a reversible process, we have

$$dS = \delta Q/T \quad \text{or} \quad \delta Q = T\, dS \tag{1.36}$$

This relation is fundamental for the interpretation of the second law and for the development of thermodynamics as a whole.

As is known [5], the efficiency η' of an irreversible Carnot engine is always smaller than the efficiency η of a reversible engine, i.e.,

$$\eta' < \eta. \tag{1.37}$$

Thus, on the basis of (1.35) and (1.37), we are led to conclude that the heat transferred from the heater in an irreversible Carnot engine is less than the heat given to the cooler. Therefore,

$$\oint (\delta Q_{\text{irrev}}/T) < 0. \tag{1.38}$$

In differential form we obtain generally

$$T\, dS > \delta Q, \tag{1.39}$$

the inequality and equality signs referring to irreversible and reversible changes, respectively. Expression (1.39) is the complete mathematical formulation of the second law in differential form; it shows that the behavior of entropy may serve as the criterion for the direction of a thermodynamic process.

Combining the first and second laws from Eqs. (1.26) and (1.39), we have, in the general case,

$$T\, dS > dU + \delta A. \tag{1.40}$$

It follows from this expression that in an irreversible process the work performed by the system is always less than in a reversible process under the same conditions, so that maximum work is obtained in a reversible process. The interposition of any irreversibility in the process leads to a decrease in the available work.

4. ENTROPY RELATIONS AND SOME EXPRESSIONS INVOLVING DERIVATIVES OF THERMODYNAMIC FUNCTIONS

In solving problems by utilizing the entropy principle, it is expedient to consider the dependence of entropy on other thermodynamic functions. From Eq. (1.40) we may write for a reversible process

$$dS = \frac{1}{T}\, dU + \frac{1}{T}\, p\, dV. \tag{1.41}$$

Expressing the internal energy in terms of the enthalpy, one has

$$dS = \frac{1}{T} dH - \frac{1}{T} V\, dp, \tag{1.42}$$

in which all differentials are perfect since S, U, H, V, and p are all functions of state. Thus, $1/T$ serves as an integrating factor.

Expressions (1.41) and (1.42) are the joint formulas describing all reversible changes of state by the two laws of thermodynamics. In other words, they pertain to infinitesimal changes of state which take the system from one equilibrium state to another infinitely close to the first. These equations are not valid for irreversible changes.

From mathematical analysis it follows that the perfect differentials should satisfy reciprocal relations. For simple systems there are only two independent vari-

ables and one reciprocal relation. By comparing the expression for the perfect differential of the internal energy in terms of the variables T and V with the equation
for the first law, we get

$$\delta Q = [\,(\partial U/\partial V)_T + p]\, dV + (\partial U/\partial T)_V\, dT. \tag{1.43}$$

Substituting this δQ into the second law of thermodynamics, we obtain

$$dS = T^{-1}\,[(\partial U/\partial V)_T + p]\, dV + T^{-1}\,(dU/dT)_V\, dT\ , \tag{1.44}$$

$$\frac{\partial}{\partial V}\,[T^{-1}\,(\partial U/\partial T)_V] = \frac{\partial}{\partial T}\,[T^{-1}\,(\partial U/\partial V)_T + p/T]. \tag{1.45}$$

After differentiating and performing some transformations, we get

$$(\partial U/\partial V)_T = T\,(\partial p/\partial T)_V - p. \tag{1.46}$$

This relation expresses the dependence of the internal energy on the volume in an
isothermal process.
    Substituting (1.46) into (1.44) we have, finally,

$$(\partial S/\partial V)_T = (\partial p/\partial T)_V \tag{1.47}$$

or, using the relation between partial derivatives of the parameters of the equation
state,

$$(\partial S/\partial V)_T = (\partial p/\partial T)_V = -(\partial V/\partial T)_p/(\partial V/\partial p)_T. \tag{1.48}$$

    Comparing the perfect differential for the enthalpy of a simple system in terms
of T and p with the first law of thermodynamics, we find

$$\delta Q = (\partial H/\partial T)_p\, dT + [(\partial H/\partial p)_T - V]\, dp. \tag{1.49}$$

Using (1.49) and (1.42), we obtain

$$dS = T^{-1}\,[(\partial H/\partial p)_T - V]\, dp + T^{-1}\,(\partial H/\partial T)_p\, dT. \tag{1.50}$$

The reciprocal relation can be written, in this case, as

$$\frac{\partial}{\partial p}[T^{-1}\,(\partial H/\partial T)_p]_T = \frac{\partial}{\partial T}\,[T^{-1}\,(\partial H/\partial p)_T - V/T]_p\,, \tag{1.51}$$

which on differentiating becomes

$$(\partial H/\partial p)_T = -T\,(\partial V/\partial T)_p + V. \tag{1.52}$$

This gives the dependence of the enthalpy on pressure in an isothermal process.
Substituting (1.52) into (1.50), we find

$$(\partial S/\partial p)_T = -(\partial V/\partial T)_p. \tag{1.53}$$

Expressions (1.46) and (1.52) are the thermodynamic equations of state, since the derivatives $(\partial U/\partial V)_T$ and $(\partial H/\partial p)_T$ impose the thermodynamic conditions to be satisfied by any empirical equation of state.

By means of Eqs. (1.46) and (1.53), together with data on the thermal properties of substances, one may determine caloric quantities like the internal energy and the enthalpy. Conversely, knowledge of the caloric quantities allows one to calculate the thermal properties of materials.

Finally, let us derive the dependence of entropy on temperature. By definition the heat capacity is

$$C = \delta Q/d\,T. \tag{1.54}$$

Substituting for δQ from (1.36) into (1.54), we have

$$C = TdS/dT. \tag{1.55}$$

This expression is valid for any infinitesimal temperature change, independently of the way it occurs. Usually heating is carried out either at constant pressure or at constant volume.

From Eq. (1.55) we obtain, at constant p,

$$dS = C_p d \ln T \tag{1.56}$$

or

$$(\partial S/\partial T)_p = C_p/T, \tag{1.57}$$

while at constant V

$$dS = C_V d \ln T \tag{1.58}$$

or

$$(\partial S/\partial T)_V = C_V/T. \tag{1.59}$$

Differentiating Eqs. (1.57) and (1.59) with respect to p and V, respectively, and using the relations

$$C_V = (\partial U/\partial T)_V \quad \text{and} \quad C_p = (\partial H/\partial T)_p, \tag{1.60}$$

one obtains

$$\partial^2 S/\partial T \partial p = T^{-1} (\partial C_p/\partial p)_T = T^{-1} (\partial^2 H/\partial T \partial p), \tag{1.61}$$

$$\partial^2 S/\partial T \partial V = T^{-1} (\partial C_V/\partial V)_T = T^{-1} (\partial^2 U/\partial T \partial V). \tag{1.62}$$

On the basis of (1.53) and (1.61) we can write

$$(\partial C_p/\partial p)_T = - T (\partial^2 V/\partial T^2)_p. \tag{1.63}$$

According to (1.62), and using the relation in (1.47), we find

$$(\partial C_V/\partial V)_T = T \, (\partial^2 p/\partial T^2)_V. \tag{1.64}$$

In the simplest cases of constant volume and constant pressure, integration of (1.56) and (1.58) leads to the following expressions for the entropy:

$$S\,(p,T) = \int (C_p/T)\,dT + \text{const}; \tag{1.65}$$

$$S\,(V,T) = \int (C_V/T)\,dT + \text{const.} \tag{1.66}$$

In order to find the entropy increase of a substance whose temperature increases from T_I to T_{II} at constant pressure one must clearly know the temperature dependence of the heat capacity.

The calculation of the entropy difference becomes extremely simple if a reversible transition between two states takes place at constant temperature, as for example in the case of phase changes. In this case we have, by the second law,

$$s_{II}^\circ - s_I^\circ = q/T, \tag{1.67}$$

where q is the latent heat of the phase transition for one mole of substance, S is the molar entropy, and the superscript "$\circ$" denotes the particular substance. This formula may be used to calculate the entropy change for reversible melting, evaporation, etc. Here, too, one has to distinguish between $V = \text{const}$ and $p = \text{const}$. At constant volume no work is performed and the latent heat equals the increase in internal energy Δu°, while at constant pressure it equals the increase in the Gibbs heat function Δh°:

$$q_V = \Delta u^\circ; \quad q_p = \Delta h^\circ. \tag{1.68}$$

For an isothermal transition between two phases, as in the liquid–vapor transition (boiling), we note that equilibrium prevails when the vapor pressure equals the externally applied pressure.

In this case, $(\partial S/\partial V)_T$ is identically equal to $\Delta s^\circ/\Delta v^\circ$, where Δs° and Δv° are the entropy and volume increase per mole of evaporating substance. We may then write, according to (1.47),

$$dp/dT = \Delta s^\circ/\Delta v^\circ \tag{1.69}$$

or, by using (1.67) and (1.68) for the liquid–vapor phase transition,

$$dp/dT = \Delta h_V^\circ \left[T_S^\circ \left(v^{\circ(G)} - v^{\circ(L)} \right) \right]^{-1}, \tag{1.70}$$

where Δh_V° is the latent heat of vaporization, $v^{\circ(G)}$ and $v^{\circ(L)}$ are the molar volumes of the vapor and liquid at the boiling temperature (G and L denote the gas and liquid phase) and T_s° is the boiling temperature of the pure substance.

Expression (1.70) is known as the Clausius–Clapeyron equation. It is, in essence, a direct consequence of the second law.

Since in the transition of liquid into vapor the volume of vapor is many times greater than that of the liquid, we may neglect $v^{\circ(L)}$ in (1.70) and obtain

$$dp/dT = \Delta h_V^{\circ} \left(v^{\circ\, G} T_S^{\circ} \right)^{-1}. \tag{1.71}$$

Expression (1.69), however, is very general and may be used for any phase transition.

For the melting transition the Clausius–Clapeyron equation has a special meaning, as it provides the slope of the crystal–liquid equilibrium curve. It is useful in this case to transcribe the equation into the following form:

$$dT/dp = T_m^{\circ} \left(v^{\circ(L)} - v^{\circ(S)} \right) / \Delta h_m^{\circ}, \tag{1.72}$$

wherein the quantity dT/dp is the pressure derivative of the melting curve of the substance and expresses the pressure dependence of the melting temperature change, while Δh_m° is the latent heat of fusion.

Using the relationship between volume and density, Eq. (1.72) may be written as

$$dT/dp = \left(d^{(S)} - d^{(L)} \right) \left(\Delta S_m^{\circ} d^{(L)} d^{(S)} \right)^{-1}, \tag{1.73}$$

where d is the density of the particular substance (the index S denotes the solid phase) and ΔS_m° is the entropy of melting.

It is important to emphasize that in melting and evaporation the latent heats, and, consequently the associated entropies of the phase transitions, are all positive, whereas the difference in the specific volumes in melting may be either positive or negative. The latter is observed in water, but as noted in a series of works [6], it is not unique to it. In fact, the negative value is quite a widespread phenomenon in nature, being inherent to semiconductors, which are characterized by loose packing in the solid state (cf. [7]).

An immediate consequence of Eqs. (1.72) and (1.73) is the rule that dp/dT is negative if $v^{\circ(L)} - v^{\circ(S)} < 0$ or if $d^{(S)} - d^{(L)} < 0$, so that a pressure increase is accompanied by a lowering of the melting temperature. When the inequalities are reversed, dp/dT is positive, so that the melting temperature increases with pressure.

5. CHARACTERISTIC THERMODYNAMIC FUNCTIONS AND THE GIBBS–HELMHOLTZ EQUATION

Direct application of both laws of thermodynamics enables one to solve a variety of realistic problems. Although the method of fictitious reversible cycles leads to the correct solution in all cases, it is not a convenient technique, as it requires purely artificial constructions and circuitous routes in solving these problems. Hence an alternative, simpler method has gained wide acceptance, namely the method of thermodynamic (or characteristic) functions, pioneered by Gibbs.

The concept of thermodynamic functions was first introduced by Masse in 1869. Gibbs gave these functions a somewhat different definition and subsequently applied them to solve a series of problems of chemical thermodynamics. Characteristic functions are functions of state whose derivatives furnish the simplest and clearest expressions for all thermodynamic properties of the system. For this purpose we take thermodynamic properties to be those that depend only on temperature, pressure (or volume), and composition. The term "characteristic functions" is not so appropriate, since it gives the impression that the characteristic feature is a property of the functions. In fact, the characteristic nature of the functions is due to the choice of independent variables (or state parameters).

The need to introduce characteristic functions was justified by Gibbs as follows. The joint equation (1.41) of the two laws of thermodynamics for reversible processes related five parameters, T, S, U, p, and V, which determine the state of a system in equilibrium. Each of these quantities may be considered to be both a state parameter and a function of state. To determine the state of the system let us for the present limit ourselves to two independent variables. This leaves three unknown parameters, and to determine these we need, in addition to (1.41), another two equations interrelating them [8].

The second equation may be taken to be the thermal equation of state, while the third equation may be constructed from a knowledge of some relation between the parameters or functions of state, derived from definite properties of the system in question. This last equation contains a new, thermodynamic function and leads to the solution of the problem.

From among the unlimited number of thermodynamic functions (among which are all functions of state) one should distinguish the characteristic functions. As we have seen, these have the property that for a specific choice of state parameters the partial derivatives of these functions with respect to the parameters will equal one of the state parameters. The number of characteristic functions is clearly not large.

Gibbs introduced four functions which are properly termed characteristic: the internal energy U, the enthalpy H, and the isochoric–isothermal and isobaric–isothermal potentials F and G, respectively. His method follows directly from the joint first and second laws in (1.41), which is conveniently written as

$$dU = T\,dS - p\,dV. \qquad (1.74)$$

Let us consider the step-by-step derivation of characteristic functions and find their partial derivatives. According to (1.74) the internal energy is a function of the independent parameters S and V. The internal energy being a function of state, it has a perfect differential, so that

$$dU = (\partial U/\partial S)_V\,dS + (\partial U/\partial V)_S\,dV. \qquad (1.75)$$

Equation (1.74) is identically satisfied for all values of dS and dV. This is possible only if

$$(\partial U/\partial S)_V = T, \qquad (1.76)$$

$$(\partial U/\partial V)_S = -p. \qquad (1.77)$$

Thus, by definition, the internal energy is a characteristic function of the variables S and V. For other choices of variables this would be impossible. If a given thermodynamic function is considered as a function of variables for which it is not a characteristic function, then the other thermodynamic quantities cannot be expressed explicitly in terms of it.

The derivation of the other functions H, F, and G proceeds by way of a Legendre transformation. The latter consists of a simultaneous variation of the independent parameters on the right-hand side of (1.74) with the function whose differential appears on the left. Subtracting $d(TS)$ from both sides of Eq. (1.74), we obtain, after some rearrangements,

$$d\,(U - TS) = -S\,dT - p\,dV. \qquad (1.78)$$

Equation (1.78) is a perfect differential of the function

$$F = U - TS\,, \qquad (1.79)$$

introduced for the first time by Masse and termed free energy by Helmholtz. It is thus called the Helmholtz free energy, or the free energy at constant volume. An alternative nomenclature is the isochoric–isothermal potential. As seen from (1.78), F is a function of T and V. Writing the perfect differential in terms of the chain rule, we get

$$dF = (\partial F/\partial T)_V\,dT + (\partial F/\partial V)_T\,dV. \qquad (1.80)$$

Comparing expressions (1.78) and (1.80), we have

$$(\partial F/\partial T)_V = -\,S, \qquad (1.81)$$

$$(\partial F/\partial V)_T = -\,p. \qquad (1.82)$$

It is thus seen that the Helmholtz free energy is a characteristic function. By means of a Legendre transformation let us change to S and p as independent parameters. Adding $d(pV)$ to both sides of (1.74), we obtain, after some manipulations,

$$d\,(U + pV) = T\,dS + V\,dp. \qquad (1.83)$$

The expression obtained is the perfect differential of the enthalpy or Gibbs heat function $H = U + pV$, so that it is a characteristic function of the two variables S and p. Since H is a function of state, one may write

$$dH = (\partial H/\partial S)_p\,dS + (\partial H/\partial p)_S\,dp. \qquad (1.84)$$

Comparison of expressions (1.83) and (1.84) gives

$$(\partial H/\partial S)_p = T, \qquad (1.85)$$

$$(\partial H/\partial p)_S = V. \qquad (1.86)$$

Subtracting from both sides of Eq. (1.74) the differential $d(TS - pV)$, and changing to independent variables T and p, we have

$$d (U - TS + pV) = -S \, dT + V \, dp. \tag{1.87}$$

The complete expression is a perfect differential of a new function G:

$$G = U - TS + pV = F + pV. \tag{1.88}$$

As seen from (1.87), G is a function of temperature and pressure. It is called the Gibbs free energy or, alternatively, the isobaric–isothermal potential and is a characteristic function of the independent parameters T and p. It, too, was introduced by Masse, although it was Gibbs who discovered its deeper meaning and applied it extensively in the development of chemical thermodynamics. It is therefore justifiably called the Gibbs free energy.

Since G is a function of state, we may write

$$dG = (\partial G/\partial T)_p \, dT + (\partial G/\partial p)_T \, dp. \tag{1.89}$$

Comparing (1.87) and (1.89), we obtain

$$(\partial G/\partial T)_p = -S, \tag{1.90}$$

$$(\partial G/\partial p)_T = V. \tag{1.91}$$

Further transformations of Eq. (1.74) of the kind already carried out are impossible, so that one has ultimately just four functions, U, H, F, and G. Taking their partial derivatives with respect to the appropriate parameters, we obtain the eight expressions given in (1.76) and (1.77), (1.81) and (1.82), (1.85) and (1.86), and (1.90) and (1.91). These are of great utility in solving various physicochemical problems, as we show later. We have already noted that there are many thermodynamic functions and that any state parameter such as entropy or volume may, in principle, play the role of a thermodynamic function. However, the state parameters taken in the form of such arbitrary functions are characterized by complicated relations between the partial derivatives. This is easily shown by the same step-by-step arguments as were used in deriving the four characteristic functions. Entropy is ranked among the characteristic functions by some authors [9].

It is of interest to consider the second derivatives of the thermodynamic potentials with respect to the conjugate independent parameters. Using them together with the eight equations mentioned above, we may obtain various relations among the thermodynamic parameters of the system. Thus we obtain

$$\partial^2 U/\partial S \, \partial V = (\partial T/\partial V)_S = -(\partial p/\partial S)_V, \tag{1.92}$$

$$\partial^2 F/\partial T \, \partial V = (\partial S/\partial V)_T = (\partial p/\partial T)_V, \tag{1.93}$$

$$\partial^2 H/\partial S \, \partial p = (\partial T/\partial p)_S = (\partial V/\partial S)_p, \tag{1.94}$$

$$\partial^2 G/\partial T \, \partial p = -(\partial S/\partial p)_T = (\partial V/\partial T)_p. \tag{1.95}$$

Relations (1.92)–(1.95) are consequences of the joint equation for the first and second laws and are to be considered as the basic differential (Maxwell) relations. The first one was obtained by Maxwell.

Differentiating Eq. (1.90) with respect to T at constant p, we get

$$(\partial^2 G/\partial T^2)_p = -(\partial S/\partial T)_p. \tag{1.96}$$

From (1.57) we have

$$(\partial^2 G/\partial T^2)_p = -C_p/T. \tag{1.97}$$

Differentiating (1.97) with respect to p at constant T and changing the order of differentiation on the left, we obtain

$$[(\partial^2/\partial T^2)(\partial G/\partial p)_T]_p = -T^{-1}(\partial C_p/\partial p)_T. \tag{1.98}$$

By using (1.91) in (1.98) we are led to a relation analogous to (1.63). Knowing the equation of state and the value of C_p at a certain pressure, we may use the equation obtained to calculate C_p at other pressures.

Similarly, from expressions (1.81) and (1.59) we obtain

$$(\partial^2 F/\partial T^2)_V = -C_V/T. \tag{1.99}$$

Differentiating this with respect to V at $T = $ const, changing the order of differentiation, and employing (1.82), we arrive at an expression similar to (1.64).

Let us consider the physical meaning of increments in the characteristic functions for certain processes. We should note that the work of a system is composed of work due to the volume increase and of work due to internal forces not associated with volume change. But Eq. (1.74) accounts only for work performed in overcoming external pressure, so that it should be generalized. Let us write all other forms of work as the product of generalized forces and of changes in generalized coordinates:

$$\sum_i X_i \, dY_i \equiv \delta A_{max}^*, \tag{1.100}$$

where δA_{max}^* is the sum of elements of effective work (apart from the work of expansion).

Using the identity (1.100), we may express the differentials of characteristic functions (1.74), (1.78), (1.83), and (1.87) in the following form:

$$dU = T \, dS - p \, dV - \sum_i X_i \, dY_i, \tag{1.101}$$

$$dF = -S \, dT - p \, dV - \sum_i X_i \, dY_i, \tag{1.102}$$

$$dH = T \, dS + V \, dp - \sum_i X_i \, dY_i, \tag{1.103}$$

$$dG = -S\,dT + V\,dp - \sum_i X_i\,dY_i. \qquad (1.104)$$

At constant generalized coordinates dY_i and volume V, we have, according to (1.101),

$$dU = T\,dS = \delta Q_{V,\,Y_i}, \qquad (1.105)$$

which equates the heat absorbed to the gain in the internal energy of the system. If in (1.101) we keep the entropy constant $(dS = 0)$, then

$$-(dU)_S = p\,dV + \delta A^{\bullet}_{max} = \delta A_{max}, \qquad (1.106)$$

so that in this case the decrease in the internal entropy equals the maximum amount of work done by the system.

At constant pressure $(dp = 0)$ and generalized coordinates $(dY_i = 0)$, we get from (1.103)

$$dH = T\,dS = \delta Q_{p,\,Y_i}. \qquad (1.107)$$

Under these conditions the enthalpy increment equals the amount of heat passed to the system.

At constant entropy $(dS = 0)$ and pressure $(dp = 0)$

$$-dH = \sum_i X_i\,dY_i = \delta A^{\bullet}_{max}, \qquad (1.108)$$

i.e., the decrease in enthalpy is equal to the maximum of work unconnected with volume changes.

The physical meaning of the Helmholtz free energy is easily understood by calculating the work performed in an isothermal process. At constant temperature $(dT = 0)$, Eq. (1.102) gives

$$-(dF)_T = p\,dV + \delta A^{\bullet}_{max} = \delta A_{max}, \qquad (1.109)$$

i.e., the maximum available work in an isothermal reversible process is equal to the decrease in free energy. At constant temperature $(dT = 0)$ and volume $(dV = 0)$, we get

$$-(dF)_{T,\,V} = \delta A^{\bullet}_{max}. \qquad (1.110)$$

Therefore, in an isochoric–isothermal process the decrease in Helmholtz free energy is equal to the maximum amount of useful work excluding that associated with volume changes.

If the isothermal process is irreversible, less work is obtained, since part of it is not utilized due to losses in the irreversible process.

At constant temperature and pressure $(dT = 0$ and $dp = 0)$, we have from (1.104)

$$-(dG)_{p,\,T} = \delta A^{\bullet}_{max}. \qquad (1.111)$$

Thus, the decrease in the Gibbs free energy determines the maximum amount of work not associated with volume change which can be obtained in an isobaric–isothermal process.

The validity of Eqs. (1.109)–(1.111) does not require the temperature to remain constant throughout the reversible process, but only that the source from (to) which the system absorbs (delivers) heat have the temperature of the system in its initial (final) state, as shown first by Gibbs [1].

By analyzing the connection between the Helmholtz free energy and the internal energy, and between the Gibbs free energy and the enthalpy, one may obtain important relations of great practical use in thermodynamic calculations.

Although the connection between the Helmholtz free energy F and the internal energy U is given by (1.79), it does not allow one to properly evaluate the relation between them, since it also involves the entropy S, itself a thermodynamic function of state. If, however, we use Eq. (1.81) and substitute it into (1.79), we shall have

$$F = U + T \, (\partial F/\partial T)_V. \tag{1.112}$$

This expression is called the Gibbs–Helmholtz equation and is of great value in isothermal processes in the course of which work is performed with practically no change in volume (such as work in galvanic cells, in elastic deformations, in various processes in solutions, in electromagnetic fields, etc.).

A similar process may be applied to the relation between the Gibbs free energy and the enthalpy:

$$G = H - TS, \tag{1.113}$$

which does not exhibit clearly the relation between G and H because of the presence of S. But if we replace S by the value provided by Eq. (1.90), we obtain

$$G = H + T \, (\partial G/\partial T)_p, \tag{1.114}$$

which is the second Gibbs–Helmholtz equation, of tremendous value in analyzing isobaric–isothermal processes (especially in thermochemistry).

By means of these properties of the F and G functions one may establish the relation between the maximum work in a reversible process and the heat in the same process carried out irreversibly, thereby obtaining another variant of the Gibbs–Helmholtz equation [10].

In calculating the integrals of the characteristic functions, we do not obtain absolute values but only differences between any two states of the system. One of the states is fixed as a reference standard, to which we ascribe an arbitrary but convenient value of the characteristic function.

We now introduce general formulas for calculating the function $G(p, T)$ at any pressure and temperature. At first we find the pressure dependence of $G(p, T)$ at $T = $ const. This problem is easily solved by integrating expression (1.91) at constant temperature:

$$G\,(p_2, \, T_2) - G\,(p_1, \, T_2) = \int_{p_1}^{p_2} V \, dp. \tag{1.115}$$

Integrating (1.97) and using (1.92), we find the temperature dependence of $G(p, T)$ at constant pressure p_1:

$$G(p_1,T_2)-G(p_1,T_1)=-S(p_1,T_1)(T_2-T_1)-\int_{T_1}^{T_2}dT\int_{T_1}^{T_2}\frac{c_p(p_1,T)}{T}dT. \qquad (1.116)$$

Using (1.115), we have, finally,

$$G(p_2,T_2)=G(p_1,T_1)-S(p_1,T_1)(T_2-T_1)$$

$$+\int_{p_1}^{p_2}V\,dp-\int_{T_1}^{T_2}dT\int_{T_1}^{T_2}\frac{c_p(p_1,T)}{T}dT. \qquad (1.117)$$

In a similar way we may obtain the equation for calculating the volume and temperature dependence of F:

$$F(V_2,T_2)=F(V_1,T_1)-S(V_1,T_1)(T_2-T_1)$$

$$-\int_{V_1}^{V_2}p\,dV-\int_{T_1}^{T_2}dT\int_{T_1}^{T_2}\frac{c_V(V_1,T)}{T}dT. \qquad (1.118)$$

Chapter 2

THEORY OF CHEMICAL POTENTIALS AND
THE GENERAL THEORY OF
THERMODYNAMIC EQUILIBRIUM

1. FUNDAMENTAL THEOREMS AND EQUATIONS.
THE GIBBS–DUHEM EQUATION

We noted in Chapter 1 that thermodynamic systems may be either closed or open. Both laws of thermodynamics were obtained for closed systems. In going from simple to more complex multicomponent systems, the possibility of material exchange with the external medium, or between different parts of the system, may eventually lead to changes in composition which should therefore be included among the state parameters of the system.

Thermodynamic relations do not distinguish in any way between open and closed systems. Gibbs ingeniously overcame the difficulty of treating composite systems by assuming that a given system is composed of an aggregate of phases, each of which is considered to be an open system [1]. Open and closed systems differ mainly by the conditions which define them and, consequently, by the opportunities for changing their compositions, since the compositions of open and closed systems are subject to different conditions of material confinement [11].

Since the main objects of our interest are heterogeneous two-component systems, let us generalize the characteristic thermodynamic functions for open systems by including composition among the basic parameters. We begin by stating the

First Theorem. For any increment of the state parameters of a thermodynamic system, the independent variables being the entropy, volume, and the number of moles of the components, the following relation should hold for the perfect differential of the internal energy:

$$dU = T\, dS - p\, dV + \sum_{i=1}^{k} (\partial U / \partial n_i)_{S,\, V,\, n_i}\, dn_i. \tag{2.1}$$

To prove the validity of this relation, let us consider the expression for the total differential of the internal energy of a composite system:

$$dU = (\partial U/\partial T)_{V,\,n}\,dT + (\partial U/\partial V)_{T,\,n}\,dV + \sum_{i=1}^{k} (\partial U/\partial n_i)_{T,\,V,\,n_j}\,dn_i. \qquad (2.1a)$$

If an open system undergoes change while the compositions are kept constant, then its change of state is equivalent to that of a homogeneous closed system. In such a situation we may write, in accordance with Eqs. (1.76) and (1.77),

$$(\partial U/\partial S)_{V,\,n} = T, \qquad (2.2)$$

$$(\partial U/\partial V)_{S,\,n} = -p. \qquad (2.3)$$

Substituting the partial derivatives from (2.2) and (2.3) into (2.1a), we obtain relation (2.1), which is therefore proved.

The derivatives $(\partial U/\partial n_i)_{S,\,V,\,n_j}$ were called by Gibbs the chemical potentials of the ith component:

$$\mu_i = (\partial U/\partial n_i)_{S,\,V,\,n_j}. \qquad (2.4)$$

The introduction of the concept of the chemical potential is of great value in chemical thermodynamics, since it plays an essential role in defining chemical and phase equilibrium.

In order to present clearly the meaning of the chemical potential, we shall consider each term of the energy in Eq. (2.1) as a product of two quantities: a generalized force and a generalized coordinate. The derivatives of the internal energy with respect to the independent parameters (at constant values of the rest of the parameters) are generalized forces. In this connection the derivatives $(\partial U/\partial n_i)_{S,\,V,\,n_j}$ may also be considered as generalized forces determining the process of redistributing the number of moles of the components of the system. Thus the chemical potential plays the same role in the mass redistribution of the components as the pressure in a volume change, or the temperature in the process of heat exchange. One may speak of the chemical potential of a component at each point of the system in exactly the same way as one speaks of concentration or mole fraction.

We may obtain perfect differentials for the functions F, H, and G the same way as for dU. The perfect differential of the function F for the variables T, V, and n_i will be

$$dF = (\partial F/\partial T)_{V,\,n}\,dT + (\partial F/\partial V)_{T,\,n}\,dV + \sum_{i=1}^{k} (\partial F/\partial n_i)_{T,\,V,\,n_j}\,dn_i. \qquad (2.5)$$

Here and below, the index n denotes holding all the compositions constant, whereas n_j signifies holding all but the ith constant. Using (1.81) and (1.82), keeping all n_i fixed, we obtain

$$dF = -S\,dT - p\,dV + \sum_{i=1}^{k} (dF/dn_i)_{T,\,V,\,n_j}\,dn_i. \tag{2.6}$$

Taking H as a function of the variables S, p, and n_i we have

$$dH = (\partial H/\partial S)_{p,\,n}\,dS + (\partial H/\partial p)_{S,\,n}\,dp + \sum_{i=1}^{k} (\partial H/\partial n_i)_{S,\,p,\,n_j}\,dn_i \tag{2.7}$$

or, using (1.85) and (1.86),

$$dH = T\,dS + V\,dp + \sum_{i=1}^{k} (dH/\partial n_i)_{S,\,p,\,n_j}\,dn_i. \tag{2.8}$$

The perfect differential of G as a function of T, p, and n_i is written in the following form:

$$dG = (\partial G/\partial T)_{p,\,n}\,dT + (\partial G/\partial p)_{T,\,n}\,dp + \sum_{i=1}^{k} (\partial G/\partial n_i)_{p,\,T,\,n_j}\,dn_i \tag{2.9}$$

or, using relations (1.90) and (1.91), keeping all n_i fixed,

$$dG = -S\,dT + V\,dp + \sum_{i=1}^{k} (\partial G/\partial n_i)_{T,\,p,\,n_j}\,dn_i. \tag{2.10}$$

Let us note that during a change of mass of the ith component the entropy can remain constant only in a specially selected temperature range, so that Eq. (2.4) is effectively useless. In practice the chemical potential is therefore always defined in terms of derivatives of other functions. We shall consider expressing the chemical potential in terms of the other characteristic functions, which leads us to the

Second Theorem. The partial derivatives of all characteristic functions with respect to the mass of the ith component, holding the remaining state parameters constant, are equal to one another. Thus we shall prove that

$$\mu_i = (\partial U/\partial n_i)_{S,\,V,\,n_j} = (\partial H/\partial n_i)_{S,\,p,\,n_j} = (\partial F/\partial n_i)_{T,\,V,\,n_j}$$

$$= (\partial G/\partial n_i)_{T,\,p,\,n_j}. \tag{2.11}$$

We differentiate the expressions for H, F, and G and substitute into the result the value of dU from Eq. (2.1). Comparing the relations thus obtained with expressions (2.6), (2.8), and (2.10), we establish the validity of (2.11). We have, finally,

$$dU = T\,dS - p\,dV + \sum_{i=1}^{k} \mu_i\,dn_i; \tag{2.12}$$

$$dH = T\,dS + V\,dp + \sum_{i=1}^{k} \mu_i\,dn_i, \tag{2.13}$$

$$dF = -\,S\,dT - p\,dV + \sum_{i=1}^{k} \mu_i\,dn_i, \tag{2.14}$$

$$dG = -\,S\,dT + V\,dp + \sum_{i=1}^{k} \mu_i\,dn_i. \tag{2.15}$$

The most widely used expression for the chemical potential is in terms of the Gibbs free energy:

$$\mu_i = (\partial G/\partial n_i)_{p,\,T,\,n_j}. \tag{2.16}$$

The reason is that processes associated with composition changes are normally studied at constant temperature and pressure. One sees from (2.16) that the chemical potential of the ith species is a partial molar quantity of the Gibbs free energy.

Gibbs called (2.12)–(2.15) the fundamental equations, in order to emphasize that they express the relation between the characteristic functions and their variables and, consequently, provide a comprehensive thermodynamic description of a k-component mixture. These equations are all equivalent, so that any one of them may be employed in describing the thermodynamic properties of a multicomponent system.

We should note that the thermodynamic potentials (U, H, F, G) are linear homogeneous functions of the corresponding extensive quantities (S, v, n_1, n_2, ..., n_k). Applying Euler's theorem on homogeneous functions, we obtain the expressions

$$U = TS - pV + \mu_1 n_1 + \mu_2 n_2 + \dots + \mu_k n_k, \tag{2.17}$$

$$H = TS + \mu_1 n_1 + \mu_2 n_2 + \dots + \mu_k n_k, \tag{2.18}$$

$$F = -\,pV + \mu_1 n_1 + \mu_2 n_2 + \dots + \mu_k n_k, \tag{2.19}$$

$$G = \mu_1 n_1 + \mu_2 n_2 + \dots + \mu_k n_k = \sum_{i=1}^{k} \mu_i n_i. \tag{2.20}$$

These formulas may also be obtained by integrating Eqs. (2.12)–(2.15) while keeping constant all intensive properties T, p, μ_1, μ_2, ..., μ_k. Since intensive quantities characterize the state of a system, integrating under these conditions means that the mass of the mixture is increased without changing the latter's state.

According to the above, the thermodynamic properties of a system are completely determined if, for example, one knows the function $G(T, p, n_1, n_2, ..., n_k)$. It follows from expression (2.20) that the thermodynamic properties of a system are uniquely determined once all k chemical potentials μ_i are known as functions of the independent variables T, p, n_1, n_2, ..., n_k. These k chemical potentials are related to one another and to the other thermodynamic parameters by relations easily derivable from the fundamental equations (2.12)–(2.15). Since G is a function of state and dG is a perfect differential, the following so-called reciprocal relations hold for any pair of components:

$$(\partial \mu_j/\partial n_i)_{T,p,\,n_{j\neq i}} = (\partial \mu_i/\partial n_j)_{T,\,p,\,n_{i\neq j}}.\qquad(2.21)$$

Using Cauchy's theorem and Eqs. (2.16) and (1.90)–(1.91) (the latter two at $n_i = $ const), it is easy to show that

$$(\partial \mu_i/\partial T)_{p,\,n} = -(\partial S/\partial n_i)_{T,\,p,\,n_j} = -\bar{s}_i,\qquad(2.22)$$

$$(\partial \mu_i/\partial p)_{T,\,n} = (\partial V/\partial n_i)_{T,\,p,\,n_j} = \bar{v}_i,\qquad(2.23)$$

where $\bar{s}_i$ and $\bar{v}_i$ are the partial molar entropy and volume of the ith species, respectively.

Differentiating Eq. (2.17) and using expression (2.12), we obtain another fundamental equation introduced by Gibbs:

$$S\,dT - V\,dp + \sum_{i=1}^{k} n_i\,d\mu_i = 0.\qquad(2.24)$$

This equation is unusual in that its independent variables are intensive parameters (temperature, pressure, and the chemical potentials $\mu_1, \mu_2, \ldots, \mu_k$) whereas the coefficients are extensive quantities (S, V, n_i).

For changes occurring at constant temperature and pressure, we may write, on account of (2.24),

$$\sum_{i=1}^{k} n_i\,d\mu_i = 0.\qquad(2.25)$$

Relation (2.25) is the Gibbs–Duhem equation. It connects the changes of chemical potentials of the components in an isobaric–isothermal process and is invaluable in the thermodynamics of solutions. It enables one to calculate one of the quantities $d\mu_i$ (e.g., $d\mu_1$) in an isobaric–isothermal process if all the others are known.

At constant T and p the perfect differential $d\mu_i$ becomes

$$d\mu_i = \sum_{j=1}^{k} (\partial \mu_i/\partial n_j)_{T,\,p,\,n_{i\neq j}}\,dn_j.\qquad(2.26)$$

This may be substituted into Eq. (2.25) to give

$$\sum_i \sum_j n_i\,(\partial \mu_i/\partial n_j)_{T,\,p,\,n_{i\neq j}}\,dn_j = 0.\qquad(2.27)$$

Since (2.27) is valid for any dn_j, it must satisfy the condition

$$\sum_i n_i \, (\partial \mu_i / \partial n_j)_{T, \, p, \, n_{i \ne j}} = 0,$$ (2.28)

which coincides with expression (1.10) if μ_i is considered as a partial molar quantity.

In addition, the reciprocal relation (2.21) enables one to rewrite (2.28) as

$$\sum_i n_i \, (\partial \mu_j / \partial n_i)_{T, \, p, \, n_j} = 0.$$ (2.29)

It should be noted that the general description of partial molar quantities developed in Chapter 1 applies to the chemical potential.

Using Eq. (2.29), we may write, for a binary mixture,

$$n_1 \, (\partial \mu_1 / \partial n_1) + n_2 \, (\partial \mu_1 / \partial n_2) = 0$$ (2.30)

(here and below the indices on the partial derivatives of the chemical potential are omitted).

Let us choose the mole fraction of the two components as independent variables. Then

$$\partial \mu_1 / \partial n_2 = x_1/n \cdot \partial \mu_1 / \partial x_2,$$ (2.31)

from which we get

$$\partial \mu_1 / \partial x_2 = n/x_1 \cdot \partial \mu_1 / \partial n_2.$$ (2.32)

Substituting expression (2.31) into Eq. (2.30), we find

$$\partial \mu_1 / \partial x_2 = -n/x_2 \cdot \partial \mu_1 / \partial n_1.$$ (2.33)

Similarly,

$$\partial \mu_2 / \partial x_2 = n/x_1 \cdot \partial \mu_2 / \partial n_2,$$ (2.34)

$$\partial \mu_2 / \partial x_2 = -n/x_2 \cdot \partial \mu_2 / \partial n_1.$$ (2.35)

On the basis of expression (2.28), and utilizing relations (2.31) and (2.34), it is easy to show that

$$x_1 \, (\partial \mu_1 / \partial x_2) + x_2 \, (\partial \mu_2 / \partial x_2) = 0.$$ (2.36)

On the other hand, the molar Gibbs free energy g may be expressed via (2.20) as

$$g = G/n = \sum_i^k x_i \mu_i.$$ (2.37)

Combining relations (2.36) and (2.37), it is simple to show that for a binary system the relation

$$\partial g/\partial x_2 = \mu_2 - \mu_1 \tag{2.38}$$

will hold. For a k-component mixture this is generalized to

$$\partial G/\partial x_i = \mu_i - \mu_k, \tag{2.39}$$

where k refers to the component whose mole fraction is a dependent variable, i.e., found from the constraint condition in (1.2).

Differentiating expression (2.38) with respect to x_2, we obtain

$$\partial^2 g/\partial x_2^2 = \partial \mu_2/\partial x_2 - \partial \mu_1/\partial x_2. \tag{2.40}$$

Substituting for $\partial \mu_1/\partial x_2$ from (2.36), we find

$$\partial^2 g/\partial x_2^2 = \partial \mu_2/\partial x_2 \left(1 + x_2/x_1\right) \tag{2.41}$$

or

$$\partial \mu_2/\partial x_2 = x_1 \, \partial^2 g/\partial x_2^2. \tag{2.42}$$

Similarly,

$$\partial \mu_1/\partial x_2 = - x_2 \, \partial^2 g/\partial x_2^2. \tag{2.43}$$

For a particular substance we write (2.23) at constant T as

$$d\mu^\circ = v^\circ \, dp. \tag{2.44}$$

Substituting for the molar value of v° from Eq. (1.4), we obtain

$$d\mu^\circ = RT \, (dp/p). \tag{2.45}$$

Integrating this at $T = $ const, we find for the chemical potential of an ideal gas

$$\mu^\circ (T, p) = \mu^{\circ *} (T) + RT \ln p, \tag{2.46}$$

where $\mu^{\circ *}(T)$ is a constant of integration depending on the nature of the gas and on temperature. Equation (2.46) shows that $\mu^{\circ *}(T)$ represents the chemical potential of the gas at a given temperature and at unit pressure.

A mixture of gases at a given volume V and temperature T is called ideal if its free energy F is equal to the sum of the partial free energies of the separate components if each were to occupy the whole volume at the same temperature:

$$F = \sum_i F (T, V, n_i) = \sum_i n_i f_i^\circ. \tag{2.47}$$

Here f_i° is the molar Helmholtz free energy (the subscript i referring to the constituent gases of the mixture).

Taking $V_1 = 1$, it follows easily from Eq. (1.118) that for one mole of the ith component of a mixture of ideal gases

$$f_i^\circ = f_i^{\circ*}(T) - RT \ln V_i^\circ, \tag{2.48}$$

where $f_i^{\circ*}(T)$ is a constant of integration which is a function of temperature but not of volume.

Substituting (2.48) into (2.47), we obtain

$$F = \sum_i n_i \left[f_i^{\circ*}(T) - RT \ln (V/n_i) \right]. \tag{2.49}$$

Differentiating expression (2.49) with respect to n_i gives the chemical potential of the constituents of a mixture of ideal gases:

$$\mu_i = (\partial F/\partial n_i)_{T,\, V,\, n_j} = f_i^{\circ*}(T) + RT + RT \ln (n_i/V). \tag{2.50}$$

These quantities may also be expressed in terms of partial pressures. A mixture of ideal gases satisfies the ideal gas equation for each component

$$p_i = n_i RT/V, \tag{2.51}$$

where the partial pressure of the ith species equals the pressure of n_i moles occupying the total volume V at the same temperature.

Combining Eqs. (2.50) and (2.51), we may write

$$\mu_i = \mu_i^*(T) + RT \ln p_i. \tag{2.52}$$

In this expression $\mu_{i}^{*}(T) = f_{i}^{\circ*}(T) + RT(1 - \ln RT)$ depends only on the temperature and is equivalent to the function $\mu^{\circ*}(T)$ given in (2.46) for a single component i. Indeed, if species i is the only component, then p_i becomes the total pressure and (2.52) becomes identical with (2.46).

Replacing p_i in relation (2.52) by px_i, we obtain

$$\mu_i = \mu_i^\circ(T, p) + RT \ln x_i, \tag{2.53}$$

where

$$\mu_i^\circ(T, p) = \mu_i^*(T) + RT \ln p, \tag{2.54}$$

i.e., $\mu^\circ(T, p)$ is the chemical potential of a pure gas at temperature T and pressure p.

For real gases we may write an expression analogous to (2.46):

$$\mu = \mu^{\circ*}(T) + RT \ln \varphi(T, p), \tag{2.55}$$

where $\mu°*(T)$ is the same function as for an ideal gas, while φ is the fugacity originating in the intermolecular interactions of a nonideal gas. The fugacity φ depends on temperature and pressure. It was introduced by Lewis in 1901.

The form of Eq. (2.55) has the advantage that the chemical potential of a real gas obeys the same equation as for an ideal gas, except that the fugacity replaces the pressure. Thus φ is an intensive parameter, like p. At sufficiently low pressures all gases tend to be ideal so that the fugacity becomes equal to the pressure (Gibbs–Dalton law), i.e.,

$$\lim_{p \to 0} (\varphi/p) = 1. \tag{2.56}$$

By analogy with expression (2.52) we may write for a mixture of real gases

$$\mu_i = \mu_i^*(T) + RT \ln \varphi_i, \tag{2.57}$$

in which μ_i^* is the same function as for ideal gases and all effects of intermolecular interactions are included in the fugacities φ_i. The latter are thus functions of the temperature and partial pressures or, equivalently, depend on $T, P, n_1, n_2, ..., n_k$.

Equation (2.57) defines the functions φ_i. This equation also preserves the ideal gas form, with the fugacities replacing the partial pressures.

As for a pure gas, the fugacities of the components in a mixture may be calculated using the equation

$$RT \ln (\varphi_i/p_i) = \lim_{p_0 \to 0} \int_{p_2}^{p} (\bar{v}_i - \bar{v}_i^{(d)}) \, dp, \tag{2.58}$$

where the superscript *id* denotes the value of $\bar{v}_i$ in an ideal mixture.

In order to calculate the fugacity of any one component at a given temperature one need know only its partial molar volume as a function of pressure. To this end one may use either the value of $\bar{v}_i$ determined directly, or the equation of state.

Extending Eq. (2.53) to real mixtures, we write

$$\mu_i = \mu_i^°(T, p) + RT \ln a_i, \tag{2.59}$$

where $\mu_i^°(T, p)$ is the same function as in Eq. (2.53), so that all effects of intermolecular interactions are included in the function $a_i(T, p, n_1, ..., n_k)$ called by Lewis the activity of the *i*th species.

As the total pressure is lowered, the activity of each component tends to its mole fraction. It should be noted that, like x_i, a_i is an intensive quantity.

Let us introduce the concept of the activity coefficient:

$$\gamma_i = a_i/x_i, \tag{2.60}$$

having the following property:

$$\lim_{p \to 0} \gamma_i = 1. \tag{2.61}$$

In a mixture of ideal gases all of these coefficients are unity, so that $(1 - \gamma_i)$ or $\ln \gamma_i$ may serve as a measure of the departure from the ideal gas laws.

The connection between the fugacity and the activity coefficients, derived from expressions (2.54), (2.59), and (2.57), is the following:

$$\gamma_i = \varphi_i / p_i. \tag{2.62}$$

The activity coefficients can be computed by the same method as the fugacity. Thus, on account of Eq. (2.62), Eq. (2.58) may be written as

$$RT \ln \gamma_i = \lim_{p_0 \to 0} \int_{p_0}^{p} (\bar{v}_i - \bar{v}_i^{id}) \, dp. \tag{2.63}$$

The preceding relations are valid not only for gaseous mixtures but for mixtures in general, including the condensed phase, although in each particular application one must choose the reference state appropriately.

The theory of chemical potentials lies at the basis of the general thermodynamic theory of heterogeneous equilibrium and constitutes the essence of Gibbs' approach to dealing with these problems in the most general form. The superiority of the chemical potential approach over other methods has often been discussed in the literature and we shall not expound on it. We would only remark that the method will always give the correct solution, as can be confirmed by applying it to several problems.

2. THE EQUILIBRIUM PRINCIPLE OF GIBBS

One of the major tasks of thermodynamics is the need to formulate explicit conditions for the thermodynamic equilibrium of various systems. This problem may be solved mathematically by the method of Gibbs, which provides a general formulation of the principle of equilibrium for material systems isolated from all external influences, in two essentially equivalent forms.

Although Gibbs did not dwell on the demonstration of the equilibrium conditions, they are in fact consequences of the two laws of thermodynamics, as formulated in Eqs. (1.41) and (2.12). Gibbs' exposition of thermodynamic equilibrium is rather laconic and quite difficult to understand.

The credit for developing further the method of Gibbs belongs to van der Waals [12]. A step-by-step, rigorous exposition and creative developments of the ideas of Gibbs and van der Waals were given in the classic work of Storonkin [13]. His method will be employed in our exposition of the equilibrium principle of Gibbs and of the stability criteria of thermodynamic systems.

Let us consider a k-component system totally isolated from its surroundings, i.e., satisfying the following conditions:

$$\begin{aligned} U &= \text{const} \\ V &= \text{const,} \\ n_1 &= \text{const,} \end{aligned} \tag{2.64}$$

$$n_2 = \text{const};$$

$$\cdot \ \cdot \ \cdot \ \cdot \ \cdot$$

$$n_k = \text{const}.$$

It then follows from expression (1.40) that

$$T \, dS \geqslant 0. \tag{2.65}$$

If the isolated system is in a nonequilibrium state then, by the second law, on the approach to equilibrium its entropy will spontaneously increase, since $T > 0$. This entropy increase of an isolated nonequilibrium system is possible only because of the irreversible redistribution of energy and matter between its various parts. If the heat and matter exchange between the parts of an isolated system proceeds reversibly, then the entropy of the system remains constant, in strict accordance with given values of the energy, volume, and mass of the system. Since $T \neq 0$, we ultimately have

$$dS = 0. \tag{2.66}$$

On the basis of the preceding discussion we may formulate the following fundamental proposition, the so-called equilibrium principle of Gibbs: an isolated system is in a state of equilibrium if for all possible changes consistent with the conditions of its isolation its entropy remains either constant or decreases, i.e.,

$$(\delta S)_{U, V, n_1, n_2, \cdots, n_k} \leqslant 0. \tag{2.67}$$

Depending on whether the processes occurring in the system proceed reversibly or irreversibly, we have the equality or inequality holding in (2.67), respectively. For isolated systems irreversible processes have only a fluctuational character.

Although the property of entropy thus formulated has a probabilistic character, it has been shown by Boltzmann that the probability of a spontaneous decrease of entropy in an isolated macroscopic system is completely negligible [14]. The idea behind this conclusion belongs to Gibbs, who stated that an uncompensated decrease of entropy is extremely improbable. This was also noted by Boltzmann.

Relation (2.67) reflects the fact that the entropy of an isolated system in a state of equilibrium has a conditional maximum. The maximum value of the entropy of an isolated system is determined by the energy, volume, and mass (i.e., number of moles) of the system.

From the combined equation for the first and second laws we have at constant volume and constant masses of components

$$(dU \leqslant T \, dS)_{V, n_1, n_2, \cdots, n_k}. \tag{2.68}$$

Since T is a positive quantity, Eq. (2.68) shows that the internal energy and the entropy change in unison in any equilibrium process. If $V \neq \text{const}$, then Eq. (1.41) holds, and the energy and entropy need not change in unison. Let us consider a system with a given volume and total number of moles of components, in which a nonequilibrium process is taking place. The entropy of such a system would increase spontaneously. In order to prevent this one must withdraw an appropriate amount of heat from the system. To maintain a constant entropy, the internal energy of the system must decrease until an equilibrium state is attained. Consequently, as the system approaches equilibrium the internal energy decreases as long as the volume, the total number of moles, as well as the entropy are maintained constant. Therefore, in an equilibrium state, the internal energy of the system is at a conditional minimum.

Based on the concept of internal energy change, the equilibrium principle may be expressed as follows: *A system is in an equilibrium state if during all processes preserving a constant value of the entropy, the volume, and the total amount of matter, the internal energy remains constant or increases.* In other words,[1]

$$(\delta U)_{S, V, n_1, n_2, \cdots, n_k} \geqslant 0. \tag{2.69}$$

We note that the first formulation of the equilibrium principle stipulates complete isolation of the system, thus including adiabatic conditions. By contrast, the second formulation assumes thermal exchange between the system and the surrounding medium.

Under different external conditions the equilibrium principle may be expressed by means of other thermodynamic potentials [13]:

$$(\delta H)_{S, p, n_1, n_2, \cdots, n_k} \geqslant 0, \tag{2.70}$$

$$(\delta F)_{T, V, n_1, n_2, \cdots, n_k} \geqslant 0. \tag{2.71}$$

$$(\delta G)_{T, p, n_1, n_2, \cdots, n_k} \geqslant 0. \tag{2.72}$$

All formulations of the equilibrium principle are equivalent in the sense that under specific assumptions one formulation gives rise to all the others. They are not identical, however, since their content and applicability vary according to the conditions prevailing over the system [13]. These conditions are expressed by fixing definite values for the thermodynamic parameters.

3. CONDITIONS FOR PHASE EQUILIBRIUM. REAL AND VIRTUAL COMPONENTS

In considering conditions for phase equilibria, we must define more precisely the terms "phase," "component," and others introduced in a preliminary way in Chapter 1 which, until now, we used in situations where an accurate definition was not essential.

Gibbs called components individual substances whose concentration increments are independent and express all possible changes in the composition of the system under consideration. These individual substances can be simple substances or chemical compounds.

---

[1]In expressions (2.67), (2.69), and those to follow we use the *variation* of the entropy and the internal energy rather than the *differential*, since we are considering virtual changes comprising a more extensive class of variations [13].

According to Gibbs, components are either real or virtual. Real components are those that are present, if only in vanishingly small amounts, in all homogeneous parts or phases of a heterogeneous system. Virtual components are substances that are part of some phases but are missing from others.

If there are no reversible chemical reactions in a system, then all species of its constituent matter will form components, since their concentrations will change independently of one another. In this instance the concepts of components and substances are synonymous, and the number of components is just the number of substances contained in the system.

If, however, reversible chemical changes do occur, then only the concentrations of those substances which enter into the composition of the system change independently. This is due to the fact that the reversible chemical reactions establish quantitative relations between the concentrations of substances. These relations are described by as many thermodynamic equations as the number of independent reversible reactions. Hence, for a system with chemical transformations, the number of independent components will be always less than the number of substances forming the system. In fact, it will be the difference between the number of particle species existing independently in the system and the number of independent reversible reactions taking place therein.

One should not confuse the number of components with the number of original substances that were mixed to produce the system. The number of components may be larger or smaller than the original number of substances, depending on the chemical nature of the system [13].

From the foregoing we see that, as emphasized by Storonkin, the choice of components is arbitrary and not unique. Thus there is no unique answer to the question as to which components are contained in a system with reversible chemical reactions. However, the number of components is absolutely determined.

It should also be pointed out that the number of components depends on the conditions prevailing in the system. Changing conditions may enhance or suppress chemical reactions in the system, thereby decreasing or increasing the number of relations imposed on the concentration changes of substances.

The concept of a "phase" was introduced by Gibbs in a precise form in the context of proving the phase rule. This vital concept in the theory of heterogeneous equilibrium was further refined by van der Waals and Storonkin [12, 13].

In proving the phase rule, Gibbs assumed that each phase satisfies a fundamental equation such as (2.24) and, accordingly, that there are as many independent equations of this form as the number of phases in the system. The physical meaning of this inference is that each phase possesses that dependence of the thermodynamic properties on the state parameters which distinguishes it from the other phases. Moreover, the phase is homogeneous and has macroscopic dimensions. This is due to the fact that the validity of (2.24) presupposes conditions under which the phase is in internal equilibrium, external fields are either absent or constant, and surface effects are negligible. Thus the essential aspect of the concept "phase" is mirrored in Eq. (2.24), whereas the surface separation between phases is only an external aspect of phases. Equations expressing the intrinsic dependence of the thermodynamic properties of a given phase on the state parameters were termed phase equations by van der Waals.

In keeping with modern concepts based on the ideas of Gibbs, a phase is considered to be an aggregate of material complexes with equal dependences of their thermodynamic properties on the state parameters, i.e., described by the same phase equation. The latter may be taken to be any one of the fundamental equations of Gibbs. By material complexes one means macroscopic homogeneous units formed by individual substances, either molecular or atomic mixtures of substances, separated from one another by surface partitions [13].

A heterogeneous system is one composed of several phases. In developing the thermodynamic analysis of heterogeneous equilibrium, a problem arises concerning the estimate of the possible number of equilibrium coexisting phases, and of the number of parameters of a given system which can have arbitrary variations without violating equilibrium.

Gibbs considered the general law to be satisfied by equilibrium systems comprising any number of phases and any number of components. The law he proved for the equilibrium of heterogeneous systems concerns the variance (i.e., the number of independent, intensive variables) and is known as the phase rule.

Let us assume that the system contains m substances and r phases at temperature T and pressure p. If there are l independent chemical reactions which can take place in the system, then the variance or number of degrees of freedom of the system is given by

$$f = (m - l) - r + 2 \text{ or } f = k - r + 2, \tag{2.73}$$

where k is the number of components.

Depending on the number of degrees of freedom, systems may be invariant ($f = 0$), monovariant ($f = 1$), divariant ($f = 2$), or polyvariant ($f \geq 3$). According to (2.73), the maximum number of coexisting phases is observed in an invariant system ($k + 2$). The maximum number of coexisting phases is 3 in a one-component system and 4 in a binary system. In an invariant system no changes are possible in the system without lowering the number of phases.

Equation (2.73) is the mathematical expression of the phase rule: The number of degrees of freedom of a heterogeneous system equals the number of components less the number of phases, plus two.

The well-known derivation of (2.73) assumes that all the components are present in every phase, i.e., that they are real. But if one lets one of the components (say, i) become virtual, so that it might be missing from one of the phases (e.g., α), this will not be reflected in the number of degrees of freedom. Indeed, in such a case one of the equations associated with (2.73) is missing, for example,

$$\mu_i^\alpha = \mu_i^\beta. \tag{2.74}$$

At the same time we must then introduce the condition for the absence of i in phase α, i.e.,

$$x_i^\alpha = 0. \tag{2.75}$$

For every condition such as (2.75), introduced for missing components, we shall lose one condition like (2.74). The total number of conditions is thus constant and the phase rule (2.73) is preserved.

The presence of the factor two in the phase rule is a consequence of the tacit assumption that the state of the system depends only on two parameters, the temperature and pressure. The phase rule may be generalized to cases where there are more than two parameters:

$$f = k - r + N,$$

where N is the number of parameters.

In many condensed systems pressure changes are insignificant and do not influence processes and transformations in the system. The phase rule in this case becomes

$$f = k - r + 1.$$

In some cases (for example, equality of the concentrations of some components in two phases), conditions arise which enable one to set up additional relations for the properties of the system. The presence of these additional relations among the system parameters results in a decrease in the variance of the system by the same amount as the increase in the number of equations characterizing the system.

The variance f determines the number of degrees of freedom of a system, not counting the masses of the phases. If the masses were included, the total number of independent variables would be given by the total variance F: $F = f + r$. Using this quantity enables one to describe completely all volume and/or surface phase phenomena.

The phase rule has been most extensively applied to various types of systems by Storonkin and his school (cf., e.g., [11, 15], etc.).

Following the method of Gibbs, let us consider the conditions for phase equilibrium. We assume that there are no intrinsic chemical reactions between the components in the case under study. Then all chemical transformations result in a transfer of components from one part of the system to another. The equilibrium conditions will accordingly refer only to the redistribution of components between the separate phases of the system. In order to derive these conditions, we shall assume that the system consists of phases containing k components. Let us consider all possible (virtual) phase changes subject to restrictions for which expression (2.69) is valid, namely, at constant total entropy, total volume, and total number of moles of the system.

We write these conditions in expanded form as follows:

$$
\begin{aligned}
\Phi_S &= S^\alpha + S^\beta + \ldots + S^r - C_S = 0, \\
\Phi_V &= V^\alpha + V^\beta + \ldots + V^r - C_V = 0; \\
\Phi_1 &= n_1^\alpha + n_1^\beta + \ldots + n_1^r - C_1 = 0, \\
&\qquad \cdot \cdot \cdot \cdot \cdot \cdot \cdot \cdot \cdot \cdot \cdot \cdot \cdot \cdot \cdot \cdot \cdot \\
\Phi_k &= n_k^\alpha + n_k^\beta + \ldots + n_k^r - C_k = 0,
\end{aligned}
\tag{2.76}
$$

where $C_S, C_V, C_1, \ldots, C_k$ are constants.

These relations impose restrictions on the parameter changes of the heterogeneous system.

According to the general conditions of (2.69), the equilibrium principle of Gibbs for a k-component, r-phase system is expressed as

$$\left(\sum_{j=1}^r \delta U^j \right)_{S, V, n_1, n_2, \ldots, n_k} \geq 0. \tag{2.77}$$

We shall use the method of Lagrange multipliers to find the conditions for which the internal energy of a system is at a conditional minimum. Let us construct the function

$$\Phi \equiv U + \lambda_S \Phi_S + \lambda_V \Phi_V + \lambda_1 \Phi_1 + \dots + \lambda_k \Phi_k, \tag{2.78}$$

where $\lambda_S, \lambda_V, \lambda_1, \dots, \lambda_k$ are constants. As we know, the function Φ has an unconditional minimum when the internal energy reaches a conditional minimum. Consequently, in a state of equilibrium the partial derivatives of the function Φ with respect to the state parameters of the system indicated in (2.76) are equal to zero.

Considering Eqs. (1.76), (1.77), and (2.4), we find from Eqs. (2.76) and (2.78)

$$\begin{aligned}
\partial \Phi / \partial S^j &= T^j + \lambda_S = 0, \\
\partial \Phi / \partial V^j &= -p^j + \lambda_V = 0, \\
\partial \Phi / \partial n_i^j &= \mu_i^j + \lambda_j = 0.
\end{aligned} \tag{2.79}$$

Solving the system of equations in (2.79) results in the following equilibrium conditions for a k-component, r-phase system:

$$\begin{aligned}
T^\alpha &= T^\beta = \dots = T^r, \\
p^\alpha &= p^\beta = \dots = p^r, \\
\mu_1^\alpha &= \mu_1^\beta = \dots = \mu_1^r, \\
&\quad\cdots\cdots\cdots\cdots \\
\mu_k^\alpha &= \mu_k^\beta = \dots = \mu_k^r.
\end{aligned} \tag{2.80}$$

The first two of these relations are the equations of thermal and mechanical equilibrium, whereas the rest characterize the conditions for the equilibrium distribution of each phase over the whole system. Then the chemical potential is the same in all phases if the component is uniformly distributed in the closed system.

The equilibrium conditions in Eq. (2.80) were derived with external fields either constant or absent. We also neglected the effect the surface separation between parts of the system has on its thermodynamic properties. The derivation also assumed that all components are present in all phases. This last assumption does not limit the equilibrium conditions of (2.80) in any way.

Let us now consider the phase equilibrium conditions for heterogeneous systems in light of the classification of components into real and virtual.

We imagine a system with an arbitrary number of phases at given temperature and pressure. Consider the process of transferring a quantity dn_1 of a real component from the α-phase to the β-phase. The total variation of the Gibbs free energy, greater than or equal to zero according to the equilibrium principle of (2.72), may be written as

$$(\delta G)_{p,\,T,\,n_1,\,n_2} = \left(\partial G / \partial n_1\right)_{p,T,\,n_1,\,n_2}^\alpha dn_1^\alpha + \left(\partial G / \partial n_1\right)_{p,T,\,n_1,n_2}^\beta dn_1^\beta \geqslant 0. \tag{2.81}$$

Alternatively, since by expression (2.16) the changes in the number of moles $dn_1{}^\alpha$ and $dn_1{}^\beta$ are equal and opposite, we have

$$\left(\mu_1^\beta - \mu_1^\alpha\right) dn_1 \geqslant 0. \tag{2.82}$$

Since by definition dn_1 is positive, we get from (2.82)

$$\mu_1^\beta - \mu_1^\alpha \geqslant 0. \tag{2.83}$$

Because the first component is real for the system as a whole, and for the β-phase in particular (i.e., it is present in this phase even if only in minute quantities), we may just as well consider the inverse process to that analyzed, consisting of the transfer of dn_1 moles from the β- into the α-phase. In that case the condition (2.81) leads to the relation

$$\mu_1^\alpha - \mu_1^\beta \geqslant 0. \tag{2.84}$$

Thus, if a component is present in both phases, the two simultaneous inequalities (2.83) and (2.84) force the conclusion that only the equality is satisfied; otherwise they are mutually exclusive. We see, therefore, that in thermodynamic equilibrium the chemical potential of any component is identical in all the phases in which this component is real. We have thus arrived at the equilibrium conditions stated in (2.80).

The foregoing considerations lead to a substantially different conclusion concerning virtual components. For example, if the first component is present only in the α-phase but is missing from the other phases, then one may consider the process of the transfer of this component from the α-phase to any other, which proceeds according to the relation (2.83). But the inverse process cannot now take place, since there is no trace of this component in any other phase. Relation (2.84) does not therefore apply, and we may not reject the inequality sign in (2.83).

Consequently, on the basis of thermodynamics we can only assert that the chemical potential of a virtual component, in any phase where it is present, is less than (or in the limiting case, equal to) the chemical potential of this component in other phases where it is not present under equilibrium conditions. In other words, if for any component the equality of chemical potentials in different phases is not maintained, then in thermodynamic equilibrium this component is concentrated in the phase (or phases) where its chemical potential is a minimum [16].

Thus the question of ignoring the existence of virtual components leads us outside the framework of thermodynamics and was, in fact, solved by us using molecular-kinetic considerations [17].

4. CONDITIONS FOR CHEMICAL EQUILIBRIUM. THE LAW OF MASS ACTION

When chemical reactions take place between the components of a system, the general state of its thermodynamic equilibrium will be specified as follows. Chemical equilibrium is a time-independent state of a system composed of components, capable of reacting chemically, and the reaction products. The equilibrium conditions, derived thermodynamically, will be affected by variations in the state parameters (such as temperature and pressure).

Following Gibbs we consider a system in which the next reaction takes place:

$$v_1 A_1 + v_2 A_2 \rightleftarrows v'_k A_k + v'_{k+1} A_{k+1}, \tag{2.85}$$

where v_i and v_i' are the stoichiometric coefficients of the initial substances and the reaction products, respectively, and A_i denotes the ith substance.

The stoichiometric equation for the reaction concerned is

$$\sum_i v_i A_i = 0. \tag{2.86}$$

By convention, the coefficients of the reaction products are given a positive sign, while for the components which are consumed v is taken as negative. In an r-phase, k-component system we may obtain a mixture of $(k-1)$ substances $A_1, A_2, ..., A_{k-1}$ since the first two enter the reversible reaction (2.85) and form two other substances.

Let $n_1{}^*, n_2, ..., n_{k-1}{}^*$ be the number of moles of the mixing substances, which are constants, since we are considering a closed system. After chemical equilibrium is established the system will consist of $(k+1)$ substances $A_1, A_2, ..., A_{k+1}$, with corresponding numbers of moles $n_1, n_2, ..., n_{k+1}$. For substances which do not participate in the reaction ($A_3, A_4, ..., A_{k-1}$) the quantities C_i and n_i are clearly identical.

We now consider the equation of constraint. In formulating such an equation for A_1 and A_2, we keep in mind that their equilibrium mole numbers n_1 and n_2 differ from their counterparts at the moment of mixing, n_1' and n_2', since part of A_1 and A_2 are expended in forming A_k and A_{k+1}, as shown by the chemical reaction equation (2.85). Hence the equations of constraint have the following form:

$$
\begin{aligned}
\Phi_S &= \sum_{j=1}^{r} S^j - C_S = 0, \\[4pt]
\Phi_V &= \sum_{j=1}^{r} V^j - C_V = 0, \\[4pt]
\Phi_1 &= \sum_{j=1}^{r} n_1^j + (v_1/v'_k) \sum_{j=1}^{r} n_k^j - C_1 = 0, \\[4pt]
\Phi_2 &= \sum_{j=1}^{r} n_2^j + (v_2/v'_k) \sum_{j=1}^{r} n_k^j - C_2 = 0, \\[4pt]
\Phi_3 &= \sum_{j=1}^{r} n_3^j - C_3 = 0, \\[4pt]
&\cdot \cdot \cdot \cdot \cdot \cdot \cdot \cdot \cdot \cdot \cdot \cdot \cdot \cdot \cdot \cdot \cdot \\[4pt]
\Phi_{k-1} &= \sum_{j=1}^{r} n_{k-1}^j - C_{k-1} = 0.
\end{aligned}
\tag{2.87}
$$

The first terms in Φ_1 and Φ_2 represent the nonreacting total number of moles of the substances A_1 and A_2, while the second terms describe the amount of A_1 and A_2 expended in forming A_k and A_{k+1}. In the absence of chemical reactions the system of equations (2.87) obviously reduces to (2.76). It follows that the equilibrium conditions of Eq. (2.80), although proved only for a system without chemical reactions, are in fact valid for systems with chemical reactions as well. But while in the former case these conditions are both necessary and sufficient, they are only necessary in the latter.

We may establish additional equilibrium conditions for a closed system with chemical reactions if we notice that the changes in the mole numbers of substances participating in the reversible reaction are interrelated according to Eq. (2.85):

$$\left|\delta n_1^j\right| : \left|\delta n_2^j\right| : \left|\delta n_k^j\right| : \left|\delta n_{k+1}^j\right| = v_1 : v_2 : v_k' : v_{k+1}', \tag{2.88}$$

where the parallel brackets denote absolute values.

Substituting Eq. (2.87) into Eq. (2.78), we differentiate the function Φ with respect to the number of moles (for example, for substance A_i in the jth phase) and equate the result to zero. After utilizing the relations in Eq. (2.88) we find, finally,

$$v_1\mu_1^j + v_2\mu_2^j = v_k'\mu_k^j + v_{k+1}'\mu_{k+1}^j. \tag{2.89}$$

Equation (2.89) expresses the condition for chemical equilibrium in the jth phase. A similar condition is found for all phases in which the chemical reaction takes place, since by Eq. (2.80) the chemical potentials of substances have identical values in all the phases.

In the general case the equilibrium condition for proper chemical reactions typified by (2.85) is given by

$$\sum_i v_i\mu_i = 0, \tag{2.90}$$

where the summation extends over all participants of the reaction. Equations (2.80) and (2.90) constitute the general conditions of heterogeneous equilibrium according to Gibbs.

The law of mass action is easily derived from Eq. (2.90) if the chemical potentials μ_i are expressed in terms of the concentrations or the partial pressures of the components. The first law of mass action was formulated by Guldberg and Waage (1867) on the basis of kinetic considerations for ideal gases. Such an approach is admissible for only the simplest reactions. The general law of mass action follows from thermodynamic considerations.

Suppose the chemical reaction (2.85) takes place in a homogeneous system containing ideal gases. Substituting the chemical potential from Eq. (2.52) into (2.90), we obtain

$$\sum_i v_i\mu_i^*(T) + \sum_i v_i RT \ln p_i = 0, \tag{2.91}$$

from which

$$\sum_i v_i \ln p_i = -\sum_i v_i\mu_i^*(T)/RT. \tag{2.92}$$

The left-hand side of (2.92) is some function of temperature only, $f(T)$. Replacing the sum of the logarithms of p_i by the logarithm of products of p_i, and writing $f(T)$ as a logarithm of some function $K_p^{id}(T)$, we may rewrite Eq. (2.92) in the form

$$\ln \prod_i \left(p_i^{v_i}\right) = \ln K_p^{id}(T), \tag{2.93}$$

where p_i are the component partial pressures at equilibrium.

The quantity $K_p{}^{id}(T)$ is a function of temperature only and does not depend on the total pressure and the partial pressures of the substances in the original mixture, i.e., on the relative amounts of original substances. We note that for equilibrium mixtures of real gases $K_p{}^{id}$ does not depend on the pressure. The quantity $K_p{}^{id}$ is called the equilibrium constant, and Eq. (2.93) is the law of mass action applied to ideal gases.

The equilibrium constant does not have an arbitrary value. Therefore, the standard state for the origin of the chemical potential is dependent on the nature of the reactants if the components can interact. Interest therefore attaches to another derivation for the law of mass action, which does not require standard states.

Following Guggenheim [18], we consider two sets of specific values of the partial pressures p_i' and p_i''. Then, independent of any choice of the standard state, we have for each component

$$\mu_i'' - \mu_i' = RT \ln (p_i''/p_i').$$ (2.94)

We now assume that the sets p_i' and p_i'' each correspond to states of chemical equilibrium at one and the same temperature. Then, according to the equation of chemical equilibrium (2.90), we may write for the most general reaction of the form (2.85)

$$\nu_1\mu_{A_1}' + \nu_2\mu_{A_2}' + \ldots = \nu_1'\mu_{A_1'}' + \nu_2'\mu_{A_2'}' + \ldots,$$

$$\nu_1\mu_{A_1}'' + \nu_2\mu_{A_2}'' + \ldots = \nu_1'\mu_{A_1'}'' + \nu_2'\mu_{A_2'}'' + \ldots.$$

Subtracting these equations and using (2.94) we obtain, after some transformations,

$$\left(p_{A_1'}''\right)^{\nu_1'} \left(p_{A_2'}''\right)^{\nu_2'} \ldots \left/ \left(p_{A_1}''\right)^{\nu_1} \left(p_{A_2}''\right)^{\nu_2} \ldots \right.$$

$$= \left(p_{A_1'}'\right)^{\nu_1'} \left(p_{A_2'}'\right)^{\nu_2'} \ldots \left/ \left(p_{A_1}'\right)^{\nu_1} \left(p_{A_2}'\right)^{\nu_2} \ldots \right.$$ (2.95)

Since a similar condition should be valid for any other set of partial pressures p_i at the same temperature, Eq. (2.95) is equivalent to expression (2.93).

The equilibrium constant is related to the partial pressures of all the substances taking part in the reaction. It is impossible to change the partial pressure of one of these substances without an accompanying change in the partial pressures of all the reactants which would restore the previous value of the equilibrium constant under the given conditions. This is in fact the content of the law of mass action.

If the partial pressures of the reactants are replaced by fugacities, then the law of mass action as formulated in Eq. (2.93) is also valid for real gases. Thus

$$\prod_i \left(\varphi_i^{\nu_i}\right) = \varphi_1'^{\nu_1'}\varphi_2'^{\nu_2'} \ldots \left/ \varphi_1^{\nu_1}\varphi_2^{\nu_2} \ldots = K_\varphi(T).\right.$$ (2.96)

Noting that $\mu_i*(T)$ is the same in both (2.52) and (2.57), we obtain

$$K_p^{id}(T) = K_\varphi(T). \tag{2.97}$$

The quantity $K_\varphi(T)$ does not depend on the pressure.

Without repeating a similar discussion we merely note that for real mixtures

$$\prod_i \left(a_i^{\nu_1}\right) = a_1^{'\nu_1'} a_2^{'\nu_2'} \ldots \bigg/ a_1^{\nu_1} a_2^{\nu_2} \ldots = K_a(T,p), \tag{2.98}$$

where

$$K_x^{id}(T,p) = K_a(T,p). \tag{2.99}$$

Formulas (2.93) and (2.96) may also be applied to situations in which, in addition to gases, solid and liquid substances also take part in the reactions (heterogeneous reactions). Then the partial pressures or fugacities of the vapors of the condensing substances will enter into the right-hand side of our formulas. If liquid or solid substances partake in a reaction without forming mixtures among themselves, then, at a given temperature, the partial pressures of these condensing reaction components will be constant depending only on temperature, and not on the partial pressures of other gases or vapors present in the system. Therefore, for convenience of calculation the vapor pressures of the condensing substances enter into the expressions of (2.93) and (2.96) via the equilibrium constant, which in this case will be determined solely by the partial pressures of the gaseous reactants.

The mechanism of a heterogeneous reaction (whether it proceeds, for example, in the gas phase or on the surface of phase separation) does not affect the derivation of the law of mass action in heterogeneous systems, because thermodynamics judges the equilibrium of a process only according to the initial and final states of a system.

5. CRITERIA OF PHASE STABILITY IN HETEROGENEOUS SYSTEMS

In states of equilibrium of thermodynamic systems the entropy and the thermodynamic potentials have extreme values. Thus for a system with constant values of the entropy, the volume and the number of moles for the components, the equilibrium condition is that the internal energy be a minimum. This is expressed by the relations $\delta U = 0$ and $\delta^2 U > 0$. To deduce the concrete equilibrium conditions, the first relation will suffice ($\delta U = 0$). But the vanishing of the first variation is only a necessary condition for an extremum and does not ensure that the internal energy is a minimum. A sufficient condition is provided by positiveness of the second variation, which also guarantees the stability of equilibrium.

Thermodynamic inequalities expressing stability conditions provide, together with the equilibrium principle, the basis for the chemical thermodynamics of heterogeneous systems. We shall present the Gibbsian deduction of the stability criteria, as systematized in the works of Storonkin [13].

5.1. Criteria for the Stability of a Given Phase with Respect to the Formation of New Phases from Within

The internal energy of a heterogeneous system containing, besides r old phases, also s new ones, is given by

$$U = \sum_{i=1}^{r} U^i + \sum_{j=1}^{s} U^j. \qquad (2.100)$$

Here and below i and j denote the old and new phases, respectively.

We shall assume that the system is prepared from a mixture of $C_1, C_2, ..., C_{k-1}$ moles of substances $A_1, A_2, ..., A_{k-1}$, and that a reversible reaction as depicted in (2.85) is taking place. The equilibrium principle will assume the form

$$\left[\sum_{i=1}^{r} \delta U^i + \sum_{j=1}^{s} \delta U^j \right]_{S, V, C_1, ..., C_{k-1}} \geqslant 0. \qquad (2.101)$$

The equations of constraint, which express the condition that the internal energy be at a conditional minimum, are written as

$$\Phi_S = \sum_{i=1}^{r} S^i + \sum_{j=1}^{s} S^j - C_S = 0, \qquad (2.102)$$

$$\Phi_V = \sum_{i=1}^{r} V^i + \sum_{j=1}^{s} V^j - C_V = 0; \qquad (2.103)$$

$$\left.\begin{aligned}
\Phi_1 &= \sum_{i=1}^{r} n_1^i + \sum_{j=1}^{s} n_1^j + (\nu_1/\nu_k') \left(\sum_{i=1}^{r} n_k^i + \sum_{j=1}^{s} n_k^j \right) - C_1 = 0, \\
\Phi_2 &= \sum_{i=1}^{r} n_2^i + \sum_{j=1}^{s} n_2^j + (\nu_2/\nu_k') \left(\sum_{i=1}^{r} n_k^i + \sum_{j=1}^{s} n_k^j \right) - C_2 = 0, \\
\Phi_3 &= \sum_{i=1}^{r} n_3^i + \sum_{j=1}^{s} n_3^j - C_3 = 0, \\
&\cdots \cdots \cdots \cdots \cdots \cdots \cdots \cdots \cdots \cdots \cdots \cdots \cdots \\
\Phi_{k-1} &= \sum_{i=1}^{r} n_{k-1}^i + \sum_{j=1}^{s} n_{k-1}^j - C_{k-1} = 0.
\end{aligned}\right\} \qquad (2.104)$$

Using expressions (2.12) and (2.80), we may also represent (2.101) in the following form:

$$\left[\sum_{j=1}^{s} \delta U_-^j + T \sum_{i=1}^{r} \delta S^i - p \sum_{i=1}^{r} \delta V^i + \mu_1 \sum_{i=1}^{r} \delta n_1^i + \dots + \mu_{k+1} \sum_{i=1}^{r} \delta n_{k+1}^i \right]_{S, V, C_1, \dots, C_{k-1}} > 0$$

$$(2.105)$$

According to (2.102)–(2.104), the following relations hold:

$$\sum_{i=1}^{r} \delta S^i = - \sum_{j=1}^{s} \delta S^j, \qquad (2.106)$$

$$\sum_{i=1}^{r} \delta V^i = - \sum_{j=1}^{s} \delta V^j; \qquad (2.107)$$

$$\left.\begin{aligned}
\sum_{i=1}^{r} \delta n_1^i &= - \sum_{j=1}^{s} \delta n_1^j - (v_1/v_k') \left(\sum_{i=1}^{r} \delta n_k^i + \sum_{j=1}^{s} \delta n_k^j \right), \\
\sum_{i=1}^{r} \delta n_2^i &= - \sum_{j=1}^{s} \delta n_2^j - (v_2/v_k') \left(\sum_{i=1}^{r} \delta n_k^i + \sum_{j=1}^{s} \delta n_k^j \right); \\
\sum_{i=1}^{r} \delta n_3^i &= - \sum_{j=1}^{s} \delta n_3^j, \\
& \cdots \cdots \cdots \cdots \cdots \cdots \cdots \cdots \\
\sum_{j=1}^{r} \delta n_{k-1}^i &= - \sum_{j=1}^{s} \delta n_{k-1}^j.
\end{aligned}\right\} \qquad (2.108)$$

On the basis of Eqs. (2.88), (2.89) as well as (2.106)–(2.108), we may transform (2.105) into the following:

$$\left[\sum_{i=1}^{s} (\delta U^j - T\delta S^j + p\delta V^j - \mu_1 \delta n_1^j - \dots \right.$$

$$\left. - \mu_{k+1} \delta n_{k+1}^j) \right]_{S, V, C_1, \dots, C_{k-1}} \geqslant 0. \qquad (2.109)$$

Bearing in mind that the changes in the extensive quantities U^j, S^j, V^j, $n_1{}^j$, ..., $n_{k+1}{}^j$ pertaining to a given new phase do not depend on the changes of the same quantities for any other phase, we can write, using (2.109),

$$\delta U^j - T\delta S^j + p\delta V^j - \mu_1 \delta n_1^j - \dots - \mu_{k+1} \delta n_{k+1}^j \geqslant 0. \qquad (2.110)$$

The inequality case of (2.110) provides an additional equilibrium condition with respect to the formation of a new phase. If at the moment of their formation the new phases are taken to be homogeneous, then by summing the left-hand side of (2.110) for the new phases over a macroscopic region we find

$$U^j - TS^j + pV^j - \mu_1 n_1^j - \dots - \mu_{k+1} n_{k+1}^j \geqslant 0. \qquad (2.111)$$

This relation is the criterion for the stability of a phase with respect to the generation of new phases within and has great practical utility. It may easily be transformed into the form

$$(U^j + U^i) - U_0^i > 0,   \qquad (2.112)$$

in which the left-hand side is the internal energy associated with the formation of a new macroscopic phase in the state U^j plus the change in the state of the old phase from U_0^i to U^i.

Equation (2.111) does not include the contribution from the work required to form the surface of separation between the old and new phases, so that it is only a sufficient criterion of phase stability with respect to finite changes of state.

If condition (2.111) applies, and the surface tension on the boundary of separation between the old and new phases is positive, namely, if work has to be expended in forming the dispersing phase so that the stricter condition (2.110) is sure to apply, then the phase in question is stable with respect to the formation of new phases (whether dispersive or nondispersive). If the surface tension is negative, then the macroscopic phase will be unstable, and the use of expression (2.111) becomes meaningless.

However, the stability criterion (2.111) does not apply to metastable phases (supersaturated solutions, supercooled melts, etc.). These are unstable with respect to the formation of any macroscopic phase from within, when surface phenomena can be disregarded. Hence, the criterion is not necessary. If one introduces into the metastable phase nuclei of a new, more stable phase, then the process of their growth to macroscopic dimensions proceeds spontaneously, and the work of their formation will be negative.

Consequently, a sufficient criterion for phase stability with respect to the formation of new phases from within may be formulated as follows: a phase is stable with respect to finite (discontinuous) changes of state if the work required for forming a new phase of macroscopic dimensions is positive [13].

This criterion allows one to distinguish between stable and metastable states. Stable and metastable states are stable with respect to the formation of a dispersive phase. Therefore, condition (2.110) is valid for stable and metastable states, and so it is a necessary and sufficient condition for the stability of a phase with respect to discontinuous (finite) changes of state.

5.2 Criterion of Phase Stability with Respect to Infinitesimal Changes of State (Fluctuational Processes)

Phases which are unstable with respect to infinitesimal changes are absolutely unstable and consequently cannot exist. Therefore, the criterion of phase stability with respect to infinitesimal changes enables one to discriminate between realizable and unrealizable states of dynamic equilibrium.

On the basis of the integral criterion of stability (2.111) any deviations from the mean phase equilibrium should satisfy the inequality

$$\Delta T \, \Delta S - \Delta p \, \Delta V + \Delta \mu_1 \, \Delta n_1 + \ldots + \Delta \mu_k \, \Delta n_k > 0, \qquad (2.113)$$

where Δ denotes the exact change of the corresponding quantity in going from a particular state to a neighboring one (infinitely close to it).

The Gibbs inequality (2.113) is a necessary and sufficient criterion of stability for two infinitely close states of phases with respect to each other.

By means of the inequality (2.113) one may establish a qualitative relation between the changes of conjugate thermodynamic parameters. The left-hand side of (2.113) contains $(k + 2)$ terms. We obtain a special stability condition by fixing the values of $(k + 1)$ parameters in each of the terms besides the one under consideration, making sure that at least one of the $(k + 1)$ fixed parameters is extensive.

The fulfillment of the first condition of fixing is necessary to ensure that the constancy of the $(k + 1)$ parameters in the inequality (2.113) gives only one nonzero term, while the second condition eliminates from consideration cases when, on account of the constancy of $(k + 1)$ intensive parameters, changes in the states of the phase would be impossible, so that (2.113) would become meaningless. We thereby include the mass of the phase among the variables.

Fixing the values of $(k + 1)$ parameters in accordance with the foregoing conditions, we obtain from (2.113)

$$(\Delta T / \Delta S)_a > 0, \tag{2.114}$$

$$(\Delta p / \Delta V)_a < 0, \tag{2.115}$$

$$(\Delta \mu_i / \Delta n_i)_a > 0, \tag{2.116}$$

where the index a indicates the method of fixing parameters determined by the nature of the process changing the state of the phase.

The inequalities (2.114)–(2.116) represent necessary and sufficient conditions for the thermal, mechanical, and chemical stability of phases with respect to infinitesimal changes of state. According to these inequalities, changes in the temperature and entropy, as well as in the chemical potentials and their associated mole numbers, have the same sign, whereas pressure and volume changes are always opposite in sign.

As is known, if the derivative of any intensive parameter with respect to the corresponding extensive quantity is positive, then the ratio of the total increments is also positive. The latter, however, may also be positive when the derivative vanishes. Hence, a sufficient criterion for the stability of phases with respect to infinitesimal changes of state may be put into the form

$$dT \, dS - dp \, dV + d\mu_1 \, dn_1 + \ldots + d\mu_k \, dn_k > 0, \tag{2.117}$$

$$(dT / dS)_a > 0, \tag{2.118}$$

$$(dp / dV)_a < 0, \tag{2.119}$$

$$(d\mu_i / dn_i)_a > 0. \tag{2.120}$$

The procedure for fixing parameters is similar to that employed to obtain equalities (2.114)–(2.116). If we introduce the equality sign in the inequalities of (2.117)–(2.120), we arrive at the necessary conditions of stability. At the stability limit one of the derivatives, (2.118) or (2.119), approaches zero, whereas their inverses do not vanish. The sufficient condition for stability in (2.113) may be written in a different form which is useful in solving certain problems:

$$\delta^2 U > 0. \tag{2.121}$$

We can set up a general form of the sufficiency condition (2.120) in terms of second derivatives of the thermodynamic potential by utilizing the systematic method applied in the work of Münster [19]. To this end we introduce generalized coordinates X and Y in the inequality (2.113) and write it in the form

$$\sum_{i=1}^{m} \Delta Y_i \, \Delta X_i > 0. \tag{2.122}$$

The internal energy as a function of the generalized coordinates Y is of the form

$$U = U(Y_1, Y_2, \ldots, Y_m), \tag{2.123}$$

and its differential is

$$dU = \sum_{i=1}^{m} (\partial U/\partial Y_i)_{j \neq i} \, dY_i = \sum_{i=1}^{m} X_i \, dY_i. \tag{2.124}$$

Using the appropriate Legendre transformation in (2.124), we obtain the general equation for the thermodynamic potentials:

$$d\Psi_k = dU - \sum_{i=1}^{k} d(X_i Y_i) = \sum_{i=1}^{m} X_i \, dY_i - \sum_{i=1}^{k} X_i \, dY_i - \sum_{i=1}^{k} Y_i \, dX_i$$

$$= -\sum_{i=1}^{k} Y_i \, dX_i + \sum_{j=k+1}^{m} X_j \, dY_j, \tag{2.125}$$

or in integral form:

$$\Psi_k = \Psi_k(X_i, Y_j), \tag{2.126}$$

where $i = 1, 2, \ldots, k; j = k + 1, \ldots, m$.

Let us now express the inequality (2.122) in terms of the second derivatives of the thermodynamic potentials Ψ_k. To do this, one should represent ΔY_s ($s = 1, 2, \ldots, k$) and ΔX_t ($t = k + 1, \ldots, m$) as functions of X_i and Y_j. Since in (2.122) we confined ourselves to second-order terms, we obtain the system of linear equations:

$$\left. \begin{array}{l} \Delta Y_s = \displaystyle\sum_{i=1}^{k} (\partial Y_s/\partial X_i) \, \Delta X_i + \sum_{j=k+1}^{m} (\partial Y_s/\partial Y_j) \, \Delta Y_j, \\[2mm] \Delta X_t = \displaystyle\sum_{i=1}^{k} (\partial X_t/\partial X_i) \, \Delta X_i + \sum_{j=k+1}^{m} (\partial X_t/\partial Y_j) \Delta Y_j. \end{array} \right\} \tag{2.127}$$

From expression (2.125) we have

$$\left. \begin{array}{l} \partial \Psi_k/\partial X_i = -Y_i, \\[1mm] \partial \Psi_k/\partial Y_j = X_j, \end{array} \right\} \tag{2.128}$$

so that

$$\partial Y_s/\partial X_i = - \partial^2 \Psi_k/\partial X_s \, \partial X_i, \tag{2.129}$$

$$\partial X_t/\partial Y_j = \partial^2 \Psi_k/\partial Y_t \, \partial Y_j. \tag{2.130}$$

Equation (2.125) also gives rise to the following Maxwell relation:

$$\partial^2 \Psi/\partial X_i \partial Y_j = - (\partial Y_s/\partial Y_j) = (\partial X_t/\partial X_i). \tag{2.131}$$

Substituting (2.127) into (2.122) and using the relations (2.129)–(2.131), we get

$$\sum_{i=1}^{k} \Delta Y_s \, \Delta X_i + \sum_{j=k+1}^{m} \Delta X_t \, \Delta Y_j = \sum_{i=1}^{k} \sum_{i=1}^{k} (\partial Y_s/\partial X_i) \, \Delta X_i^2$$
$$+ \sum_{i=1}^{k} \sum_{j=k+1}^{m} (\partial Y_s/\partial Y_j) \, \Delta Y_j \, \Delta X_i + \sum_{j=k+1}^{m} \sum_{i=1}^{k} (\partial X_t/\partial X_i) \, \Delta X_i \, \Delta Y_j$$
$$+ \sum_{j=k+1}^{m} \sum_{j=k+1}^{m} (\partial X_t/\partial Y_j) \, \Delta Y_j^2 \equiv [\delta^2 \Psi_k (Y_j)]_{X_i} - [\delta^2 \Psi_k (X_i)]_{Y_j} > 0. \tag{2.132}$$

The original expression is therefore decomposed into two quadratic forms, the first of which depends only on extensive and the second only on intensive parameters.

The inequality (2.132) is generally fulfilled when the first quadratic form is positive while the second is negative. If we exclude variations which only change the system's mass and take Y_m = const, we arrive at the sufficient condition for stability with respect to infinitesimal (continuous) changes of state in a form valid for all thermodynamic potentials:

$$|\delta^2 \Psi_k (X_1, \ldots, X_k)]_{Y_j} < 0, \tag{2.133}$$

$$[\delta^2 \Psi_k (Y_{k+1}, \ldots, Y_{m-1})]_{Y_m, \, X_i} > 0. \tag{2.134}$$

Bearing in mind the signs of the quadratic forms, an equivalent and practically more important formulation may be stated in the following form:

for odd-order principal minors:

$$|\partial^2 \Psi_k/\partial X_i \, \partial X_s| < 0, \tag{2.135}$$

for even-order principal minors:

$$|\partial^2 \Psi_k/\partial X_i \, \partial X_s| > 0, \tag{2.136}$$

for all principal minors:

$$|\partial^2 \Psi_k/\partial Y_j \, \partial Y_t| > 0. \tag{2.137}$$

Schottky [20] was the first to derive the conditions in the form (2.133) and (2.134).

When the quantities T, p, and $\Sigma_i n_i$ are constant, we obtain an expression for the Gibbs free energy which is equivalent to (2.134):

$$[\delta^2 g (x_1, \ldots, x_{k-1})]_{p, T} > 0 , \tag{2.138}$$

or, for all principal minors:

$$\sum_{i=1}^{k-1} \sum_{j=1}^{k-1} \left| \partial^2 g / \partial x_i \, \partial x_j \right| > 0, \quad (i, j = 1, 2, \ldots, k-1). \tag{2.139}$$

For a binary system (2.139) reduces to the condition

$$\left(\partial^2 g / \partial x_i^2 \right)_{p, T} > 0. \tag{2.140}$$

On account of (2.39) and (2.43), the stability condition may be reexpressed as

$$(\partial \mu_2 / \partial x_2)_{p, T} > 0. \tag{2.141}$$

In general the stability condition demands that the Gibbs free energy surface be convex with respect to the coordinate axes of concentrations. The usual formulations of the stability condition are at the same time restrictions on the magnitudes of the thermodynamic potentials [cf., for example, the two inequalities (2.121) and (2.139)]. Restrictions on the partial molar thermodynamic functions of mixtures have also been deduced from the stability condition in the form of differential relations [21]. On the basis of the rigorous relations [23] for the activities and activity coefficients of the components resulting from the condition of stability with respect to continuous changes of state, Toikka and Susarev [22] obtained, without further assumptions, bounds in integral form on the free energy of formation of solutions.

5.3. Necessary Criteria for the Stability of Heterogeneous Systems with Respect to Continuous Changes of State

So far we have discussed stability criteria for uniform systems and for separate phases of heterogeneous systems. We now show that a heterogeneous system as a whole is in stable equilibrium if the stability criterion with respect to continuous changes of state applies to each constituent phase. We restrict consideration to a one-component system containing two phases α and β of macroscopic dimensions. We assume that in an infinitesimal change of state the number of moles in both phases n^α and n^β remains unchanged, i.e.,

$$n^j = \text{const} \ (j = \alpha, \beta). \tag{2.142}$$

We consider possible changes in ΔU subject to the following additional conditions:

$$\left.
\begin{aligned}
\Phi_S &= n^\alpha s^\alpha + n^\beta s^\beta - C_S = 0, \\
\Phi_V &= n^\alpha v^\alpha + n^\beta v^\beta - C_V = 0, \\
\Phi_n &= n^\alpha + n^\beta - C_n = 0 .
\end{aligned}
\right\} \tag{2.143}$$

Here s and v are molar entropies and volume defined by

$$s = S/n; \quad v = V/n. \tag{2.144}$$

Combining Eqs. (2.143) and (2.142), we find

$$\left.\begin{array}{l} n^\alpha \delta s^\alpha + n^\beta \delta s^\beta = 0, \\ n^\alpha \delta v^\alpha + n^\beta \delta v^\beta = 0, \\ \delta n^\alpha = \delta n^\beta = 0. \end{array}\right\} \tag{2.145}$$

In equilibrium the first-order variation must vanish for all possible perturbations. As a result we obtain the stability condition in the following form:

$$\delta^2 U = \frac{1}{2}\left[\frac{\partial^2 U}{\partial s^{(\alpha)2}}(\delta s^\alpha)^2 + 2\frac{\partial^2 U}{\partial s^\alpha \partial v^\alpha}\delta s^\alpha \delta v^\alpha + \frac{\partial^2 U}{\partial v^{(\alpha)2}}(\delta v^\alpha)^2 \right] > 0. \tag{2.146}$$

Using the relation $U = n^\alpha u^\alpha + n^\beta u^\beta$ (where u is the molar internal energy), together with (2.145), the inequality in (2.146) is written in expanded form as follows:

$$\delta^2 U = \frac{1}{2}n^{(\alpha)2}\left[\left(\frac{1}{n^\alpha}\frac{\partial^2 u^\alpha}{\partial s^{(\alpha)2}} + \frac{1}{n^\beta}\frac{\partial^2 u^\beta}{\partial s^{\beta2}} \right)(\delta s^\alpha)^2 \right.$$

$$+ 2\left(\frac{1}{n^\alpha}\frac{\partial^2 u^\alpha}{\partial s^\alpha \partial v^\alpha} + \frac{1}{n^\beta}\frac{\partial^2 u^\beta}{\partial s^\beta \partial v^\beta} \right)\delta s^\alpha \delta v^\alpha$$

$$\left. + \left(\frac{1}{n^\alpha}\frac{\partial^2 u^\alpha}{\partial v^{(\alpha)2}} + \frac{1}{n^\beta}\frac{\partial^2 u^\beta}{\partial v^{(\beta)2}} \right)(\delta v^\alpha)^2 \right] > 0. \tag{2.147}$$

Because this condition must be valid for any value of n^α and n^β, we may set $n^\alpha = n^\beta = 1$ and obtain the stability criterion in explicit form:

$$\partial^2 u^\alpha/\partial s^{(\alpha)2} + \partial^2 u^\beta/\partial s^{(\beta)2} > 0, \tag{2.148}$$

$$\partial^2 u^\alpha/\partial v^{(\alpha)2} + \partial^2 u^\beta/\partial v^{(\beta)2} > 0, \tag{2.149}$$

$$\left(\frac{\partial^2 u^\alpha}{\partial s^{(\alpha)2}} + \frac{\partial^2 u^\beta}{\partial s^{(\beta)2}} \right)\left(\frac{\partial^2 u^\alpha}{\partial v^{(\alpha)2}} + \frac{\partial^2 u^\beta}{\partial v^{(\beta)2}} \right) - \left(\frac{\partial^2 u^\alpha}{\partial s^\alpha \partial v^\alpha} + \frac{\partial^2 u^\beta}{\partial s^\beta \partial v^\beta} \right)^2 > 0. \tag{2.150}$$

The condition for the stability of a system as a whole is the stability of its separate phases. Consequently, the sufficient criterion of phase stability with respect to infinitesimal changes of state implies the satisfaction of the following inequalities:

$$\left.\begin{array}{ll} \partial^2 u^\alpha/\partial s^{(\alpha)2} > 0, & \partial^2 u^\beta/\partial s^{(\beta)2} > 0, \\ \partial^2 u^\alpha/\partial v^{(\alpha)2} > 0, & \partial^2 u^\beta/\partial v^{(\beta)2} > 0; \end{array}\right\} \tag{2.151}$$

$$\frac{\partial^2 u^\alpha}{\partial s^{(\alpha)2}} \frac{\partial^2 u^\alpha}{\partial v^{(\alpha)2}} - \left(\frac{\partial^2 u^\alpha}{\partial s^\alpha \partial v^\alpha}\right)^2 > 0,$$
$$\frac{\partial^2 u^\beta}{\partial s^{(\beta)2}} \frac{\partial^2 u^\beta}{\partial v^{(\beta)2}} - \left(\frac{\partial^2 u^\beta}{\partial s^\beta \partial v^\beta}\right)^2 > 0. \qquad (2.152)$$

It is obvious that (2.148) and (2.149) follow from (2.151), while (2.151) and (2.152) also give rise to the inequality in (2.150).

In the theory of heterogeneous systems it is of great interest to discuss the relation between the varying conjugate intensive and extensive parameters referring to the heterogeneous system as a whole, rather than to the separate phases. As shown by Storonkin [13], this relation is a consequence of the necessary condition of stability for heterogeneous systems:

$$dT\, dS^g - dp\, dV^g + d\mu_1\, dn_1^g + \cdots + d\mu_k\, dn_k^g > 0. \qquad (2.153)$$

This inequality imposes a restriction on the variation of intensive parameters (T, p, μ_1, ..., μ_k) referring, due to the phase stability condition, both to separate phases and to the heterogeneous system as a whole, and on the extensive parameters (S^g, V^g, n_1^g, ..., n_k^g), referring only to the heterogeneous system as a whole.

According to condition (2.153) the following inequality is valid for conjugate thermodynamic parameters (except the pressure and the volume):

$$(\partial X_i / \partial Y_i^g)_a > 0, \qquad (2.154)$$

where Y_i^g is an extensive property of the heterogeneous system.

For the pressure and volume the inequality is

$$(\partial p / \partial V^g)_a < 0. \qquad (2.155)$$

We see from expressions (2.154) and (2.155) that changes always proceed in the same direction for all conjugate intensive and extensive parameters of a heterogeneous system, except for pressure and volume, for which they are in the opposite direction. This statement is valid if the system is in stable equilibrium, if the processes occurring in the system cause changes in the states of phases, and, finally, if in each pair of conjugate parameters (save one that is changing) one of the parameters is a constant.

The inequalities of (2.153)–(2.155) are necessary but not sufficient conditions of stability with respect to infinitesimal changes of state, since one may imagine situations in which these inequalities are satisfied yet the state of a heterogeneous system is unstable. On the other hand, if the heterogeneous system is in a state of stable equilibrium, and if the processes occurring in it result in changes of phase states, then condition (2.153) and its consequences (2.154) and (2.155) are necessarily satisfied.

A special but important case of condition (2.154) is the inequality

$$(\partial \mu_i / \partial n_i^g)_{T, p, n_1^g, ..., n_{i-1}^g, n_{i+1}^g, ..., n_k^g} > 0, \qquad (2.156)$$

which is valid for a heterogeneous system in which the number of components is larger than the number of phases ($k > r$). This inequality may be expressed in terms of mole fractions as

$$\left(\partial\mu_i/\partial x_i^g\right)_{T,\ p,\ x_1^g/x_k^g,\ \ldots,\ x_{i-1}^g/x_k^g,\ x_{i+1}^g/x_k^g,\ \ldots,\ x_{k-1}^g/x_k^g} > 0. \qquad (2.157)$$

It is seen from (2.157) that the chemical potential and the total mole fraction of any component vary in the same direction when temperature, pressure, and the other mole fractions are kept constant. Physically the conditions imposed in (2.157) are of greater significance than those in (2.156), since the condition of constant number of moles is considerably less restrictive than the condition of constant mole fractions of the components. Obviously, if the inequality (2.157) applies, then, as a consequence, the analogous inequality of (2.156) is also satisfied.

In conclusion we remark that calculations involving the equation of the curve of monovariant equilibrium usually require the adoption of some model assumption (absence of solubility, ideal, or regular solutions) which will inevitably reduce the usefulness of the results. The application of the stability conditions to the analysis of the equation of phase equilibria enables one to exhibit, in some cases without further assumptions, the regime of possible values of the state parameters (see, for example, the work of Toikka and Susarev [23], in which the stability condition was applied to delineate the concentration distribution regions of a eutectic and the points of minimum crystallization temperatures in solid–liquid systems).

6. THE PRINCIPLE OF EQUILIBRIUM DISPLACEMENT (THE GIBBS–LE CHATELIER PRINCIPLE)

One of the cardinal questions in chemical thermodynamics is the direction in which processes proceed when a system is under external influences, since on the basis of this direction one can determine which variant of this or that process is most likely to occur.

The simplest formulation of the principle of equilibrium displacement is that given by Le Chatelier (1888). This says, in essence, that any system in equilibrium reacts to an external effect in such a manner as to reduce the effect. Such a formulation actually encompasses the stability principle. In the simplest cases the Le Chatelier principle is equivalent to the stability condition and practically duplicates the principle of least action already formulated in the works of d'Alembert and Gauss. However, the inexact and limited nature of this formulation for the purposes of chemical thermodynamics can often lead to completely incorrect conclusions.

Braun (1887) tried to give a more precise formulation for the principle of Le Chatelier on the basis of the equilibrium stability criterion but, practically speaking, did not achieve any essential improvement.

We should point out that the completely general formulation of this principle is extremely complex. For this reason Prigogine and Defay [24] elected to discuss this problem more logically in terms of an appropriate moderation principle, which can be deduced from considerations based on de Donder's concept [25] of chemical affinity, and which provided unambiguous answers. In this sense, de Donder's inequality ($A'\nu > 0$) represents the most general formulation of Le Chatelier's principle [24].

The analysis of processes which served as the basis for proving the Le Chatelier–Braun principle led Epstein [26] to the conclusion that they are related to two completely different types of equilibrium displacement processes, and that the principle should accordingly be expanded into at least two different independent rules. One group of processes is associated with transformations and transfers of substances (chemical and phase changes), the second group with the so-called secondary forces.

A detailed analysis of both types of processes is contained in Storonkin's monograph [13]. A systematic consideration of the displacements of chemical and phase processes leads to the following conclusions.

The mathematical principle of the displacement of chemical equilibrium may be represented in the most general form by the following inequality:

$$(\partial \ln K/\partial X_i)_{X_{l \neq i}} \gtrless 0 \quad \text{if} \quad \Delta Y_i^\circ \gtrless 0, \tag{2.158}$$

where K is the equilibrium constant for a reaction of the type (2.85).

At constant values of the pressure and of the intensity of external (electric and magnetic) force fields, we have

$$(\partial \ln K/\partial T)_{p, \, x_t} \gtrless 0 \quad \text{if} \quad \Delta H^\circ \gtrless 0, \tag{2.159}$$

where ΔH° is the standard heat of reaction at constant T and p.

The inequalities (2.158) and (2.159) are valid for both homogeneous and inhomogeneous chemical reactions.

The content of the inequality (2.158) is that a change in any generalized force X_i (except $X_i = T$), while the other forces are fixed, causes a shift in chemical equilibrium in such a direction that the change in the conjugate generalized coordinate ΔY_i° will oppose that in the generalized force. Inequality (2.159) stipulates that if the temperature is changed while the other generalized forces are kept constant, the chemical reaction is always shifted in such a direction that the reaction-generated thermal effect will oppose the original temperature change (van't Hoff's theorem). Bearing in mind the contradictory nature of pressure and volume changes, the pressure as a generalized force should be accorded a minus sign ($X_i = -p$). For the particular case of pressure, inequality (2.158) assumes the form

$$(\partial \ln K/\partial p)_{T, \, x_t} \gtrless 0 \quad \text{if} \quad \Delta V^\circ \lessgtr 0. \tag{2.160}$$

This says that when the pressure is increased the chemical equilibrium is shifted toward a volume decrease, and vice versa.

In phase equilibria the equilibrium constant K may be taken as the coefficient of distribution of the ith substance among the phases. In this case we find an inequality which characterizes the equilibrium distribution of the separate substances among the coexisting phases under the action of external forces:

$$(\partial \ln K_i/\partial X_j)_{X_{k \neq j}} \gtrless 0 \quad \text{if} \quad \Delta \bar{y}_{ji}^\circ \gtrless 0, \tag{2.161}$$

$$(\partial \ln K_i/\partial T)_{p, \, x_k} \gtrless 0 \quad \text{if} \quad \Delta \bar{h}_i^\bullet \gtrless 0, \tag{2.162}$$

where $\Delta \bar{y}_{ji}{}^0$ are the standard partial molar quantities.

In their physical interpretation the inequalities (2.161) and (2.162) for phase equilibria are completely analogous to (2.158) and (2.159) for chemical equilibria. It is therefore possible to give a general formulation of the principle of equilibrium displacement associated with the transfer and transformation of substances resulting from chemical or phase interactions. Changing a generalized force X_i while keeping the other constant gives rise to a chemical or phase equilibrium displacement in such a direction that the conjugate generalized coordinate Y_i experiences a change opposing the variation of the generalized force [13]. The principle of displacement thus formulated is valid for intrinsic dynamic equilibrium but does not apply to so-called "quenched" or pseudoequilibrium. In addition, it is valid only under the action of any one force while all the others are fixed, and when one takes into account the possible changes of sign of thermal, volume, and other effects accompanying the relevant processes.

It was Gibbs [1] who first considered processes occurring in the presence of so-called secondary forces, namely forces induced under the action of the primary forces coupling directly into the system, and related them to the stability conditions with respect to continuous changes of state. Following the logic of Gibbs' exposition [1] and expressing the masses of the components as defined in this work in terms of the number of moles, we obtain the following inequality:

$$(\partial \mu_k / \partial n_k)_{H, V, \mu_1, \ldots, \mu_{k-1}} \geqslant (\partial \mu_k / \partial n_k)_{T, V, \mu_1, \ldots, \mu_{k-1}} \geqslant 0. \qquad (2.163)$$

This is an expression of the so-called "reduced" Le Chatelier–Braun principle involving the parameters μ_k and n_k for homogeneous processes occurring in the presence of secondary forces [21].

In the literature the formulation of the Le Chatelier–Braun principle is often based on the result of Ehrenfest [27], according to which external forces driving a system from equilibrium stimulate within it processes tending to attenuate the result of this action. Emphasizing the inadequacy of this formulation, in that it makes no reference to any condition for the occurrence of the equilibrium displacement process, Storonkin noted [13] that it loses the generality of the Gibbs formulation expressed in (2.163).

The role of Gibbs in developing the displacement principle was well known to Le Chatelier himself, who made the astute remark that the disparity between the various individual approaches and the general method of the Gibbs solution is akin to that between the solutions of the extremum problem by the ancient Greeks and by differential calculus. Yet, while in the former case the general method was created before the emergence of particular solutions, in the latter case it was worked out by generalizing particular solutions (cf. [29]).

From this observation of Le Chatelier there follows not just the fact of his awareness of the role of Gibbs in solving the problem under discussion, but a clear understanding of the relative generality of his and Gibbs' formulations.

These observations, plus the fact that the more general formulation of Gibbs appeared at least ten years before the work of Le Chatelier, justify calling the principle of equilibrium displacement the Gibbs–Le Chatelier principle.

A systematic analysis of the Gibbs inequality enabled Storonkin [13] and Rusanov and Schultz [28] to obtain the following generalized expression for the principle of equilibrium displacements for homogeneous processes in the presence of secondary forces:

$$(\partial Y_i/\partial X_i)_{a_1} \geqslant (\partial Y_i/\partial X_i)_{a_2} \geqslant \cdots \geqslant (\partial Y_i/\partial X_i)_{a_{t-1}} \geqslant (\partial Y_i/\partial X_i)_{a_t}, \qquad (2.164)$$

where

$$a_1 = Y_1, \ X_2, \ \ldots, \ X_{i-1}, \ X_{i+1}, \ \ldots, \ X_t,$$
$$a_2 = Y_1, \ Y_2, \ X_3, \ \ldots, \ X_{i-1}, \ X_{i+1}, \ \ldots, \ X_t; \ a_{i-1} = Y_1, \ \ldots, \ Y_{i-1},$$
$$Y_{i+1}, \ \ldots, \ Y_{t-1}, \ X_t; \ a_t = Y_1, \ \ldots, \ Y_{i-1}, \ Y_{i+1}, \ \ldots, \ Y_t.$$

It follows from inequality (2.164) that, commensurate with the extent of replacing the condition of constant generalized forces by conditions of constant respective conjugate derivatives dY_i/dX_i – because this is accompanied by the increase in the number of secondary forces – the counteracting influence of the primary forces on the equilibrium state is diminished. The secondary forces are clearly functions of the primary forces X_i, since all derivatives in the chain (2.164) describe monovariant processes of equilibrium displacements. Consequently, for monovariant processes of equilibrium displacements, the effect of a primary force on its conjugate parameter decreases in proportion to the reduction in the number of fixed other generalized forces from among the $(t-1)$ fixed parameters, where t is the number of independent parameters.

Since the principle of equilibrium displacements under the participation of secondary forces is a direct consequence of the stability criteria, which, as we have remarked, apply to a heterogeneous system, the equilibrium displacement of such a system ought to be considered by analogy with homogeneous systems on the basis indicated.

It can easily be shown that the Gibbs thermodynamic potential G^g applied to a heterogeneous system as a whole possesses the same properties with respect to the variables S^g, V^g, and n_i^g of the heterogeneous system as the corresponding function for the homogeneous system considered by Gibbs. Therefore, the derivations of inequalities for the equilibrium displacement principle of a heterogeneous system with secondary forces and for a homogeneous system are completely analogous.

Nevertheless, for a heterogeneous system we must keep in mind the requirement of choosing the fixed quantities so as to make the system monovariant [13]. In keeping with the Gibbs phase rule, this will be fulfilled whenever the number of phases is less than the number of components of the heterogeneous system. Taking into account this requirement, and exploiting the analogy with homogeneous systems, we find for a heterogeneous system with secondary forces the chain of inequalities

$$(\partial Y_t^g/\partial X_i)_{a_1^g} \geqslant (\partial Y_t^g/\partial X_i)_{a_2^g} \geqslant \cdots$$
$$\geqslant (\partial Y_t^g/\partial X_i)_{a_{t-1}^g} \geqslant (\partial Y_t^g/\partial X_i)_{a_t^g}, \qquad (2.165)$$

where the parameters a_1^g, a_2^g, etc., are interpreted as for (2.164) but refer to the heterogeneous system as a whole (indicated by the index g).

The coordinates $Y_i^{(g)}$ in (2.165) are global quantities also pertaining to the system as a whole. According to the restrictive condition whereby the number of phases is smaller than the number of components, the chain in (2.165) is shorter than in (2.164).

The cases of equilibrium shifts we have considered illustrate the quite complex nature of the Gibbs–Le Chatelier principle. At the same time they provide the basis for a detailed analysis of various systems. For heterogeneous phase equilibria a further elaboration of the principle is feasible. In particular, it is of interest to consider the displacement principle along lines representing phase equilibria in two-component systems.

7. DISPLACEMENT ALONG EQUILIBRIUM LINES. GENERALIZED VAN DER WAALS DIFFERENTIAL EQUATIONS

A displacement along a phase equilibrium line in a heterogeneous system may be described in various ways. Particular expressions of the displacement principle along a phase equilibrium line are the laws of Gibbs–Konowalow [1, 30], Vrevskii [31], and others. The most general form has been given by van der Waals [12], who obtained a differential equation of two-phase equilibrium for a two-component system. This has been generalized subsequently by Storonkin to multicomponent systems [13, 32].

The van der Waals equation in combination with the stability conditions introduced by Gibbs provides an exhaustive characterization of the thermodynamic properties of two-component systems [32]. On its basis one may also analyze phase diagrams. For this reason we shall devote more space to the thorough foundations of this equation.

It is simple and convenient to derive the van der Waals equation from the fundamental Gibbs equations for the phases a and β:

$$s^{\alpha}\, dT - v^{\alpha}\, dp + d\mu_1^{\alpha} + x_2^{\alpha}\left(d\mu_2^{\alpha} - d\mu_1^{\alpha}\right) = 0, \qquad (2.166)$$

$$s^{\beta}\, dT - v^{\beta}\, dp + d\mu_1^{\beta} + x_2^{\beta}\left(d\mu_2^{\beta} - d\mu_1^{\beta}\right) = 0. \qquad (2.167)$$

Subtracting the first from the second and remembering that according to Eq. (2.39)

$$\mu_2 - \mu_1 = (\partial g/\partial x_2)_{p,\, T}, \qquad (2.168)$$

we obtain

$$\left(s^{\beta} - s^{\alpha}\right) dT - \left(v^{\beta} - v^{\alpha}\right) dp + \left(x_2^{\beta} - x_2^{\alpha}\right) d\,(\partial g/\partial x_2)_{p,\, T} = 0. \qquad (2.169)$$

Extending the discussion to multicomponent two-phase systems, Storonkin obtained similar relations:

$$\left(v^{\beta} - v^{\alpha}\right) dp = \left(s^{\beta} - s^{\alpha}\right) dT + \sum_{i=1}^{k-1}\left(x_i^{\beta} - x_i^{\alpha}\right) d\,(\partial g/\partial x_i)_{p,\, T,\, x_{j\neq i}}. \qquad (2.170)$$

According to expressions (2.169) and (2.170), the differential $d(\partial g/\partial x_i)_{p,\,T,\,x_i\neq j}$ can be taken both in the first and in the second phase. In writing the differentials in an expanded form, one should differentiate with respect to the state variables of the appropriate phases. Because the partial derivatives $(\partial g/\partial x_i)_{p,\,T,\,x_j\neq i}$ for each phase have a functional dependence on the state parameters of the phases, two independent differential equations follow from formulas (2.169) and (2.170). The equilibrium conditions for a multicomponent system will therefore be expressed by the following system of differential equations:

$$(v^\beta - v^\alpha)\, dp = (s^\beta - s^\alpha)\, dT + \sum_{i=1}^{k-1} (x_i^\beta - x_i^\alpha)\, d\,(\partial g/\partial x_i)^\alpha,$$

$$(v^\beta - v^\alpha)\, dp = (s^\beta - s^\alpha)\, dT + \sum_{i=1}^{k-1} (x_i^\beta - x_i^\alpha)\, d\,(\partial g/\partial x_i)^\beta, \qquad (2.171)$$

$$d\,(\partial g/\partial x_i)^\alpha_{p,\,T,\,x_{j\neq i}} = d\,(\partial g/\partial x_i)^\beta_{p,\,T,\,x_{j\neq i}},$$

where the index i ranges from 1 to $(k-1)$, excluding $i=s$, s being the number of redundant equations. It is easy to show [10, 13, 32] that if the equation

$$d\,(\partial g/\partial x_s)^\alpha_{p,\,T,\,x_{j\neq s}} = d\,(\partial g/\partial x_s)^\beta_{p,\,T,\,x_{j\neq s}} \qquad (2.172)$$

is redundant then it should follow from the system (2.171).

Expanding the differentials $d(\partial g/\partial x_i)$ of the last $(k-2)$ equations of the system (2.172), one may write the expanded form as follows:

$$[(\partial s/\partial x_i)^\beta - (\partial s/\partial x_i)^\alpha]\, dT - [(\partial v/\partial x_i)^\beta$$

$$- (\partial v/\partial x_i)^\alpha]\, dp - D\,(\partial g/\partial x_i)^\beta + D\,(\partial g/\partial x_i)^\alpha = 0, \qquad (2.173)$$

where

$$D\,(\partial g/\partial x_i) \equiv \sum_{j=1}^{k-1} (\partial^2 g/\partial x_i\,\partial x_j)\, dx_j \qquad (2.174)$$

is the differential with respect to the composition.

Utilizing the representations in terms of partial molar quantities (cf. Chapter 1), one may write

$$(\partial s/\partial x_i)_{T,\,p,\,x_{j\neq i}} = \bar{s}_i - \bar{s}_k, \qquad (2.175)$$

whence

$$(\partial s/\partial x_i)^\beta - (\partial s/\partial x_i)^\alpha = (\bar{s}_i^\beta - \bar{s}_i^\alpha) - (\bar{s}_k^\beta - \bar{s}_k^\alpha) = [\Delta\bar{h}_i^{(\alpha\to\beta)} - \Delta\bar{h}_k^{(\alpha\to\beta)}]/T. \qquad (2.176)$$

Here $\Delta\bar{h}_i^{(\alpha\to\beta)}$ and $\Delta\bar{h}_k^{(\alpha\to\beta)}$ are the partial molar heats of transition of the ith and kth components from the α– to the β-phase.

Similarly, one may easily prove that

$$(\partial v/\partial x_i)_{T,\,p,\,x_{j \neq i}} = \bar{v}_i - \bar{v}_k. \qquad (2.177)$$

Furthermore,

$$(\partial v/\partial x_i)^\beta - (\partial v/\partial x_i)^\alpha = \Delta \bar{v}_i^{(\alpha \to \beta)} - \Delta \bar{v}_k^{(\alpha \to \beta)}, \qquad (2.178)$$

where

$$\left. \begin{aligned} \Delta \bar{v}_i^{(\alpha \to \beta)} &\equiv \bar{v}_i^\beta - \bar{v}_i^\alpha, \\ \Delta \bar{v}_k^{(\alpha \to \beta)} &\equiv \bar{v}_k^\beta - \bar{v}_k^\alpha. \end{aligned} \right\} \qquad (2.179)$$

Here $\overline{\Delta v_i}^{(\alpha \to \beta)}$ is the increment in the partial molar volume of the ith component going from the α- to the β-phase.

On account of relations (2.173)–(2.179) the system of differential equations (2.171) may be written as

$$\left[v^\beta - v^\alpha - \sum_{i=1}^{k-1} (x_i^\beta - x_i^\alpha)(\partial v/\partial x_i)^\alpha \right] dp$$

$$= \left[s^\beta - s^\alpha - \sum_{i=1}^{k-1} (x_i^\beta - x_i^\alpha)(\partial s/\partial x_i)^\alpha \right] dT$$

$$+ \sum_{i=1}^{k-1} \sum_{j=1}^{k-1} (x_i^\beta - x_i^\alpha)[\partial^2 g/(\partial x_i\, \partial x_j)]\, dx_j^\alpha;$$

$$\left[v^\beta - v^\alpha - \sum_{i=1}^{k-1} (x_i^\beta - x_i^\alpha)(\partial v/\partial x_i)^\beta \right] dp$$

$$= \left[s^\beta - s^\alpha - \sum_{i=1}^{k-1} (x_i^\beta - x_i^\alpha)(\partial s/\partial x_i)^\beta \right] dT \qquad (2.180)$$

$$+ \sum_{i=1}^{k-1} \sum_{j=1}^{k-1} (x_i^\beta - x_i^\alpha)[\partial^2 g/(\partial x_i\, \partial x_j)]\, dx_j^\beta,$$

$$[(\Delta \bar{h}_i^{(\alpha \to \beta)} - \Delta \bar{h}_k^{(\alpha \to \beta)})/T]\, dT + (\Delta \bar{v}_k^{(\alpha \to \beta)} - \Delta \bar{v}_i^{(\alpha \to \beta)})\, dp$$

$$- D\,(\partial g/\partial x_i)^\beta + D\,(\partial g/\partial x_i)^\alpha = 0,$$

where $i = 1, \ldots, s-1, s+1, \ldots, k-1$.

The first two equations in (2.180) were called by Storonkin the generalized van der Waals differential equations, while the last $(k-2)$ were termed the supplementary equilibrium conditions. The first two equations are more general and allow a better exposition of the thermodynamics of heterogeneous systems than the supplementary conditions. From the first two equations of (2.180) it follows that for a one-component, two-phase system

$$(v^\beta - v^\alpha)\, dp = (s^\beta - s^\alpha)\, dT; \qquad (2.181)$$

i.e., we have obtained the well-known Clausius–Clapeyron equation. For a two-component system we find an equation first introduced by van der Waals:

$$[v^\beta - v^\alpha - (x^\beta - x^\alpha)(\partial v/\partial x)]\,dp = [s^\beta - s^\alpha - (x^\beta - x^\alpha)$$
$$\times (\partial s/\partial x)]\,dT + (\partial^2 g/\partial x^2)(x^\beta - x^\alpha)\,dx. \tag{2.182}$$

The derivatives $\partial v/\partial x$, $\partial s/\partial x$, $\partial^2 g/\partial x^2$, and the differential dx pertain either to the a- or to the β-phase.

The physical meaning of the quantities Δv and Δs with dp and dT given by the generalized van der Waals equation is that they describe the change in volume and entropy of a two-component system for an isobaric–isothermal formation of one mole of the β-phase from an infinitely large a-phase, i.e., they are characteristic of a process which is incompatible with the conditions of equilibrium between phases [12, 13, 32].

Zharov and Korobov [33] have given another clear interpretation of the physical meaning of the coefficient Δs, associated with the equilibrium conversion of the a-phase into the β-phase. According to their expression, Δs is determined by the following experimentally measurable quantities: first by the differential thermal effect of the equilibrium isobaric formation of the β-phase in a system containing initially only one mole of the a-phase, and, second, by the differential effect of the phase process and the molar heat capacity of the a-phase.

The physical content of the coefficients dx_j in the first two equations of (2.180) resides in the difference of the phase effects for the jth and kth components, characterizing the influence of isobaric–isothermal changes of phase compositions along a conoidal path induced by phase processes on the chemical potentials of components of a given phase. The derivatives of the chemical potentials with respect to composition characterize the effects of diffusion and separation (or salting-in and salting-out, as observed by Storonkin [13]) when phase processes (along a conoid) cause changes in the composition of phases.

The thermal effects associated with the phase transformations $\alpha \to \beta$ and $\beta \to \alpha$ may be represented by

$$\bar{q}^{(\alpha \to \beta)} \equiv T\bar{s}^{(\alpha \to \beta)}, \quad \bar{q}^{(\beta \to \alpha)} \equiv T\bar{s}^{(\beta \to \alpha)}, \tag{2.183}$$

where $\overline{q}^{(a \to \beta)}$ and $\overline{q}^{(\beta \to a)}$ are differential molar heats of formation of the β-phase from the a-phase and vice versa, respectively.

These quantities, like the corresponding volume changes $v^{(\alpha \to \beta)}$ and $v^{(\beta \to \alpha)}$, are not equal to each other, since they pertain to processes bearing the same initial but different final states, which are therefore not reversible with respect to one another [13].

One may employ the usual van der Waals equation in calculations if, on the basis of model considerations, one has knowledge of the dependence of the Gibbs free energy of phases on the state parameters. It may also be applied whenever the coefficients of Eq. (2.182) can be expanded in a power series in the vicinity of some point in the state diagram, so that the calculation proceeds by way of local equations in various approximations [34–37].

If a two-phase multicomponent system satisfies only the condition of partial equilibrium (whereby in the phases present there are nonidentical amounts of substances, or when for whatever reason some of the interphase equilibrium conditions do not hold), then the application of the generalized van der Waals differential equation becomes somewhat difficult. Marinicheva and Storonkin [38] have obtained a modified van der Waals equation enabling one to describe not only a heterogeneous system in a state of partial equilibrium, but also a system consisting of, for example, two parts, separated by a semipermeable membrane. Their equation is thus suitable for describing various types of systems. In complete thermodynamic equilibrium, and in the presence of identical amounts of substances in the existing phases, it reduces to the usual generalized van der Waals differential equation.

8. THE ADEQUACY OF VARIOUS DEDUCTIONS OF THE PRINCIPLE OF DISPLACEMENTS ALONG EQUILIBRIUM LINES

Two main lines of approach may be clearly distinguished in proving the principle of equilibrium displacements in thermodynamics. One approach, developed in the works of Gibbs [1], Planck [2], and van der Walls [12], as well as Storonkin [32], is based on the Gibbs equilibrium principle as applied to heterogeneous systems with unchanging component masses.

The other approach to deducing the equilibrium displacement principle is contained in the works of Schottky [20] and Prigogine and Defay [24] and is based on the concept of the chemical affinity function or something analogous, which does not require strict material isolation of the system and may accordingly be applied to open systems as well (in which mass exchange with the surroundings is possible).

Semenchenko's work [39] can be classed here, too, though his derivation uses the idea of chemical affinity only implicitly.

Van der Waals [12] and Storonkin [32], applying the methodology of Gibbs, obtained differential equations for two-phase equilibrium in two-component and multicomponent systems.

Equations different in form but similar in content have been advanced in [20, 24, 39]. Since their forms differ essentially from the van der Waals equation, the question arises as to the interrelation between the two approaches.

It was pointed out in Semenchenko's paper [39] that the mathematical formulation of the principle of equilibrium displacements is tantamount to finding the equilibrium condition, namely to varying one of the thermodynamic potentials keeping all thermodynamic forces constant, and then differentiating it with respect to these forces (this being an expression of an equilibrium displacement). In contrast, condition (2.182), derived by van der Waals, contains a twofold differentiation with respect to temperature and pressure. On the strength of this it was concluded in [39] that Eq. (2.182), as well as its generalization to any number of forces and components, is incorrect.

In the following we shall compare the two approaches on the basis of a recent work of ours [40]. We show that in spite of external differences the results of derivations of the principle of equilibrium displacements using both approaches are essentially correct. For clarity we discuss a two-component system.

In the second approach, according to Prigogine and Defay [24], differential affinities may be expressed by the system of equations

$$dA_1' = d\mu_1^\alpha - d\mu_1^\beta, \ dA_2' = d\mu_2^\alpha - d\mu_2^\beta. \tag{2.184}$$

Since $\mu_i = f(T, p, x_2)$, we may write for the total differential of the chemical potential

$$d\mu_i = -\bar{s}_i \, dT + \bar{v}_i \, dp + (\partial\mu_i/\partial x_2)_{p,T} \, dx_2, \tag{2.185}$$

where we used (2.22) and (2.23).

Substituting (2.185) into (2.184), using (2.168) and the Gibbs–Duhem equation (2.36), as well as the fact that the affinity vanishes along an equilibrium line, we obtain

$$
\begin{aligned}
&\Delta\bar{s}_1^{(\alpha\rightarrow\beta)} \, dT - \Delta\bar{v}_1^{(\alpha\rightarrow\beta)} \, dp - x_2^\alpha \left(\frac{\partial^2 g}{\partial x_2^2}\right)_{p,T}^\alpha dx_2^\alpha \\
&+ x_2^\beta \left(\frac{\partial^2 g}{\partial x_2^2}\right)_{p,T}^\beta dx_2^\beta = 0, \\
&\Delta\bar{s}_2^{(\alpha\rightarrow\beta)} \, dT - \Delta\bar{v}_2^{(\alpha\rightarrow\beta)} \, dp + \left(1 - x_2^\alpha\right) \left(\frac{\partial^2 g}{\partial x_2^2}\right)_{p,T}^\alpha dx_2^\alpha \\
&- \left(1 - x_2^\beta\right) \left(\frac{\partial^2 g}{\partial x_2^2}\right)_{p,T}^\beta dx_2^\beta = 0,
\end{aligned}
\tag{2.186}
$$

where

$$\Delta\bar{s}_i^{(\alpha\rightarrow\beta)} = \bar{s}_i^\beta - \bar{s}_i^\alpha, \quad \Delta\bar{v}_i^{(\alpha\rightarrow\beta)} = \bar{v}_i^\beta - \bar{v}_i^\alpha. \tag{2.187}$$

Multiplying the first equation in the system of equations (2.186) by x_1^j and the second by x_2^j, one may eliminate the variables dx_2^β or dx_2^α. As a result, we shall have

$$
\begin{aligned}
&\left(x_1^\beta \, \Delta\bar{s}_1^{(\alpha\rightarrow\beta)} + x_2^\beta \, \Delta\bar{s}_2^{(\alpha\rightarrow\beta)}\right) dT - \left(x_1^\beta \, \Delta\bar{v}_1^{(\alpha\rightarrow\beta)} + x_2^\beta \, \Delta\bar{v}_2^{(\alpha\rightarrow\beta)}\right) dp \\
&- \left(x_2^\alpha - x_2^\beta\right) (\partial^2 g / \partial x_2^2)^\alpha \, dx_2^\alpha = 0, \\
&\left(x_1^\alpha \, \Delta\bar{s}_1^{(\alpha\rightarrow\beta)} + x_2^\alpha \, \Delta\bar{s}_2^{(\alpha\rightarrow\beta)}\right) dT - \left(x_1^\alpha \, \Delta\bar{v}_1^{(\alpha\rightarrow\beta)} \right. \\
&\left. + x_2^\alpha \, \Delta\bar{v}_2^{(\alpha\rightarrow\beta)}\right) dp - \left(x_2^\alpha - x_2^\beta\right) (\partial^2 g/\partial x_2^2)^\beta \, dx_2^\beta = 0.
\end{aligned}
\tag{2.188}
$$

We see that Eq. (2.188) contains a complete thermodynamic description of a two-component two-phase system.

On the basis of similar considerations, Semenchenko [39] obtained the following equation[1]:

$$
\left(m_1^\beta \, \Delta\bar{s}_1 + m_2^\beta \, \Delta\bar{s}_2\right) dT - \left(m_1^\beta \, \Delta\bar{v}_1 + m_2^\beta \, \Delta\bar{v}_2\right) dp + \left(m_1^\beta m_2^\alpha - m_1^\alpha m_2^\beta\right) \varphi^\alpha \, dc^\alpha = 0,
\tag{2.189}
$$

where m is the number of moles in the ith component and

$$c^\alpha = m_1^\alpha/m_2^\alpha; \quad \varphi^\alpha = (\partial\mu_1/\partial m_1)_{p,T,m_2^\alpha}^\alpha. \tag{2.190}$$

Equation (2.189) also provides a complete thermodynamic description of two-phase equilibrium in a two-component system. To show the adequacy of both derivations of the principle of equilibrium displacements, we prove that Eq. (2.189) is equivalent to the first equation in the system of equations (2.188).

---

[1]In Eq. (2.189) and those related to it we have preserved the nomenclature of [39].

We change from the number of moles to mole fractions as composition variables and express $(\partial \mu_1/dm_1)^\alpha_{p,T}$ in terms of $(\partial \mu_1/\partial x_1)^\alpha_{p,T}$. By analogy with (2.32), we have for our case

$$(\partial \mu_1/\partial m_1)^\alpha_{p,T} = (\partial \mu_1/\partial x_1)^\alpha [m_2^\alpha/(m_1^\alpha + m_2^\alpha)^2]. \tag{2.191}$$

Substituting expression (2.191) into Eq. (2.189), we obtain

$$\begin{aligned}
&(m_1^\beta \Delta \bar{s}_1 + m_2^\beta \Delta \bar{s}_2)\, dT - (m_1^\beta \Delta \bar{v}_1 + m_2^\beta \Delta \bar{v}_2)\, dp \\
&+ (\partial \mu_1/\partial x_1)^\alpha_{p,T}\, x_2 (m_1^\beta x_2^\alpha - m_2^\beta x_1^\alpha)\, d\,(x_1/x_2)^\alpha = 0.
\end{aligned} \tag{2.192}$$

Dividing (2.192) by $(m_1{}^\beta + m_2{}^\beta)$, and bearing in mind that

$$d\,(x_1/x_2)^\alpha = -\left(1/x_2^{(\alpha)}\right)^2 dx_2^\alpha, \tag{2.193}$$

we find

$$\begin{aligned}
&(x_1^\beta \Delta \bar{s}_1 + x_2^\beta \Delta \bar{s}_2)\, dT - (x_1^\beta \Delta \bar{v}_1 + x_2^\beta \Delta \bar{v}_2)\, dp \\
&- (1/x_2^\alpha)(\partial \mu_1/\partial x_1)^\alpha_{p,T} (x_1^\beta x_2^\alpha - x_2^\beta x_1^\alpha)\, dx_2^\alpha = 0.
\end{aligned} \tag{2.194}$$

On the basis of expression (2.168) we obtain

$$(\partial^2 g/\partial x_1^2)_{p,T} = (\partial \mu_1/\partial x_1)_{p,T} - (\partial \mu_2/\partial x_1)_{p,T}, \tag{2.195}$$

whereas the Gibbs–Duhem equation gives

$$(\partial \mu_2/\partial x_1)_{p,T} = [-(1-x_2)/x_2] \cdot (\partial \mu_1/\partial x_1)_{p,T}. \tag{2.196}$$

After substituting (2.196) into (2.195) we get

$$(\partial \mu_1/\partial x_1)_{p,T} = -x_2 (\partial^2 g/\partial x_1^2)_{p,T} \tag{2.197}$$

or

$$(\partial \mu_1/\partial x_1)_{p,T} = x_2 (\partial^2 g/\partial x_2^2)_{p,T}. \tag{2.198}$$

Substituting (2.198) into (2.194) and noting that $x_1^\beta x_2^\alpha - x_2^\beta x_1^\alpha = x_2^\alpha - x_2^\beta$, we are led to the first of the system of equations (2.188). At this point one may ascertain the identity of the approaches and final results of [20, 24, 39]. In order to reduce Eq. (2.188) to the van der Waals equation (2.182), we transform the coefficients of dT and dp.

We use $\Delta \bar{s}_1^{(\alpha \to \beta)}$ and $\Delta \bar{s}_2^{(\alpha \to \beta)}$ from (2.187) and the relation (1.14) in order to write the coefficient of dT in the first of the system of equations (2.188) as follows:

$$\left(x_1^\beta \Delta \bar{s}_1^{(\alpha \to \beta)} + x_2^\beta \Delta \bar{s}_2^{(\alpha \to \beta)}\right) = s^\beta - x_1^\beta \bar{s}_1^\alpha - x_2^\beta \bar{s}_2^\alpha. \tag{2.199}$$

Adding and subtracting from (2.199) the quantity

$$s^\alpha = x_1^\alpha \bar{s}_1^\alpha + x_2^\alpha \bar{s}_2^\alpha, \tag{2.200}$$

we obtain, after some manipulations,

$$\left(x_1^\beta \, \Delta \bar{s}_1^{(\alpha \to \beta)} + x_2^\beta \, \Delta \bar{s}_2^{(\alpha \to \beta)}\right) = s^\beta - s^\alpha - \left(x_2^\beta - x_2^\alpha\right)\left(\bar{s}_2^\alpha - \bar{s}_1^\alpha\right). \qquad (2.201)$$

But according to Eq. (2.175)

$$\left(\bar{s}_2^\alpha - \bar{s}_1^\alpha\right) = (\partial s/\partial x_z)_{p,\,T}^\alpha. \qquad (2.202)$$

Consequently, the coefficient of dT becomes

$$\left(x_1^\beta \, \Delta \bar{s}_1^{(\alpha \to \beta)} + x_2^\beta \, \Delta \bar{s}_2^{(\alpha \to \beta)}\right) = s^\beta - s^\alpha - \left(x_2^\beta - x_2^\alpha\right)(\partial s/\partial x_2)_{p,\,T}^\alpha. \qquad (2.203)$$

It can easily be seen that this value corresponds to the coefficient of dT in the van der Waals equation (2.182). The coefficient of dp in (2.188) $(x_1{}^\beta \Delta \overline{\nu}_1{}^{(\alpha \to \beta)} \times x_2{}^\beta \Delta \overline{\nu}_2{}^{(\alpha \to \beta)})$ is similarly transformed. To sum up, it has been shown that Eq. (2.189) reduces completely to the van der Waals equation (2.188). On account of this equivalence the apparently different equation of Semenchenko (2.189) may also be reduced to the van der Waals equation (2.182). We see, therefore, that despite the differences noted in the two approaches for deriving the principle of equilibrium displacements the final results in the two groups of works are completely equivalent. This serves to confirm that there is no difference between open and closed systems as far as the thermodynamic relations are concerned. They differ only in the possibility of changing their compositions as a result of the different conditions of isolation prevailing for open and closed systems [11, 13].

As far as multicomponent two-phase systems are concerned, the van der Waals approach and the method based on affinities lead to identical results. Nevertheless, the van der Waals method in the form developed by Storonkin [13, 32] is definitely more elegant and simpler to use, as we shall now demonstrate by the example of a three-component two-phase system.

The differentials of the chemical potentials for a three-component system have the form

$$d\mu_i = -\bar{s}_i \, dT + \bar{v}_i dp + (\partial \mu_i/\partial x_1)_{p,\,T,\,x_2} \, dx_1 + (\partial \mu_i/\partial x_2)_{p,\,T,\,x_1} \, dx_2, \qquad (2.204)$$

while the differential affinities are

$$dA_i' = d\mu_i^\alpha - d\mu_i^\beta. \qquad (2.205)$$

Along the equilibrium lines of the two coexisting phases the affinity is zero, and these lines satisfy equations of the form (2.205) provided their right-hand sides vanish. The equilibrium of a three-component two-phase system will therefore be described by the following differential conditions:

$$d\mu_1^\alpha - d\mu_1^\beta = \Delta \bar{s}_1^{(\alpha \to \beta)} \, dT - \Delta \bar{v}_1^{(\alpha \to \beta)} \, dp + (\partial \mu_1/\partial x_1)_{p,\,T,\,x_2^\alpha}^\alpha \, dx_1^\alpha$$

$$+ (\partial \mu_1/\partial x_2)_{p,\,T,\,x_1^\alpha}^\alpha \, dx_2^\alpha - (\partial \mu_1/\partial x_1)_{p,\,T,\,x_2^\beta}^\beta \, dx_1^\beta - (\partial \mu_1/\partial x_2)_{p,\,T,\,x_1^\beta}^\beta \, dx_2^\beta;$$

$$d\mu_2^\alpha - d\mu_2^\beta = \Delta\bar{s}_2^{(\alpha\to\beta)}\, dT - \Delta\bar{v}_2^{(\alpha\to\beta)}\, dp \qquad (2.206)$$
$$+ (\partial\mu_2/\partial x_1)_{p,T,x_2^\alpha}^\alpha\, dx_1^\alpha + (\partial\mu_2/\partial x_2)_{p,T,x_1^\alpha}^\alpha\, dx_2^\alpha$$
$$- (\partial\mu_2/\partial x_1)_{p,T,x_2^\beta}^\beta\, dx_1^\beta - (\partial\mu_2/\partial x_2)_{p,T,x_1^\beta}^\beta\, dx_2^\beta,$$
$$d\mu_3^\alpha - d\mu_3^\beta = \Delta\bar{s}_3^{(\alpha\to\beta)}\, dT - \Delta\bar{v}_3^{(\alpha\to\beta)}\, dp + (\partial\mu_3/\partial x_1)_{p,T,x_2^\alpha}^\alpha\, dx_1^\alpha$$
$$+ (\partial\mu_3/\partial x_2)_{p,T,x_1^\alpha}^\alpha\, dx_2^\alpha - (\partial\mu_3/\partial x_1)_{p,T,x_2^\beta}^\beta\, dx_1^\beta - (\partial\mu_3/\partial x_2)_{p,T,x_1^\beta}^\beta\, dx_2^\beta.$$

These relations may be simplified by employing the molar Gibbs free energy g. Thus, by Eq. (2.39),

$$(\partial g/\partial x_1) = \mu_1 - \mu_3; \quad (\partial g/\partial x_2) = \mu_2 - \mu_3. \qquad (2.207)$$

Using this together with the Gibbs–Duhem relation

$$x_1\, d\mu_1 + x_2\, d\mu_2 + (1 - x_1 - x_2)\, d\mu_3 = 0, \qquad (2.208)$$

it is easily verified that

$$(\partial\mu_1/\partial x_1)_{p,T,x_2} = (1 - x_1)(\partial^2 g/\partial x_1^2)_{p,T,x_2} - x_2(\partial^2 g/\partial x_1\,\partial x_2)_{p,T},$$
$$(\partial\mu_1/\partial x_2)_{p,T,x_1} = (1 - x_1)(\partial^2 g/\partial x_1\,\partial x_2)_{p,T} - x_2(\partial^2 g/\partial x_2^2)_{p,T,x_1};$$
$$(\partial\mu_2/\partial x_1)_{p,T,x_2} = (1 - x_2)(\partial^2 g/\partial x_1\,\partial x_2)_{p,T} - x_1(\partial^2 g/\partial x_1^2)_{p,T,x_2}.$$
$$\qquad (2.209)$$
$$(\partial\mu_2/\partial x_2)_{p,T,x_1} = (1 - x_2)(\partial^2 g/\partial x_2^2)_{p,T,x_1} - x_1(\partial^2 g/\partial x_1\,\partial x_2)_{p,T},$$
$$(\partial\mu_3/\partial x_1)_{p,T,x_2} = -x_1(\partial^2 g/\partial x_1^2)_{p,T,x_2} - x_2(\partial^2 g/\partial x_1\,\partial x_2)_{p,T},$$
$$(\partial\mu_3/\partial x_2)_{p,T,x_1} = -x_2(\partial^2 g/\partial x_2^2)_{p,T,x_1} - x_1(\partial^2 g/\partial x_1\,\partial x_2)_{p,T}.$$

Substituting the expressions of (2.209) into (2.206), we arrive at the following system of equations:

$$\partial\mu_1^\alpha - \partial\mu_1^\beta = \Delta\bar{s}_1^{(\alpha\to\beta)}\, dT - \Delta\bar{v}_1^{(\alpha\to\beta)}\, dp + [(1 - x_1^\alpha)(\partial^2 g/\partial x_1^2)^\alpha$$
$$- x_2^\alpha(\partial^2 g/\partial x_1\partial x_2)^\alpha]\, dx_1^\alpha + [(1 - x_1^\alpha)(\partial^2 g/\partial x_1\,\partial x_2)^\alpha$$
$$- x_2^\alpha(\partial^2 g/\partial x_2^2)^\alpha]\, dx_2^\alpha - [(1 - x_1^\beta)(\partial^2 g/\partial x_1^2)^\beta - x_2^\beta$$
$$\times (\partial^2 g/\partial x_1\,\partial x_2)^\beta]\, dx_1^\beta - [(1 - x_1^\beta)(\partial^2 g/\partial x_1\,\partial x_2)^\beta$$
$$- x_2^\beta(\partial^2 g/\partial x_2^2)^\beta]\, dx_2^\beta = 0,$$
$$d\mu_2^\alpha - d\mu_2^\beta = \Delta\bar{s}_2^{(\alpha\to\beta)}\, dT - \Delta\bar{v}_2^{(\alpha\to\beta)}\, dp + [(1 - x_2^\alpha)(\partial^2 g/\partial x_1\,\partial x_2)^\alpha$$
$$- x_1^\alpha(\partial^2 g/\partial x_1^2)^\alpha]\, dx_1^\alpha + [(1 - x_2^\alpha)(\partial^2 g/\partial x_2^2)^\alpha - x_1^\alpha(\partial^2 g/\partial x_1\,\partial x_2)^\alpha]$$
$$\times dx_2^\alpha - [(1 - x_2^\beta)(\partial^2 g/\partial x_1\,\partial x_2)^\beta - x_1^\beta(\partial^2 g/\partial x_1^2)^\beta]\, dx_1^\beta$$
$$\qquad (2.210)$$
$$- [(1 - x_2^\beta)(\partial^2 g/\partial x_2^2)^\beta - x_1^\beta(\partial^2 g/\partial x_1\,\partial x_2)^\beta]\, dx_2^\beta = 0,$$
$$d\mu_3^\alpha - d\mu_3^\beta = \Delta\bar{s}_3^{(\alpha\to\beta)}\, dT - \Delta\bar{v}_3^{(\alpha\to\beta)}\, dp + [x_1^\alpha(\partial^2 g/\partial x_1^2)^\alpha - x_2^\alpha$$
$$\times (\partial^2 g/\partial x_1\,\partial x_2)^\alpha]\, dx_1^\alpha + [x_1^\alpha(\partial^2 g/\partial x_1\,\partial x_2)^\alpha - x_2^\alpha(\partial^2 g/\partial x_2^2)^\alpha] \times$$

$$\times \, dx_2^\alpha + [x_1^\beta \, (\partial^2 g/\partial x_1^2)^\beta + x_2^\beta \, (\partial^2 g/\partial x_1 \, \partial x_2)^\beta] \, dx_1^\beta$$
$$+ [x_2^\beta \, (\partial^2 g/\partial x_1 \, \partial x_2)^\beta + x_2^\beta \, (\partial^2 g/\partial x_2^2)^\beta] \, dx_2^\beta = 0.$$

Subtracting the third equation in (2.210) from the first and second, we obtain

$$(\Delta \bar{s}_1^{(\alpha \to \beta)} - \Delta \bar{s}_3^{(\alpha \to \beta)}) \, dT - (\Delta \bar{v}_1^{(\alpha \to \beta)} - \Delta \bar{v}_3^{(\alpha \to \beta)}) \, dp$$
$$+ (\partial^2 g/\partial x_1^2)^\alpha \, dx_1^{(\alpha)} + (\partial^2 g/\partial x_1 \, \partial x_2)^\alpha \, dx_2^\alpha - (\partial^2 g/\partial x_1^2)^\beta \, dx_1^\beta$$
$$- (\partial^2 g/\partial x_1 \, \partial x_2)^\beta \, dx_2^\beta = 0; \tag{2.211}$$

$$(\Delta \bar{s}_2^{(\alpha \to \beta)} - \Delta \bar{s}_3^{(\alpha \to \beta)}) \, dT - (\Delta \bar{v}_2^{(\alpha \to \beta)} - \Delta \bar{v}_3^{(\alpha \to \beta)}) \, dp$$
$$+ (\partial^2 g/\partial x_1 \, \partial x_2)^\alpha \, dx_1^\alpha + (\partial^2 g/\partial x_2^2)^\alpha \, dx_2^\alpha - (\partial^2 g/\partial x_1 \, \partial x_2)^\beta \, dx_1^\beta$$
$$- (\partial^2 g/\partial x_2^2)^\beta \, dx_2^\beta = 0. \tag{2.112}$$

Multiplying Eq. (2.211) by x_1^β and Eq. (2.212) by x_2^β, and combining with the third equation in (2.210), we find, finally,

$$[x_1^\beta (\Delta \bar{s}_1^{(\alpha \to \beta)} - \Delta \bar{s}_3^{(\alpha \to \beta)}) + x_2^\beta (\Delta \bar{s}_2^{(\alpha \to \beta)} - \Delta \bar{s}_3^{(\alpha \to \beta)}) + \Delta \bar{s}_3^{(\alpha \to \beta)}] \, dT$$
$$- [x_1^\beta (\Delta \bar{v}_1^{(\alpha \to \beta)} - \Delta \bar{v}_2^{(\alpha \to \beta)}) + x_2^\beta (\Delta \bar{v}_2^{(\alpha \to \beta)} - \Delta \bar{v}_3^{(\alpha \to \beta)})$$
$$+ \Delta \bar{v}_3^{(\alpha \to \beta)}] \, dp + (x_1^\beta - x_1^\alpha) \, [- (\partial^2 g/\partial x_1 \, \partial x_2)^\alpha$$
$$+ (\partial^2 g/\partial x_1^2)^\alpha] \, dx_1^\alpha$$
$$+ (x_2^\beta - x_2^\alpha) \, [(\partial^2 g/\partial x_2^2)^\alpha - (\partial^2 g/\partial x_1 \, \partial x_2)^\alpha] \, dx_2^\alpha = 0. \tag{2.213}$$

We appreciate from the foregoing that to obtain the final result by using the affinity function requires quite complex computations even for a three-component system. Equation (2.213) also provides a complete thermodynamic description of a two-phase three-component system.

Meanwhile we may utilize the generalized van der Waals–Storonkin equation to obtain relations containing analogous information. Indeed, taking the number of components equal to three in (2.180), we shall have

$$[s^\beta - s^\alpha - (x_1^\beta - x_1^\alpha) \, (\partial s/\partial x_1)^\alpha - (x_2^\beta - x_2^\alpha) \, (\partial s/\partial x_2)^\alpha] \, dT$$
$$- [v^\beta - v^\alpha - (x_1^\beta - x_1^\alpha) \, (\partial v/\partial x_1)^\alpha - (x_2^\beta - x_2^\alpha) \, (\partial v/\partial x_2)^\alpha] \, dp$$
$$+ (x_1^\beta - x_1^\alpha) \, [(\partial^2 g/\partial x_1^2)^\alpha - (\partial^2 g/\partial x_1 \, \partial x_2)^\alpha] \, dx_1^\alpha + (x_2^\beta - x_2^\alpha)$$
$$\times [(\partial^2 g/\partial x_2^2)^\alpha - (\partial^2 g/\partial x_1/\partial x_2)^\alpha] \, dx_2^\alpha = 0. \tag{2.214}$$

Equations (2.213) and (2.214) are completely equivalent, since (2.213) reduces to the corresponding van der Waals–Storonkin equation (2.214). We show how it was derived for a two-component system and transform the coefficients of dT and dp. Then we find

$$
\begin{aligned}
& x_1^\beta \left(\bar{s}_1^\beta - \bar{s}_1^\alpha \right) - x_1^\beta \left(\bar{s}_3^\beta - \bar{s}_3^\alpha \right) + x_2^\beta \left(\bar{s}_2^\beta - \bar{s}_2^\alpha \right) \\
& - x_2^\beta \left(\bar{s}_3^\beta - \bar{s}_3^\alpha \right) + \bar{s}_3^\beta - \bar{s}_3^\alpha = s^\beta - s^\alpha - \left(x_1^\beta - x_1^\alpha \right) \\
& \times (\partial s / \partial x_1)^\alpha - \left(x_2^\beta - x_2^\alpha \right) (\partial s / \partial x_2)^\alpha.
\end{aligned}
\tag{2.215}
$$

The coefficient of dp may also be transformed in this manner. On the basis of Eqs. (2.213), (2.215), and an expression such as (2.215) for the coefficient of dp, we finally obtain just Eq. (2.214), as we set out to do.

The conclusion, therefore, is that the approach developed by van der Waals and Storonkin [12, 13, 32] is entirely correct. Their method has proved convenient and universal, providing real possibilities for analyzing heterogeneous systems.

Chapter 3

PHASE EQUILIBRIA IN
TWO-COMPONENT SYSTEMS

1. GENERAL CONSIDERATIONS ON STATE DIAGRAMS

In analyzing phase equilibria in heterogeneous systems, the final aim is the establishment of precise relations between parameters characterizing the state of the system. Knowing the mutual dependence of the state parameters, one may determine not just the equilibrium state of a heterogeneous system but, in addition, predict the nature of phase transitions proceeding in a definite direction under varying temperature, pressure, and concentration.

As is known, starting with the work of Gibbs in the thermodynamics of heterogeneous systems, geometrical methods have been widely used side by side with the analytical method. The graphical interpretation of heterogeneous processes and equilibrium is based on strict rules, is distinguished by great visualizability, and enables one to encompass the whole topic under study. This method gave rise to an independent development of thermodynamics, the so-called geometrical thermodynamics [41–43].

The graphical interpretation of the changes of characteristic thermodynamic functions with state parameters, and the consequent establishment of graphical rules and relations between them, are the basis of studying the state diagrams of heterogeneous systems. Phase equilibrium or state diagrams are therefore the graphical means of representing relations between state parameters. Each point on the state diagram, termed a figurative point, defines the numerical values of the parameters characterizing a given state of the system.

As we know, it is difficult to establish the form of the equation of state (especially for condensed systems). Without additional data it is, in any case, insufficient to provide a complete characterization of the thermodynamic state of a system [13]. For this reason the experimental and theoretical construction of state diagrams is an important, if not unique, means of solving the problem of describing heterogeneous equilibrium in real systems. Phase diagrams are an important working tool for the technologist trying to determine the direction of processes associated with phase transitions, for choosing the regime of heat treatment of materials, for optimizing the composition of alloys for special purposes, etc. Phase diagrams are of great value in such important areas as materials science, metallurgy, molten-salt technology, petroleum refining, chemical technology, etc. Defining the role of state diagrams, Kurnakov noted [44] that all details of chemical interaction processes, as for example the appearance of new phases and specific compounds, and the formation of liquid and solid solutions, are precisely and unambiguously reflected in the geometrical complex of surfaces, lines, and points which constitute the chemical diagrams.

By revealing the general connections arising in combining the separate elements of the diagrams, a well-structured language of chemical diagrams may be developed, whose use will resolve the apparent complexities associated with the study of chemical interactions between components [45].

State diagrams also contain a wealth of information which may be extracted by a combination of graphical and analytical methods of chemical thermodynamics. Unlike other diagrams illustrating the dependence between certain quantities by means of lines or surfaces, each point in a state diagram has a physical meaning, no matter where it is situated, since it reflects a definite state of the system [43].

The phase diagram provides an answer to the question of whether, and which, concrete phases form in the system at given values of the state parameters.

Since temperature, pressure, and concentration have been chosen as the basic state parameters of a heterogeneous system, the unambiguous choice of the basic thermodynamic function characterizing it is the Gibbs free energy. In a one-component system the Gibbs free energy is clearly only a function of two state parameters, the temperature and the pressure:

$$G = \varphi\,(T,\ p). \tag{3.1}$$

Assuming a two-phase equilibrium (for example equilibrium at the melting point), we have, among others, the following equilibrium condition:

$$G^{(S)} = G^{(L)}. \tag{3.2}$$

On the basis of (3.1) we obtain

$$G^L = \varphi\,(T,\ p);\ G^S = \varphi^{\cdot}\,(T,\ p). \tag{3.3}$$

Solving these together and using (3.2), we find

$$\Phi\,(T,\ p) = 0. \tag{3.4}$$

Equation (3.4) represents a cylindrical surface which contains the curve of intersection a_1–a_2 with the surface of $G^{(L)}$ and $G^{(S)}$ (Fig. 1). The intersection of this surface with the T–p plane gives the line a_1'–a_2' whose position in the G–T–p coordinate system is determined by

$$\Phi\,(T,\ p) = 0;\ G = 0. \tag{3.5}$$

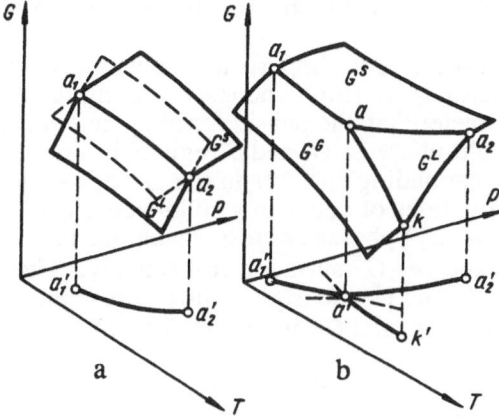

Fig. 1. Gibbs free energy surfaces: a) for solid–liquid phase equilibrium of a one-component system; b) for possible monovariant equilibria in a one-component system and for three-phase invariant equilibrium.

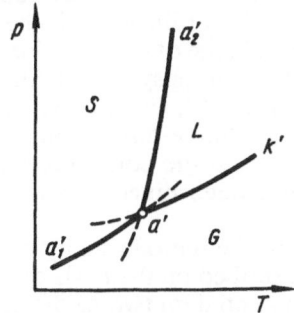

Fig. 2. State diagram of a one-component system.

The position of the same line in $T–p$ coordinates may be represented by the equation

$$p = f(T). \tag{3.6}$$

Consequently, the line $a_1'–a_2'$ is a graphical representation both of Eq. (3.6) and of the projection of the line of intersection of the $G^{(S)}$ and $G^{(L)}$ surface with the $p–T$ plane. It may be constructed from the diagram of the isobaric–isothermal potential (Gibbs free energy).

For a three-phase one-component system we obtain (Fig. 1b) an intersection of three surfaces for isobaric potentials, corresponding to the liquid, solid, and vapor phases, whose traces are projected onto the $p–T$ plane.

If we now represent all the results of the above construction in a single state diagram (Fig. 2), we discern that it describes the interrelation of phases for various values of the state parameters. Although this diagram does not describe precisely the changes in the Gibbs function accompanying the changes of state of the system, it can obviously be clearly related to the isobaric–isothermal potential via the above construction.

The phase diagram therefore imparts a connection between state parameters (temperature and pressure) by simultaneously relating them with the changes of functions of state characterizing the phases forming the system and their reciprocal transitions. The single-phase L, G, and S regions in the phase diagrams are responsible for the corresponding surfaces in the isobaric–isothermal potential diagram (Fig. 1b), and the lines of coexisting solid–vapor (a_1'–a'), liquid–vapor (a'–k'), and liquid–solid (a'–a_2') phases are responsible for the lines of intersection of the G^S–G^G, G^L–G^G, and G^L–G^S surfaces, respectively. The point of coexistence of the three phases L, S, and G (the triple point) corresponds to the common point of the three surfaces G^S, G^L, and G^G in Fig. 1b for which

$$G^S = G^L = G^G. \tag{3.7}$$

In going to multicomponent systems the concentrations join temperature and pressure as state parameters. In principle the picture is unchanged, though it becomes more complicated. Just as for a single-component system, here, too, one may go from a three-dimensional diagram describing the isobaric–isothermal potential as a function of any two state parameters (the rest being constant) to a planar state diagram in the T–x, p–x, and p–T coordinates (x is the concentration of one of the components). The fundamental scheme of construction and the transition to the T–x diagram for a two-component two-phase system at constant pressure was given in the works of Petrov [45], Zlomanov [46], and others. A two-component system should, in principle, be represented in a three-dimensional space, the variables used being the temperature, the pressure, and the concentration of one of the components, since the second component is determined by a constraint equation such as (1.2).

The three-dimensional figure may be reproduced by means of successive planar sections obtained by the method described on the basis of an analysis of the dependence of the isobaric–isothermal potential on two of the appropriate state parameters, keeping the third constant. It is therefore important to know this dependence on temperature, pressure, and concentration separately.

2 . THE ISOBARIC–ISOTHERMAL POTENTIAL AS A FUNCTION OF TEMPERATURE, PRESSURE, AND CONCENTRATION

The dependence of the isobaric–isothermal potential (Gibbs free energy) on temperature and pressure is found from the appropriate derivatives in Eqs. (1.90) and (1.91). It follows from these relations that the free energy increases with temperature, and its rate of decrease on heating the system at constant pressure and concentration should equal the entropy. A rise in pressure causes an increase in the Gibbs free energy, the rate of increase at constant temperature and concentration being equal to the volume. The curvature of the temperature-dependent free energy curve (the G–T curve) is determined by the sign of the second derivative. From expression (1.90) we have

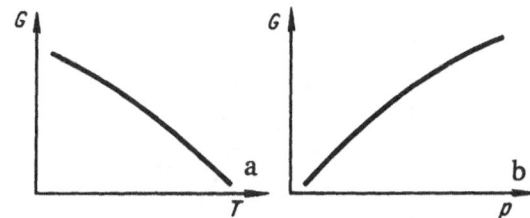

Fig. 3. The Gibbs free energy as a function of temperature at $p = $ const (a), and as a function of pressure at $T = $ const (b).

$$(\partial^2 G/\partial T^2)_{p,\,x_i} = -(\partial S/\partial T)_{p,\,x_i}. \tag{3.8}$$

Since heating always causes an entropy increase, we obtain

$$(\partial^2 G/\partial T^2)_{p,\,x_i} < 0. \tag{3.9}$$

The G–T curve is therefore concave with respect to the abscissa. In exactly the same way, differentiating (1.91), we find

$$(\partial^2 G/\partial p^2)_{T,\,x_i} = (\partial V/\partial p)_{T,\,x_i}. \tag{3.10}$$

Since a pressure increase is always accompanied by a volume decrease, $(\partial V/\partial p)_{T,\,x_i}$ is negative, so that

$$(\partial^2 G/\partial p^2)_{T,\,x_i} < 0. \tag{3.11}$$

Thus the G–p curve is also concave with respect to the abscissa.

Figure 3 illustrates the properties of G. The temperature dependence of G is especially important in the theory of state diagrams, since it provides a direct and clear representation of phase transitions among various phases of a given system.

Returning to a one-component system, an analysis of the relative disposition of the Gibbs free energy curves for the solid and liquid phases provides, first, a clear idea of the melting temperature and, second, a graphically transparent and unambiguous illustration of the equilibrium condition in the form (3.2).

Indeed, below the melting temperature the solid phase is stable in accordance with the stability criterion, so that the solid's Gibbs free energy is below that of the liquid phase. The situation with respect to G is reversed on heating above the melting temperature. Consequently, at the melting point itself the curves $G^L = f(T)$ and $G^S = f(T)$ should intersect (Fig. 4).

The preceding features of the G–T curves are given unambiguously by the sign of the second derivative in (3.9). Nevertheless, in many handbooks and monographs on materials science, the curves $G = f(T)$ or $F = f(T)$ are often presented incorrectly, convex with respect to the T axis. Because this error is so widespread, Novikov has discussed the problem thoroughly in his work [47], although a correct exposition of the temperature dependence of the Gibbs and Helmholtz free energies is contained in many handbooks on chemical thermodynamics and phase transitions (see, for example, [44, 48, 49]).

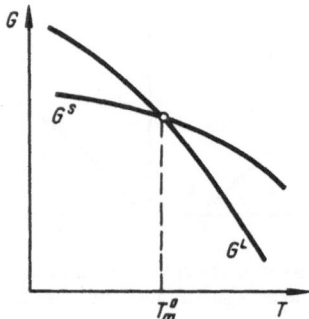

Fig. 4. The relative position of Gibbs free-energy curves in the solid and liquid phases of a one-component system.

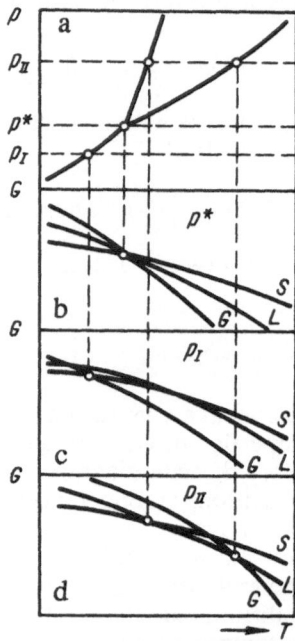

Fig. 5. Determination of the fixed position of the temperature of two- and three-phase equilibria in a one-component system.

We shall now illustrate the fixing of the temperatures of two- and three-phase equilibria by means of the G–T curves. Figure 5a shows the state diagram of a one-component system. In accordance with the equilibrium condition at the triple point described by (3.7), while taking into account the stability of phases in practical temperature regions and at a pressure p^* corresponding to the equilibrium pressure of three coexisting phases, we obtain a picture indicated in Fig. 5b. For pressures p_I and p_{II}, respectively lower and higher than p^*, Figs. 5c and 5d show the relative positions of the G–T curves for the S, L, and G phases.

According to the conditions for the monovariant equilibria $S \rightleftharpoons G$, $S \rightleftharpoons L$, and $L \rightleftharpoons G$, the curves of the Gibbs free energy intersect each other at temperatures corresponding to the phase transition temperatures at pressures p_I and p_{II}, respectively. Such a picture emerges from considering the three-dimensional dependence of the Gibbs free energy on the state parameters, as already explained, and emphasizes once more the active interrelation between state diagrams and free-energy diagrams.

Let us turn to the concentration dependence of the Gibbs free energy. The nature of this dependence follows directly from the stability criteria with respect to continuous changes of state. The physical picture of the stability of a given state of a phase with respect to adjoining states has been described in a general way in Chapter 2. To establish the general nature of the concentration dependence of the Gibbs free energy, we must specify the expression for the stability criterion.

From the analysis of the criterion of stability of a phase with respect to the formation of new phases within it in a form which is analogous to (2.111), and bearing in mind Eq. (2.20), we arrive at the stability condition of a given state with respect to all possible adjoining states:

$$(G'' - G') + (T'' - T') S'' - (p'' - p') V'' - \mu_1' (n_1'' - n_1') - \cdots$$
$$- \mu_k' (n_k'' - n_k') > 0.$$

A single prime denotes the state whose stability is being analyzed; a double prime refers to an adjoining state. At constant temperature and pressure the above inequality becomes

$$\left(\Delta G - \sum_{i=1}^{k} \mu_i' \Delta n_i \right) > 0. \qquad (3.12)$$

However,

$$\Delta G = \sum_{i=1}^{k} \mu_i' \Delta n_i + 1/2! \, (\delta^2 G)_{p, T} + \cdots. \qquad (3.13)$$

Substituting (3.13) into (3.12), we obtain a relation similar to (2.138):

$$(\delta^2 G)_{p, T} > 0. \qquad (3.14)$$

By considering the Gibbs free energy to be a function of the molar fractions of the components, the inequality (3.14) may be written in the following form:

$$\left[\sum_{i=1}^{k-1} \sum_{j=1}^{k-1} (\partial^2 G / \partial x_i \, \partial x_j) \, \delta x_i \, \delta x_j \right]_{p, T} > 0. \qquad (3.15)$$

By the theorem of Sylvester we obtain from the inequality (3.15) the following expression:

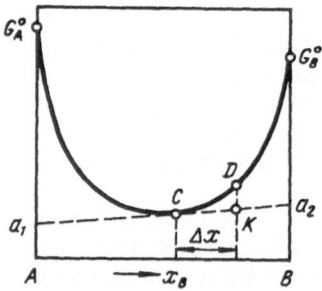

Fig. 6. Establishing the nature of the concentration dependence of the Gibbs free energy of a homogeneous phase in a two-component system at constant T and p.

$$\Delta_{k-1} \equiv \begin{vmatrix} G_{11} & G_{12} & \cdots & G_{1,\,k-1} \\ G_{21} & G_{22} & \cdots & G_{2,\,k-1} \\ \cdot\cdot\cdot\cdot\cdot\cdot\cdot\cdot\cdot\cdot\cdot\cdot\cdot\cdot\cdot\cdot\cdot\cdot\cdot\cdot \\ G_{k-1,\,1} & G_{k-1,\,2} & \cdots & G_{k-1,\,k-1} \end{vmatrix} > 0. \tag{3.16}$$

This gives for a ternary system

$$\Delta_2 \equiv \begin{vmatrix} G_{11} & G_{12} \\ G_{21} & G_{22} \end{vmatrix} > 0 \tag{3.17}$$

and for a binary system

$$\Delta_1 \equiv G_{11} > 0. \tag{3.18}$$

In expressions (3.16)–(3.18), G_{ij} is

$$G_{ij} \equiv \partial^2 G/\partial x_i\, \partial x_j. \tag{3.19}$$

Since the numbering of the components is arbitrary, we may write

$$G_{ii} > 0. \tag{3.20}$$

From the foregoing, and in particular on the basis of the stability criterion (3.15), one may generally conclude that the Gibbs free energy surface should be convex with respect to the concentration axis. If, therefore, we fix the state parameters T and p of a two-component homogeneous system, we may easily establish the general nature of the concentration dependence of the Gibbs free energy (Fig. 6). To complete the characterization of this important G–x curve, we must establish its curvature and the way it touches the ordinates for components A and B.

Let C and D (Fig. 6) represent two neighboring states separated along the concentration axis by Δx. The tangent to the $G = f(x)$ curve at the point C is $a_1 a_2$. Then the segment DK may be found from the equation

$$\overline{DK} = (G + \Delta G) - [G + (\partial G/\partial x)\, \Delta x], \tag{3.21}$$

$$\overline{DK} = 1/2! \cdot (\partial^2 G / \partial x^2)(\Delta x)^2 + \cdots \qquad (3.22)$$

The stability condition (3.15) dictates that the segment DK be positive. It follows from this that the G–x curve may become convex with respect to the concentration axis in the vicinity of the point C if C is stable with respect to continuous changes of state. Such a conclusion is in complete accord with the general consequences of the stability criterion in (3.15).

Here it is important to stress that both of the neighboring states C and D are stable. By considering the changes in the Gibbs free energy, similar conclusions may be drawn even in cases when phase homogeneity does not hold. Indeed, if the various components of a system are not miscible, so that they form a mechanical mixture, then the Gibbs free energy of this mixture is given by an additivity rule. If these components form a homogeneous stable phase (defined by given conditions), then any departure from homogeneity will lead to an increase in the Gibbs free energy. Consequently, all curves should become convex with respect to the composition axes.

The change in the inclination of the tangent to the curve $G = f(x)$ as one approaches the pure components indicates the way the curve joins the A and B ordinates. According to the definition given in Chapter 2 [Eq. (2.39)], it is easy to show that the angle of inclination of the tangent to the Gibbs free energy curve (see Fig. 6) will be given by the difference in the chemical potentials between the second and first components in a phase of given composition, i.e.,

$$\tan \alpha = (\partial G / \partial x_B)_{p,\,T} = \mu_B - \mu_A = \Delta\mu. \qquad (3.23)$$

From the expression for the chemical potential of a solution in the simplest case

$$\Delta\mu = \Delta\mu^\circ + RT \ln (x_B / x_A), \qquad (3.24)$$

when

$$x_B \to 0; \ \ln (x_B / x_A) \to -\infty; \ \Delta\mu \to -\infty. \qquad (3.25)$$

Therefore, by Eqs. (3.24) and (3.23),

$$\tan \alpha \to -\infty; \ \alpha \to 90°. \qquad (3.26)$$

This means that the G–x curve tends to coalesce with the ordinate as A is approached, and at the point of contact the Gibbs free energy is that due to the pure component A at the given pressure and temperature. A similar discussion and conclusion hold for the component B.

We have therefore established quite clearly the nature of the concentration dependence of the Gibbs free energy. When combined with the temperature dependence G (see Fig. 3), this enables one to predict the various types of phase transitions in two-component systems.

The relation between the angle of inclination of the tangent to the free energy curve and the chemical potentials (3.23) serves as the basis for a geometric interpretation of phase equilibrium conditions in a two-component system.

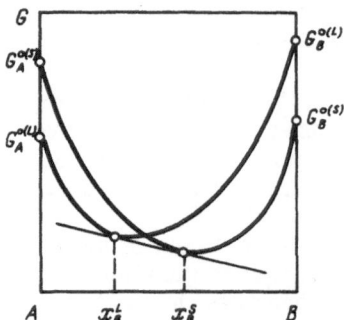

Fig. 7. The establishment of two-phase equilibrium in a two-component system $(T = T_1)$ by means of the concentration-dependent Gibbs free energy curves of the solid and liquid phases.

We show that by constructing a common tangent to the G–x curves one may establish the composition of phases which are in equilibrium. We consider a two-component system A–B in which the liquid L and solid S phases are in equilibrium at some temperature T_1. Clearly each phase is characterized by its own composition-dependent Gibbs free energy curve (Fig. 7). The conditions for constructing a common tangent to the two curves are, first, that the slopes at the points of contact be equal, and, second, that these points lie on the same straight line. From the previous considerations the first condition may be written in the form

$$(\partial G/\partial x_B)^L_{p,\,T} = (\partial G/\partial x_B)^S_{p,\,T}. \tag{3.27}$$

This condition is necessary but not sufficient, since tangents to two curves may have identical inclinations but be at different levels, i.e., the points of contact need not lie on the same straight line. Therefore, apart from the equality of slopes we must write the same straight-line condition as well, which is clearly (see Fig. 7)

$$G^L = G^S - (x_B^S - x_B^L)\,(\partial G/\partial x_B)_{p,\,T}. \tag{3.28}$$

Using (3.27), we may manipulate this into

$$G^L - x_B^L\,(\partial G/\partial x_B)^L_{p,\,T} = G^S - x_B^S\,(\partial G/\partial x_B)^S_{p,\,T}. \tag{3.29}$$

This means, on account of (2.37) and (2.38), that

$$\mu_A^L = \mu_A^S. \tag{3.30}$$

Choosing the A concentration as an independent variable and repeating the previous argument, we may write

$$G^L - x_A^L\,(\partial G/\partial x_A)^L_{p,\,T} = G^S - x_A^S\,(\partial G/\partial x_A)^S_{p,\,T} \tag{3.31}$$

or

$$\mu_B^L = \mu_B^S. \tag{3.32}$$

Equations (3.30) and (3.32) are a reflection of the equilibrium condition of the solid and liquid phases at a given temperature and pressure. The construction of a common tangent to the free energy curves is thus a geometrical interpretation of the phase equilibrium conditions and enables one to fix uniquely the compositions of the corresponding phases. This geometrical condition for equilibrium is widely used in practice to fix the possible forms of equilibrium in two-component systems. The segment of the common tangent included between the concentrations of the equilibrium phases is a conoid, and its projection on the $T–x$ plane is transformed into a straight segment joining the coordinates of the nominal points, symbolizing the coexistence of phases under the given conditions.

If the simultaneous composition and temperature dependence of the Gibbs free energy of a two-component system is considered for each phase separately, then clearly it will be represented by a trough-shaped surface, convex with respect to the $T–x$ plane. We note that the lines of intersection of this trough with the boundary $G–T$ planes of the A and B components are the curves of the temperature dependence of the Gibbs free energy of these components, similar to the lines indicated in Fig. 3.

By considering the two curves of the concentration-dependent Gibbs free energy in an isothermal section, we obtain the picture displayed in Fig. 7. Constructing the common tangent, and noting that the value of the Gibbs free energy within the concentration interval between the coexisting phases is everywhere lower than that of either phase, we may conclude that within that interval (at given T and p) a mixture of equilibrium phases is stable and the free energy of the mixture is determined by an additivity rule. We have noted already that in condensed systems the role of pressure is minimal and may be neglected within well-defined limits, so that in two-component systems only temperature and concentration remain as state parameters governing the planar phase diagrams. If one of these is fixed, we can establish the explicit dependence of the Gibbs free energy on the other parameter. Analysis of the relative positions of these dependences for different phases enables one to ascertain the nature of phase equilibria in the system.

3. THERMODYNAMIC DERIVATION OF BASIC TYPES OF STATE DIAGRAMS IN TWO-COMPONENT SYSTEMS

3.1. Two-Phase Equilibrium

The principle of establishing the nature of phase equilibria via the temperature and concentration-dependent Gibbs free energy curves amounts to the establishment of fixed positions for nominal points having coordinates of equilibrium-coexisting phases on the $T–x$ plane.

We note that to a certain extent the thermodynamic method is *a posteriori*, whereby we are not talking about performing a precise calculation, but rather constructing a specific scheme reflecting the phase interrelations in a system, assuming that the general features of the interactions in the system are in principle known.

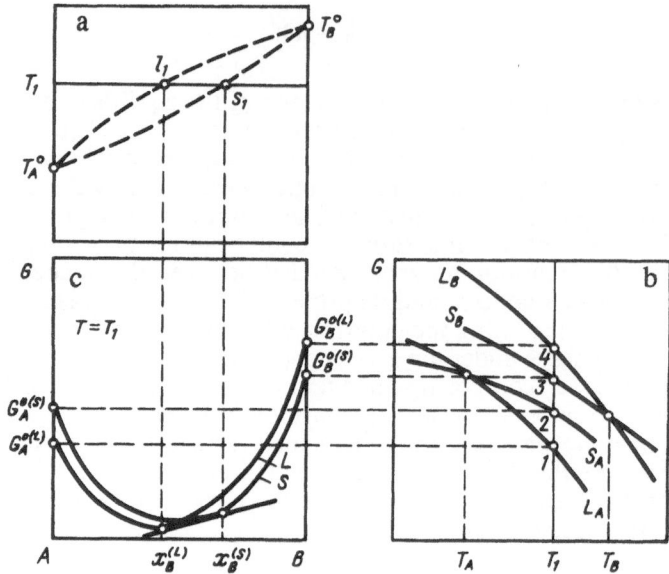

Fig. 8. Specific scheme for establishing the fixed position coordinates of nominal points corresponding to the equilibrium coexistence of phases in a two-component two-phase system.

Figure 8 presents a method for establishing the fixed positions of the coordinates of nominal points corresponding to the equilibrium coexistence of phases in a two-component, two-phase system. Consequently, one knows in advance that the system forms only two phases. In addition, one knows the melting temperatures of components A and B.

Figure 8a denotes the T–x plane. We mark on this plane what is known in advance, notably the melting temperatures of the components T_A° and T_B°. We assume that $T_B^\circ > T_A^\circ$. We determine the compositions of the equilibrium phases at some temperature T_1 intermediate between T_A° and T_B°.

We mark by a vertical line the position of this temperature on the graph of the temperature dependence of the Gibbs free energy of the solid and liquid phases for components A and B (Fig. 8b). The intersection of the curves S_A, S_B, L_A, L_B with the vertical T_1 line fixes the points 1–4, which determine the Gibbs free energies of the liquid and solid phases of the A and B components. Transposing these values to the corresponding coordinates of the G–x curve (Fig. 8c), we thereby determine the edge points $G_A^{\circ(L)}$, $G_A^{\circ(S)}$ and $G_B^{\circ(S)}$, $G_B^{\circ(L)}$ of this graph.

Furthermore, noting the general features of the concentration dependence of the Gibbs free energy according to Fig. 6, we join by the corresponding curves the pair of points $G_A^{\circ(L)}$ and $G_B^{\circ(L)}$, as well as $G_A^{\circ(S)}$ and $G_B^{\circ(S)}$. We finally have two intersecting curves for the concentration dependence of the Gibbs free energy for the liquid and solid phases, respectively, and by constructing the common tangent to these we obtain the compositions x_B^L and x_B^S of the liquid and solid phases in equilibrium at the temperature T_1. In order to obtain the corresponding conjugate points in the T–x state diagram, these compositions must be projected on the horizontal T_1, indicated in Fig. 8a. Thus the fixed positions of the nominal points l_1

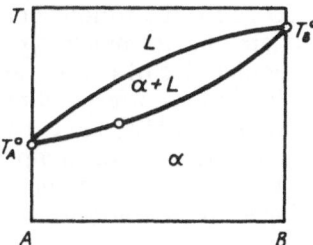

Fig. 9. Phase diagram of a two-component system with a continuous series of solid and liquid mixtures.

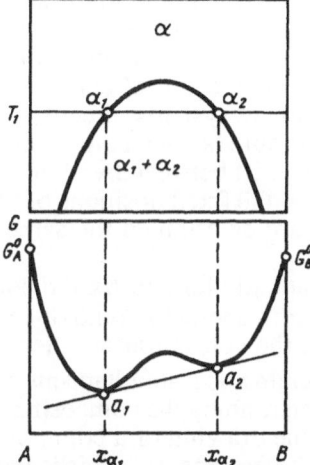

Fig. 10. The concentration dependence of the Gibbs free energy of a two-component system for dissociation of a solid solution or demixing on melting.

and s_1 in Fig. 8a symbolize the states of the conjugate liquid and solid phases which are in equilibrium at the temperature T_1.

The pair of conjugate points corresponding to any other temperature in the interval between $T_A°$ and $T_B°$ may be found by a similar process. Joining the aggregate of curves in all l_i and s_i, and closing them at points $T_A°$ and $T_B°$, we obtain the simplest variant of a phase diagram, describing equilibrium between only two phases (Fig. 9). If the equilibrium phases are liquid (L) and solid (S), then the state diagram describes an equilibrium system within which a continuous series of solid and liquid mixtures may form. A similar picture holds for liquid–vapor systems.

The curve which joins the points l_i and corresponds to the temperature of the initial formation of the solid phase from the liquid (crystallization) on cooling is called the liquidus curve. The corresponding curve joining all points s_i in Fig. 8a and appropriate to the temperature of the end of the process of transforming the liquid into the solid phase on cooling, or of the beginning of the formation of the liquid phase from the solid (melting) on heating, is called the solidus curve.

The foregoing formulation is completely rigorous for the system considered. But the system presented in Figs. 8 and 9 is simple and is characterized by a single branch for the liquidus as well as for the solidus curves. In applications to more complex systems to be treated later, the formulation presented is incomplete, since it has nothing to say about the boundaries of curves, and it is between these boundaries that the limits of application of the corresponding particular phase equation are defined. In this context we note that [13] gave a precise definition of liquidus curves as being the branches of diagrams of the temperature of initial crystallization which are responsible for the various particular phase equations. Such a definition is general and is intimately linked to the rigorous thermodynamic definition of the concept of phase.

The variant of the phase diagram considered in Fig. 9 features a typical two-phase equilibrium whose nature is intrinsic to all other systems, and in which it figures as an essential element. In the following we shall relate this case of phase equilibrium to the transition from the liquid to the solid state, and vice versa, since we are talking about condensing systems. Nevertheless, one can extend the discussion of the formation or thermodynamic construction of various types of phase equilibrium diagrams to other forms of phase equilibria.

If one talks about condensing systems, then it is obvious that, apart from the most important form of equilibrium between the liquid and solid phases already described, equilibrium may also exist between two liquids or two solids in the case of a miscibility gap. We note that the fixed positions of the coordinates of phase equilibrium on the $T-x$ diagram are determined the same way as for the liquid–solid equilibrium.

Figure 10 shows schematically how to fix the coordinates of the equilibrium phases α_1 and α_2 which may correspond to liquid or solid phases. The instability of the homogeneous mixture in the concentration interval $x_{\alpha_1}-x_{\alpha_2}$ of the $G-x$ curve has associated with it a maximum and a simultaneous double minimum. In this case the fundamental method of determining the composition of the equilibrium phases at a temperature T_1 reduces to the drawing of a common tangent to the curves of the concentration-dependent Gibbs free energy. This is the graphical application of the condition of thermodynamic equilibrium of a heterogeneous system in terms of the equality of the chemical potentials of the given compounds in each of the coexisting phases.

We have thus dealt with the basic forms of two-phase equilibrium in heterogeneous condensed systems. If one considers the variants of phase equilibria which may be constructed thermodynamically in a two-component system, these may in principle be quite numerous, especially when one or both components undergo several polymorphic transformations. Such variants have been treated in detail in [45], but they contribute no additional principles beyond those already discussed for two-phase equilibrium diagrams.

The type of phase diagram in Fig. 9 shows that the addition of component B to component A increases the latter's melting temperature, whereas adding A to B achieves the opposite effect of lowering the melting temperature.

The source of changes in the melting temperature due to admixing of one component with another must be sought in the properties of interatomic interactions at the electronic level. Nevertheless, if one has a clear idea of the mutual influence of the components on their respective melting temperature shifts, and if one knows that one deals with a two-phase equilibrium, then, on the basis of the foregoing method, one may derive quite clearly the character of the phase diagram.

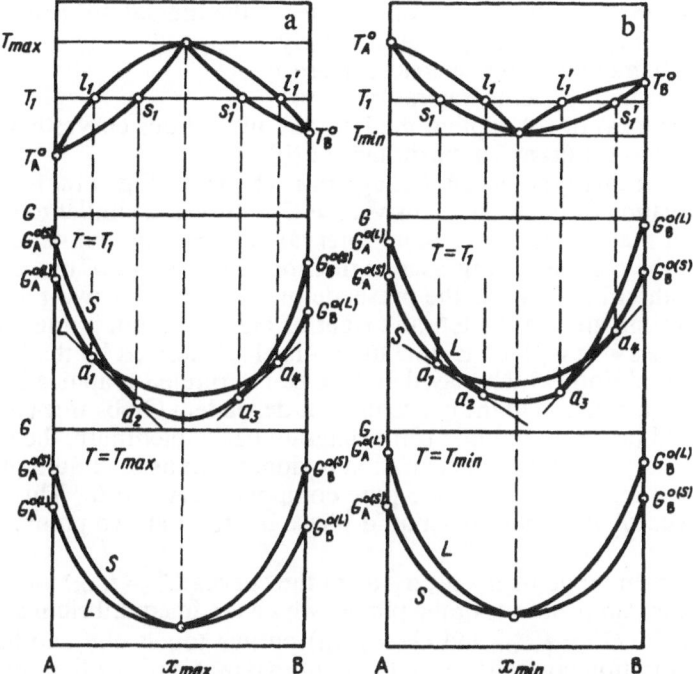

Fig. 11. Construction of the phase diagram of two-component systems with a continuous series of solid and liquid mixtures in the presence of a common maximum (a) or minimum (b) of the liquidus and solidus curves.

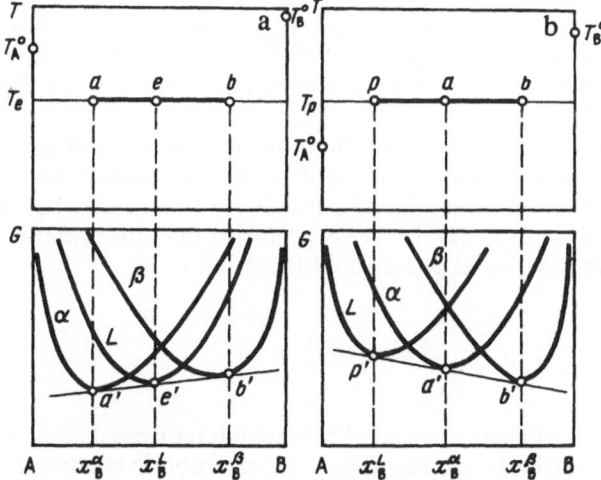

Fig. 12. The establishment of eutectic (a) and peritectic (b) three-phase equilibrium in a two-component system with the aid of concentration-dependent Gibbs free energy curves.

It is of interest to consider two other cases of the mutual influence of admixing components on the melting temperature, namely:

a) when the melting temperatures decrease;

b) when the melting temperatures increase.

When either (a) or (b) applies, the liquidus and solidus curves have a common point, a minimum for (a) and a maximum for (b).

The form of this type of phase diagram is shown in Fig. 11a, b. The thermodynamic derivation of these diagrams via the $G–x$ curves is similar to that presented above, except that in the temperature interval between the points of maxima (or minima) and the corresponding nearest melting temperature of one of the components, the isotherm T_1 reveals the presence of two different two-phase equilibria concentrations situated to the left and right of the extremum points. Each pair of conjugate phases – at a given temperature – is characterized by the intrinsic phase equilibrium condition, which may therefore be interpreted graphically by the common tangent construction to the concentration-dependent Gibbs free energy curves.

We should bear in mind that if the diagram has a maximum, the solid phase is stable at some intermediate concentration region, whereas the liquid phase is stable at composition regions close to the pure components A and B. This requires that the $G–x$ curves of the solid and liquid phase intersect at two points (cf. Fig. 11a, $T = T_1$).

The common tangents $a_1–a_2$, $a_3–a_4$ to the curves $G^L = f(x_B)$ and $G^S = f(x_B)$ in Fig. 11a fix the compositions of the phases which are in equilibrium at T_1.

At T_{max} the $G^L = f(x_B)$ and $G^S = f(x_B)$ curves touch one another at a point whose concentration corresponds to the joint maximum of the liquidus and solidus curves. Thus the G^L curve is necessarily below the G^S curve in the whole concentration interval, emphasizing the stability of the liquid and the instability of the solid at this temperature.

If the diagram has a minimum the liquid becomes stable in the intermediate range of concentration, whereas in regions close to the ordinates of the components A and B a solid solution is stable. This also entails a double intersection of the $G^L = f(x_B)$ and $G^S = f(x_B)$ curves (cf. Fig. 11b, $T = T_1$). By constructing the common tangents $a_1–a_2$ and $a_3–a_4$ to the $G–x$ curves of the solid and liquid phases, we again fix the compositions of the equilibrium phases determined by the endpoints of the conoids $s_1–l_1$ and $s_1'–l_1'$.

At the temperature of the minimum of the liquidus and solidus the $G^L = f(x_B)$ and $G^S = f(x_B)$ curves touch at a point whose concentration corresponds to the composition at the minimum, the G^S curve being necessarily below G^L in the whole relevant concentration interval. At the points of maxima and minima the compositions of the liquid and solid phases are identical, i.e.,

$$x_B^L = x_B^S. \tag{3.33}$$

Equation (3.33) is an extra condition which reduces the variance of the system by one at these particular points. Thus, for an isobaric section through the system, equilibrium becomes invariant at the maxima and minima, and the alloys corresponding to the compositions at the extremal points of Fig. 11 are, in fact, single-component systems.

3.2. Three-Phase Equilibrium

The maximum number of phases in simultaneous equilibrium in a two-component system is three, as shown by the Gibbs phase rule. This limits the number of possible three-phase equilibria in the following way. In any phase transformation in an r-phase system the number of transforming phases may be $1, 2, ..., r - 1$, and the number being generated correspondingly $r - 1, r - 2, ..., 2, 1$. Thus, in a three-phase system there can be clearly at most two transforming phases to which corresponds one phase being generated. Correspondingly there can be at most two phases forming with which one can associate only one transforming phase. In a two-component system in the simplest case the possible phases are, first, the solid solutions α and β formed from components A and B, respectively and, second, the liquid mixture L. In accordance with what has been said about the possible number and kind of three-phase equilibria in a two-component system, we may write

$$L \rightleftarrows \alpha + \beta, \tag{3.34}$$

$$L + \beta \rightleftarrows \alpha. \tag{3.35}$$

The phase equilibria of (3.34) and (3.35) are termed eutectic and peritectic equilibria. Due to the invariant nature of three-phase equilibria in two-component isobaric systems, the equilibria of (3.34) and (3.35) are observed at constant eutectic or peritectic temperatures T_e and T_p, respectively. For the eutectic temperature the concentration limits of these transformations are governed by the extent of the homogeneity regions with respect to components A and B, or by the stability limits of the liquid phase. For the peritectic temperature the limit is set by the extent of the solid solution region based on one of the components.

In order to fix the position of eutectic or peritectic horizontals on the T–x diagram of phase equilibrium, one may employ the same method of geometrical thermodynamics as for two-phase equilibria.

From the general condition of phase equilibrium for eutectic and peritectic transformations in the A–B system, we have

$$\mu_A^\alpha = \mu_A^\beta = \mu_A^L; \; \mu_B^\alpha = \mu_B^\beta = \mu_B^L. \tag{3.36}$$

In accordance with expressions (2.80) and (3.29) and on account of (3.27), we can write

$$G^\alpha - x_B^\alpha (\partial G/\partial x_B)_{p,\,T} = G^\beta - x_B^\beta (\partial G/\partial x_B)_{p,\,T}$$
$$= G^L - x_B^L (\partial G/\partial x_B)_{p,\,T}, \tag{3.37}$$

$$G^\alpha - x_A^\alpha (\partial G/\partial x_A)_{p,\,T} = G^\beta - x_A^\beta (\partial G/\partial x_A)_{p,\,T}$$
$$= G^L - x_A^L (\partial G/\partial x_A)_{p,\,T}. \tag{3.38}$$

These relations are expanded forms of (3.36). Furthermore, as already indicated in the treatment of two-phase equilibria, they are an expression of the common-tangent condition to the G–x curves of coexisting phases. The geometrical expression of the condition of three-phase equilibrium in a two-component system is therefore the common tangent $a'e'b'$ or $p'a'b'$ to the curves $G = f(x_B)$ for the corresponding three phases. The compositions x_B^α, x_B^β, and x_B^L of the equilibrium phases are thus fixed precisely, and the projection of these compositions on the T–x diagram at temperatures T_e or T_p (Fig. 12) enables one to establish the fixed positions of the eutectic (aeb in Fig. 12a) or peritectic (pab, Fig. 12b) horizontals. We note that the unambiguous position of eutectic and peritectic horizontals with respect to the melting temperatures of component A and B already follows from expressions (3.34) and (3.35) for a condensed system. Indeed, since in a eutectic transformation the transforming phase in the cooling process is the liquid, while the phases generated are the two solid phases α and β, one concludes that the left branch of the $G^\alpha = f(x_B)$ curve will always be below that of the $G^{(L)} = f(x_B)$ curve, which means that

$$G_A^S < G_A^L. \tag{3.39}$$

In the same way we can show that the right branch of $G^\beta = f(x_B)$ should always be below that of $G^L = f(x_B)$, from which it follows that

$$G_B^S < G_B^L. \tag{3.40}$$

From a comparison with the temperature-dependent Gibbs free energy curves (cf. Fig. 4) for a one-component system, we see from expressions (3.39) and (3.40) that

$$T_A^\circ > T_e; \; T_B^\circ > T_e. \tag{3.41}$$

These results show that the slope of the right and left branches of the G–x curves are not arbitrary but are completely fixed. Both branches are steepest for the liquid. The left branch of phase α is steeper than for β, since both branches should meet at a common point pertaining to the value of the Gibbs free energy of A at the eutectic temperature, i.e., at the point $G_A^{\circ(S)}$. Similarly, the right branch is steeper for β than α, because both branches must meet at the common point $G_B^{\circ(S)}$ at a given temperature. One, therefore, has a bona fide way of establishing the fixed positions of the $G = f(x_B)$ curves for all three phases on the basis of the Gibbs free energies of the solid and liquid phases of the A and B components at a particular temperature. The relative positions of these energies at the ordinates of the components are established from $G = f(T)$ graphs by using relation (3.41).

A similar discussion for a peritectic transformation of the position of the point p in Fig. 12b leads one to

$$G_A^L < G_A^S, \tag{3.42}$$

$$G_B^L > G_B^S, \tag{3.43}$$

from which (cf. Fig. 4)

$$T_A^\circ < T_p < T_B^\circ. \tag{3.44}$$

In general, the G–x curves at temperatures of three-phase equilibrium lie at different levels, but one can still construct a single common tangent to all three. As we saw, in setting the conditions for three-phase equilibrium, one need not invoke any further ideas on the nature of two-phase equilibria, or decide which of the components is the solvent or the solute. The important thing is to fix the parameters correctly, and to decide which of the concentrations are the dependent and independent parameters.

Our derivation of the condition of three-phase equilibrium in a two-component system is therefore quite general and no additional consequences or limitations are imposed by it on the Gibbs free energies of the coexisting phases. Hence the assertion [50] that the conditions (3.37) and (3.38) for invariant equilibrium – which are a direct consequence of the Gibbs equilibrium condition and are, in fact, its geometrical representation – are necessary but not sufficient, is in our opinion without foundation. The conclusion of the authors of [50], that in three-phase equilibrium of a two-component system the derivatives of the free energies of the coexisting phases with respect to concentrations are not only equal but vanish, while the actual free energy values are the same, is based on a treatment of two-phase equilibrium in the liquidus–solidus region for each of the three phases and in the region of demixing of the solid solution. Their error consisted in equating the concentration derivatives of all the components in the separation region, which is tantamount to equating the chemical potential differences between components B and A in the α-phase, and between A and B in the β-phase. At this point the conclusion of these authors is in contradiction to the fundamental conditions of phase equilibria obtained completely rigorously by Gibbs, without any model assumptions of extrapolations of the phase equilibrium line. Their conclusions on the equality of free energies in three-phase invariant equilibrium of a two-component system elevates this situation to a special position *vis-a-vis* two-phase equilibrium without any justification, since the latter can also occur at the given fixed temperature, so that the variance of the system should play no role. In addition, this conclusion gives rise to the equality of the chemical potentials of both compounds. Although such equality is not excluded thermodynamically, nevertheless it is obvious that it is realized physically only for a very special combination of the properties of the components and phases, so that it corresponds to a particular case of phase equilibrium.

The eutectic and peritectic three-phase equilibria we have discussed are quite general in that they characterize the processes of the transformation of a phase into the formation of two others, or the converse – one of transformation of two phases accompanied by generation of a third. However, depending on the states of aggregation (still in a condensed system) of the interacting phases, other forms of three-phase equilibria are possible in a two-component system.

If all three phases are solid (which is feasible even if one of the components has only one polymorphic transformation), eutectic and peritectic-like three-phase equilibria are possible. They are termed eutectoidal and peritectoidal equilibria. They may be represented as follows:

$$\gamma \rightleftarrows \alpha + \beta, \tag{3.45}$$

$$\gamma + \alpha \rightleftarrows \beta. \tag{3.46}$$

If in some temperature and concentration range one has two liquids and a solid, then so-called monotectic and syntectic three-phase equilibria may occur, corresponding to the phase reactions

$$L_1 \rightleftarrows L_2 + \alpha, \tag{3.47}$$

$$L_1 + L_2 \rightleftarrows \alpha. \tag{3.48}$$

It is easy to see that the nature of the monotectic (3.47) resembles the eutectic (3.34) or eutectoid (3.45), while the syntectic (3.48) is similar to the peritectic (3.35) or the peritectoid (3.46) three-phase equilibria. The state of aggregation of a phase obviously places no restriction whatever on the condition of three-phase equilibria, and, therefore, the thermodynamic basis of the three-phase equilibria in (3.45)–(3.48) should not differ in principle from the corresponding derivations of the eutectic and peritectic equilibria (cf. Fig. 12).

A combination of the several forms of two- and three-phase equilibria leads to various phase relations in the system and, consequently, to various state diagrams. Many modifications of the nature of the monovariant equilibrium curves are possible by combining various phase regions within the boundaries of a single phase equilibrium diagram. We shall now consider the derivation of some basic varieties of state diagrams of two-component systems, based on the principles already developed of treating two- and three-phase equilibria by means of the concentration-dependent Gibbs free energy curves.

3.3. Establishment of Basic Types of State Diagrams in Two-Component Systems

Let us consider systems with a three-phase eutectic equilibrium. According to the above, we have an initial situation in which the reciprocal addition of components A and B leads to a reduction of their melting temperatures. We draw the eutectic horizontal line aeb as well as the melting temperatures of A and B on the $T–x$ state diagram (Fig. 13a). Let us show how the relative positions of the $G–x$ curves of the three phases coexisting at the eutectic temperature T_e change as the temperature is raised or lowered relative to T_e.

Analyzing the physicochemical content of three-phase eutectic equilibrium, it is easily shown that it is a combination of three two-phase equilibria, namely $L \rightleftarrows \alpha$, $L \rightleftarrows \beta$, and $\alpha \rightleftarrows \beta$. This conclusion follows from a consideration of the common tangent to the curves of the concentration-dependent Gibbs free energies of the three phases at the eutectic temperature.

Indeed, section $a'e'$ (cf. Fig. 12) of the tangent refers to the first of the three two-phase equilibria mentioned, sections $e'b'$ and $a'b'$ to the second and third, respectively. Consequently, the eutectic horizontal may be considered to be the sum of the conoids, symbolizing the presence of the three two-phase equilibria referred to.

Looking at the relative positions of the melting temperature lines and the eutectic horizontal aeb, we conclude that on increasing the temperature the first of the two two-phase equilibria is stabilized, whereas on lowering it the third is stabilized. In the former case the $G–x$ curve of the liquid phase is shifted downward, so that its

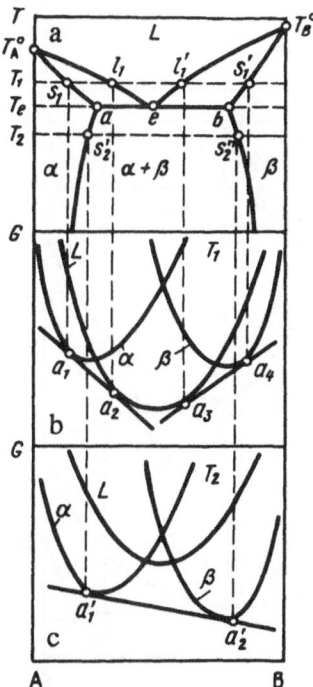

Fig. 13. Construction of a eutectic state diagram by means of concentration-dependent Gibbs free energy curves.

minimum is situated below the minimum of at least one of the curves of the other two phases (Fig. 13b). This excludes the possibility of constructing a common tangent to the Gibbs free energy curves of the α and β phases, missing the curve L. In the second case the curve L is raised and the common tangent construction becomes possible (Fig. 13c). Remembering our discussion on the relative positions of the $G = f(x_B)$ curves of the three phases in a system with a eutectic three-phase equilibrium for temperatures $T_1 > T_e$, and drawing the pairwise tangents a_1a_2 and a_3a_4 to the G–x curves, we may locate the compositions of the equilibrium phase forming at $T_1 > T_e$ corresponding to the two-phase equilibria. Projecting these compositions onto the T_1 line on the T–x diagram, we determine the fixed positions of the nominal points l_1, s_1, and l_1', s_1' for the phases forming the two-phase equilibria $\alpha \rightleftharpoons L$ and $L \rightleftharpoons \beta$.

Similarly, for other temperatures we reproduce entirely the liquidus and solidus curves. For temperatures $T_2 < T_e$ the relative positions of the G–x curves of the three phases under discussion allow one to draw the tangents $a_1'a_2'$ to the $G^\alpha = f(x_B)$ and $G^\beta = f(x_B)$ curves, and thereby determine the compositions of the α and β phases participating in the third of the two-phase equilibria (Fig. 13c). Projecting these compositions on the temperature T_2 in the T–x diagram we obtain the fixed positions of the nominal points of the α and β phases at T_2 in the T–x diagram. The same discussion for temperatures below T_e allows one to trace out curves delineating the regions of solid solutions based on components A and B. These curves are

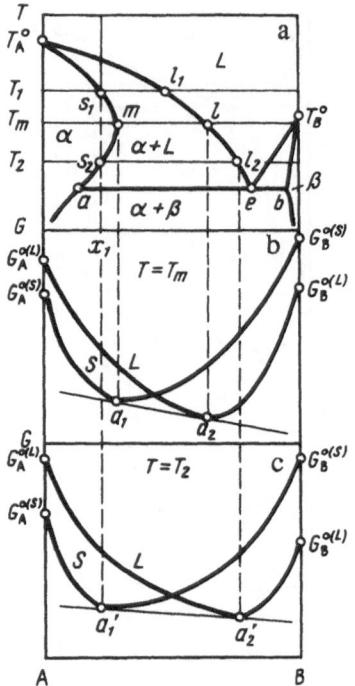

Fig. 14. Retrograde solidus in a eutectic system and the relative position of $G-x$ curves of the solid and liquid phases at various temperatures.

called lines of stability limits. A eutectic state diagram describes completely the behavior of a mixture and provides a clear idea of the phase composition in various temperature and concentration regimes. It is easy to see that sections of the diagram referring to two-phase equilibrium between solid solutions of A and B on the one hand, and liquid solutions on the other, represent parts of the diagram with a continuous series of solid and liquid solutions (cf. Fig. 9), whereas the sections enclosed by stability limit lines may be treated as a truncated immiscibility hump, shown in Fig. 10. Such descriptions of transitions from one phase region to another, as well as processes of crystallization or melting of alloys of various compositions, are outside the scope of our work. Details may be found in many handbooks on heterogeneous equilibrium [45, 51–58]. The actual mechanism of eutectic crystallization as a complex phenomenon of the generation of two crystalline phases from a liquid has been described in the classic work of Bochvar [59], who treated it as a step-by-step segregation of driving and driven phases, commensurate with the accompanying crystallization, during the changing composition of the surrounding solution and its supersaturation in favor of one or the other phase.

Of particular interest is the joining of the solidus and the solubility limit lines. Figure 13 shows that this joining occurs at the point of maximum extension of the solid solution based on A and B. In the literature such points are called saturation points.

We note, however, that thermodynamics does not impose any limitation on the direction of the solidus and solubility curves and, therefore, the maximum extent of the homogeneous region need not be located precisely at the eutectic temperature. In principle the maximum solubility may lie both above and below the eutectic temperature. Nevertheless, the picture presented covers a sufficiently common case, when each curve of monovariant equilibrium changes in the way depicted in Fig. 13a, and, therefore, it is natural that the maximum solubility will occur at their melting point. But, of course, however common, this is a special case. The maximum extent of the region of homogeneity on the basis of one of the components below the eutectic (or other three-phase equilibria) line is seldom met. Much more common is the case of maximum solubility above the temperature of three-phase equilibrium, which has acquired the name "retrograde solidus."

Figure 14 depicts a section of a state diagram with a retrograde solidus in three-phase eutectic equilibrium. It is apparent that the maximum extent of the region of solid solution based on A, denoted on the solidus curve by the letter m, corresponds to a temperature T_m which is higher than the temperature of three-phase eutectic equilibrium. The picture shown is unusual in that, if one considers, on cooling from a liquid, an alloy of composition x_1, then it transpires that after complete solidification of the liquid of composition l_1, at temperature T_1 at the point s_1, it melts on further cooling to a lower temperature at point s_2 and forms liquids of composition l_2 and s_2 at temperature T_2.

Since it is difficult to explain physically the melting of a substance in the process of cooling, one meets statements in the literature to the effect that a retrograde solidus does not exist (cf., for example, [60]). However, the possibility of its existence was already proved thermodynamically by van Laar [61] in 1908. The problem of retrograde melting has been further discussed in [62–65]. Experimental studies based on a germanium–silicon system have shown [66] that in the majority of cases the maximum extent of the solid solution is situated above the temperature of three-phase equilibrium and, consequently, that the solidus is retrograde. Since retrograde melting is of great interest both theoretically and experimentally, we shall consider its thermodynamic justification and its possible analytic description in Chapter 5. Here we only remark that no objections are raised in principle from the point of view of geometrical thermodynamics against the existence of a retrograde solidus. Figure 14b, c shows the concentration-dependent Gibbs free energy curves of the solid and liquid phases at temperatures T_m and T_2 (where $T_m > T_2 > T_e$).

We can see that the position of the nominal points of the equilibrium phases in the T–x diagram is uniquely fixed by drawing the common tangents a_1a_2 and $a_1'a_2'$ to the $G = f(x_B)$ curves of the phases which are in equilibrium at each temperature. Lowering the temperature from T_m to T_2 only shifts the corresponding curves in the region of higher values of the Gibbs free energy due to the general trend of its temperature dependence, and increases the concentration interval between the points of contact with the curves $G^\alpha = f(x_B)$ and $G^L = f(x_B)$. However, one cannot provide a tangible cause for the shifts of the Gibbs free energy curves along the concentration axis on the basis of geometrical thermodynamics alone. It should be pointed out that a retrograde solidus is possible not only for eutectics, but for other possible three-phase equilibria in a two-component system as well, since the nature of three-phase equilibrium is, in principle, not relevant to the essence of the phenomenon under discussion. The only important feature is that the region of two-phase equilibrium between the solid and liquid solutions should be descending.

If a system has a three-phase peritectic transformation, one may estimate the position of its temperature relative to the melting temperatures of the components A and B according to the inequality (3.44) above. The relative magnitudes of the melting temperatures of the component may naturally be reversed, but the temperature of the three-phase peritectic transformation should still be intermediate between the two melting temperatures of A and B.

Consequently, in discussing phase changes such as the three-phase peritectic transformation in a two-component system, we may plot in the T–x diagram the position of the peritectic horizontal and mark the melting temperatures of A and B according to (3.44). The position of the peritectic horizontal is determined with the aid of the G–x curves in Fig. 12b. Starting from the picture presented in Fig. 12b, we show, first, the changes in the relative positions of concentration-dependent Gibbs free energy curves of each of the phases forming the peritectic equilibrium, when the temperature is either above or below the peritectic horizontal.

Analysis of the physicochemical content of three-phase peritectic equilibrium supports the assertion that, like the eutectic three-phase counterpart, it, too, is a combination of three two-phase equilibria: $L \rightleftharpoons \alpha$, $L \rightleftharpoons \beta$, and $\alpha \rightleftharpoons \beta$. Whereas in the eutectic the phase at intermediate composition is liquid, in the present case it is the α-phase. Of course the role of the α-phase in the peritectic transformation is essentially different, since the liquids in the eutectic are being transformed; here, the α-phase is being generated according to (3.34) and (3.35). Therefore, in cooling below the temperature of the peritectic transformation, the α-phase is stabilized, whereas on cooling below the eutectic temperature the liquid phase disappears. This feature is essentially due to the nature of the shift of the G–x curve for the α-phase when the temperature is changed with respect to the peritectic horizontal.

From the above considerations, the peritectic horizontal pab may be considered to be made up of the three conoids pb, pa, and ab, as seen quite clearly from common tangents to the G–x curves of all three phases in Fig. 12b.

One may conclude, therefore, that on increasing the temperature above T_p the two-phase equilibrium $L \rightleftharpoons \beta$ will be stabilized, whereas decreasing it below T_p results in the stabilization of the other two phase equilibria $L \rightleftharpoons \alpha$ and $\alpha \rightleftharpoons \beta$ in the appropriate concentration regimes.

In the Gibbs free energy diagram the composition will be expressed via the shifts of the G–x curves of each phase under consideration. Thus, since the α-phase is formed on cooling, the curve of the Gibbs free energy of this phase moves downward, lowering the temperature to T_2, say (Fig. 15a), so that the minimum of this curve is lower than at least one of the concentration-dependent G curves for the other two phases (Fig. 15b). If the temperature increases to, say, T_1 (Fig. 15a), then the $G^\alpha = f(x_B)$ curve moves upward, as shown in Fig. 15c. On account of the above discussion of the positions of the G curves for all three phases constituting peritectic equilibrium, the composition of the equilibrium phases may be determined by the usual common tangent constructions. From an analysis of Fig. 15c for $T_1 > T_p$ one infers that in the concentration region between the tangent points a_1' and a_2' the two-phase equilibrium $L \rightleftharpoons \beta$ is the stable one, since the Gibbs free energy of the mixture of two of these phases, determined at each composition by the position on the common tangent a_1'–a_2', is always below the Gibbs free energy of any one separate phase. Projecting the compositions pertaining to the points of contact a_1' and a_2' onto the temperature line T_1 in the T–x diagram, one obtains the fixed positions of the conoid $l_1 s_1$ defining the conjugate points on the liquidus and solidus at a

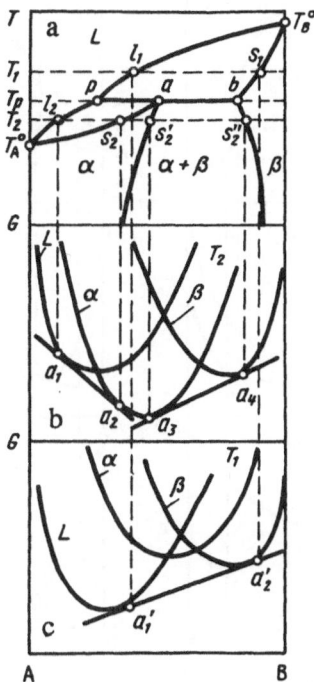

Fig. 15. Construction of a peritectic phase diagram by means of concentration-dependent Gibbs free energy curves.

given temperature. Using a similar construction for other temperatures in the interval T_p–T_B°, one may reconstruct the whole of the liquidus and solidus curves which delineate the region of the two-phase equilibrium $L \rightleftharpoons \beta$.

At temperatures $T_2 < T_p$ the relative positions of the G curves are such (Fig. 15b) that common tangents may be drawn to them, namely a_1–a_2 and a_3–a_4. This is caused by the downward shift of the intermediate G^α curve because of the stability of the α-phase on cooling. Obviously, the position of the common tangent a_1–a_2 limits the concentration interval l_2–s_2 within which a mixture of the α and L phases is stable at a given temperature. The position of the tangent a_3–a_4 limits, at T_2, the stability regime of the two-phase equilibrium $\alpha \rightleftharpoons \beta$. The projection of the common tangent a_3–a_4 onto the temperature T_2 in the T–x diagram will determine the compositions of the equilibrium phases α and β at a given temperature, through fixing the position of the conoid s_2'–s_2''. Changing the temperature within the limits T_p and T_A, one may describe, through the reconstructed liquidus and solidus curves, the stability limit of the two-phase $\alpha \rightleftharpoons L$, and the corresponding portions of the solubility limit curves limiting the position of the region of two-phase equilibrium $\alpha \rightleftharpoons \beta$.

At temperature T_2 and concentrations between a_2 and a_3, the G–x curve of the α-phase for each composition determines the lowest possible value of the G function which specifies the stable single-phase state at this concentration range and

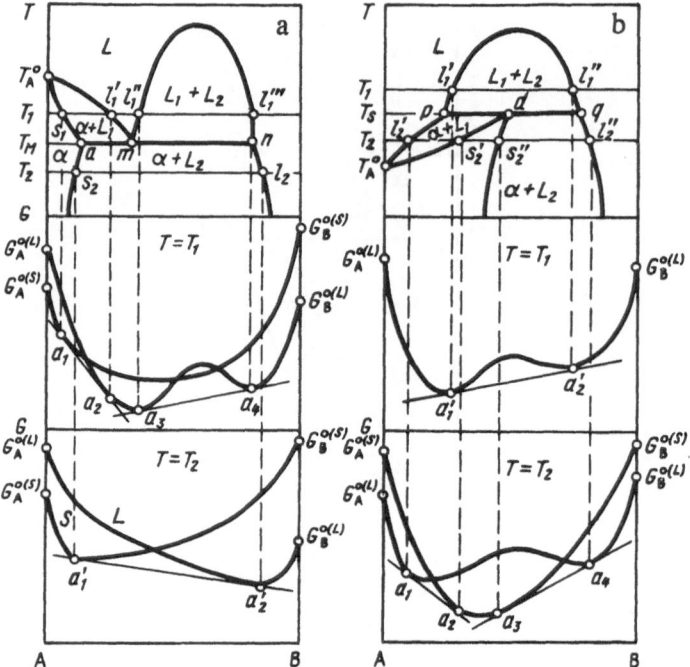

Fig. 16. Phase equilibria in the vicinity of three-phase monotectic (a) and syntectic (b) equilibria, in accordance with the relative positions of the composition-dependent Gibbs free energy curves above and below the temperature of invariant transformation.

temperature. Below the melting temperature of A the curve $G^\alpha = f(x_B)$ is always lower than $G^L = f(x_B)$, which in the end is responsible for determining the region in which only the two-phase equilibrium $\alpha \rightleftharpoons \beta$ is stable. As a result of the foregoing construction in the $T–x$ plane, one can build up the peritectic state diagram (Fig. 15a), which joins the eutectic as an important type, since it illustrates one of the many ways of combining three-phase equilibria with the corresponding two-phase equilibria.

In concluding the treatment of eutectic and peritectic state diagrams, we note that, in constructing these diagrams by means of geometrical thermodynamics, we have assumed, in both cases, the maintenance of the two crystalline α and β phases, differing from one another by their structure and their properties and, accordingly, characterized by independent equations of state. Therefore, in contrast to Mlodzeevskii [42], who used only a single curve of the concentration-dependent Gibbs free energy for the solid phase in establishing these diagrams, in the way shown in Fig. 10, our more rigorous derivation employed two independent curves $G^\alpha = f(x_B)$ and $G^\beta = f(x_B)$. The final result turns out to be the same.

If one does not take into account the state of aggregation of the phases, then the eutectic and peritectic equilibria, in principle, completely encompass the possible combinations of three- and two-phase equilibria. Hence, from the point of view of geometrical thermodynamics, a consideration of possible variants of equilibrium, wherein one takes account of other states of aggregation of phases in two-component systems, contains practically nothing new in principle. This point is illustrated clearly by the example of three-phase equilibrium with two participating liquid phases. In accordance with (3.47) and (3.48) one or two liquid phases may play the role of a transforming phase. Then the generated phases may either be a liquid in combination with a solid, or one solid. In the first case, as already remarked above, the three-phase equilibrium is monotectic; in the second case it is syntectic. The presence of two liquid phases in a two-component system is coupled to the solubility gap in a definitive concentration regime and to the appearance of the two-phase equilibrium $L_1 \rightleftharpoons L_2$, which is described in the T–x phase diagram by the immiscibility hump shown in Fig. 10. The hump may join the horizontal three-phase equililbrium line in two ways. In the first case, one liquid is the transforming phase, and one solid and the other liquid the generated phases, thereby creating a monotectic three-phase equilibrium (Fig. 16a). In the second case, corresponding to three-phase syntectic equilibrium, the two liquids are the transforming phases and the solid, the generated phase. For the monotectic equilibrium the liquid phase L_1 occupies an intermediate position of composition between the solid solution α (or β) and the liquid L_2. In considering variants for which the melting temperature of A is higher than of B, the temperature of three-phase monotectic equilibrium T_m is clearly between T_A° and T_B°.

We may consider the horizontal of monotectic three-phase equilibrium, *amn*, in the same way as for a eutectic, to be a combination of three conoids, and derive a picture of the relative positions of the G–x curves of the three equilibrium phases much as in Fig. 12a. We find that when the monotectic temperature is exceeded, both two phase equilibria $L \rightleftharpoons \alpha$ and $L_1 \rightleftharpoons L_2$ become stabilized. Dropping below the monotectic horizontal should facilitate the stabilization of the equilibrium $\alpha \rightleftharpoons L_2$. The changes in the relative positions of the G–x curves of the three phases will, in principle, be similar to those illustrated in Fig. 13b, c for eutectic systems. However, keeping in mind immiscibility in the liquid and the fact that for specific superheating the two liquids become one, it will be sufficient to consider a single concentration-dependent F curve for the liquid, as was done in Fig. 10, rather than two curves as was done for a eutectic. The relative shift of the corresponding sections of this curve and the curve of $G^\alpha = f(x_B)$ as the temperature changes is completely analogous to that observed in a eutectic system.

Therefore, from the point of view of geometrical thermodynamics, three-phase and eutectic equilibria are analogous.

Let us consider syntectic three-phase transformations in which the liquids L_1 and L_2 are the transforming phases and the solid solution α the phase generated. We are led to conclude, first of all, that the horizontal of the three-phase syntectic equilibrium *pdq* lies above the melting temperature of the more refractory component (e.g., A). Second, we find that heating above the temperature T_s of syntectic three-phase equilibrium leads to the stabilization of the two-phase equilibrium $L_1 \rightleftharpoons L_2$, while cooling below T_s stabilizes the two-phase equilibria $L_1 \rightleftharpoons \alpha$ and $L_2 \rightleftharpoons \alpha$.

The relative positions of the curve $G^\alpha = f(x_B)$ and the corresponding sections of the single curve $G^L = f(x_B)$ are similar in this case to those indicated in Fig. 12b at the syntectic temperature, whereas above and below it the qualitative picture is as represented in Fig. 15b, c for a peritectic system. From the point of view of geometrical thermodynamics the syntectic three-phase equilibrium is analogous to the

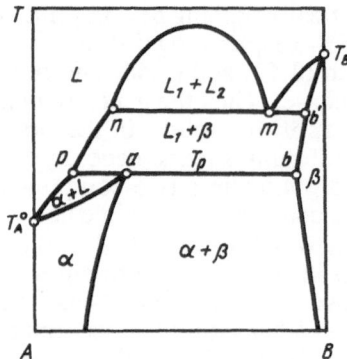

Fig. 17. Diagram of phase equilibria of a two-component system in the presence of monotectic and peritectic three-phase equilibria.

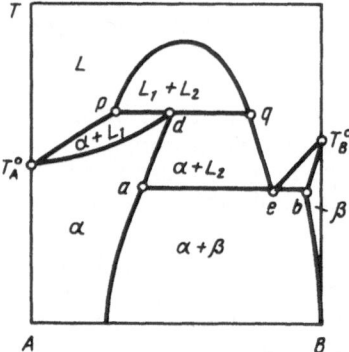

Fig. 18. Diagram of phase equilibria of a two-component system for syntectic and eutectic three-phase equilibria.

peritectic case. The difference lies only in the state of aggregation of one of the three phases participating in equilibrium.

To sum up, the general picture of equilibrium in the vicinity of the syntectic horizontal *pdq* may be represented as shown in Fig. 16b.

We note that the fixed position of the line of monovariant equilibrium in systems represented by Fig. 16 is established in principle in the same manner as for eutectic and peritectic types, namely with the aid of the Gibbs free energy versus concentration curves at various temperatures.

The general picture of the relative positions of the $G^a = f(x_B)$ and $G^L = f(x_B)$ curves above and below the monotectic (syntectic) horizontal is illustrated in Fig. 16.

Possible variants of phase diagrams in the presence of a dissociation hump, and for three-phase monotectic and syntectic equilibrium are indicated in Figs. 17 and 18.

Figure 17 illustrates the existence of two three-phase equilibria in the system: monotectic and peritectic. A possible variant is when eutectic equilibrium replaces peritectic equilibrium. The same holds for the system represented in Fig. 18, namely, instead of eutectic, peritectic equilibrium may occur.

As can be seen, the appearance of even one extra liquid phase in the system fundamentally complicates the picture of phase equilibrium in a two-component system. Obviously, the formation of intermediate solid phases should also be reflected in the state diagram. As a rule, intermediate solid phases are formed from definite chemical compounds [67], which may undergo congruent melting or may break up as a result of peritectic transformation.

In treating systems with intermediate phases the concentration-dependent Gibbs free energies of these phases are of interest. The treatment of this question in [41–43, 49, 52] is not always presented in a thermodynamically rigorous fashion and leads, therefore, to known ambiguities in the explanation of the problem. One must first of all provide a thermodynamically strict definition of the concept of "phase." The presence of definite compounds means that we have to define whether the solutions based on them relate to different phases or the same one. We shall adhere strictly to the systematic thermodynamic exposition of this problem presented in [13].

As is known, the nature of a phase is fixed by the properties of the intermolecular interaction, which in turn is specified by the species of particles forming the phase. The nature of the particles forming a given phase determines the magnitude and character of the force of the exchange interaction responsible for the chemical bond. If the solutions and phases differ by the species of their constituent particles, then their chemical compositions (we are talking about pure components) are qualitatively different and the thermodynamic characteristics of these phases are described by different fundamental equations [13]. This very important observation leads to the conclusion that such solutions, even in the limit of a single homogeneous system, should be treated as independent phases. The differences between the dependences on the state parameters of the properties of the solutions having qualitatively different chemical compositions should appear, if not in the actual forms of the functions, at least in the values of the constants featured in the equations for these functions, and reflecting specific interparticle interactions. If the composition of solutions or phases is variable, then to a given qualitative composition, in other words to a given set of particle species, there corresponds a finite interval of composition magnitudes in the system, and only within this region does there exist a strictly defined form for the state-parameter dependence of thermodynamic and other properties. The proposition that the nature of the dependence of the properties on the state parameters is determined by the qualitative chemical composition is most fundamental, and has been coined by Storonkin [13] "the principle of qualitative uniqueness of the specific chemical compounds." In essence this principle allows one to assign clearly the various solutions to the independent phases on the basis of specific compound content and, consequently, permits a more rigorous application of geometrical thermodynamics to establishing the general character of phase transformations in systems with intermediate phases.

Thus, if a nondissociating compound A_mB_n is formed in a system, then obviously the solutions that are formed on addition of the A component to A_mB_n and of B to A_mB_n should be assigned to different phases, independently of the state of aggregation of the system. We now reach the important conclusion that, as a result of the interaction between components A and B, a chemical compound A_mB_n has been formed with its own crystal structure, if it is a solid, which as a rule is totally different from that of A and B, and with its own short-range order if it is a liquid. Therefore, stable chemical compounds should be treated as independent components, separating the system into independent subsystems.

It should be noted that this important conclusion is at the basis of all further treatment of the phase relations in systems having intermediate phases based on well-defined chemical compounds. It is remarkable that this conclusion had already been reached by Mendeleev [68] when he considered the possible formation of well-defined chemical compounds in solutions. He pointed out that solutions will break up, separating into chemical compounds, if the latter can occur in the solutions. Wittorf [52] also pointed out that well-defined chemical compounds should be treated as independent components. The systematic use of this concept, together with the above-mentioned assigning of solutions on the basis of compounds to independent phases, fundamentally alters the approach to establishing phase relations by means of the concentration-dependent Gibbs free energy curves of the phases. The use of the gross (overall) compositions of alloys as state parameters for each part of the system forces one to introduce corrections.

In view of the foregoing we consider a liquid solution in which a definite nondissociative compound A_mB_n is formed. Since this compound is considered as a component, it will be characterized at each temperature by independent values of the Gibbs free energy. Consequently, the mixtures $A + A_mB_n$ and $A_mB_n + B$ are independent phases. On this basis the variation of the Gibbs free energy with the gross composition of the liquid solution is determined by the curve shown in Fig. 19. It follows that each part of the system having its own liquid phase is characterized by its own curve of $G = f(x)$ which is everywhere convex with respect to the composition axes. Here one can obviously dispense with the need to discuss the construction of the common tangent a_1a_2 to two of these curves, as is done in [52], since they belong to essentially different systems, and only for the sake of clear presentation of the characteristics of the system $A–B$ as a whole was the Gibbs free energy given as a function of the gross composition, i.e., of the molar fraction of the B component.

Even an insignificant dissociation of the compound A_mB_n radically changes the situation. A qualitative molecular composition to the left of the ordinate of the compound rearranges the basically different composition to its right, so that one finally has a single liquid phase. Since one is talking about a single phase, the concentration dependence of its Gibbs free energy should be given by curves such as those in Fig. 6.

The above discussion also applies to systems in the solid state, since for this purpose there is no difference in principle between liquid and solid solutions. We consider two fundamentally different types of state diagrams with chemical compounds, namely those which 1) are stable right up to the melting temperature and turn into a liquid at the same composition via congruent melting, and 2) participate in a three-phase peritectic transformation as an intermediate phase, via incongruent melting.

In the absence of dissociation into solid and liquid states the congruently melting compounds separate into two systems $A–B$, in accordance with the number of

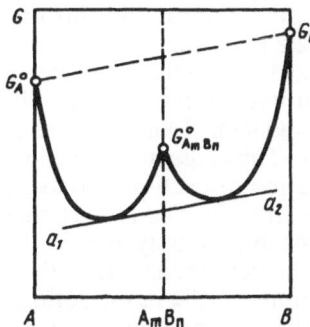

Fig. 19. Concentration-dependent Gibbs free energy in a system with a stable nondissociating compound A_mB_n in the liquid phase.

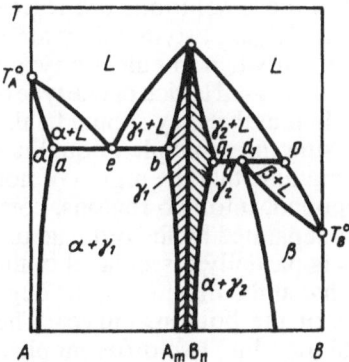

Fig. 20. State diagram of a system with congruent melting of nondissociating compounds.

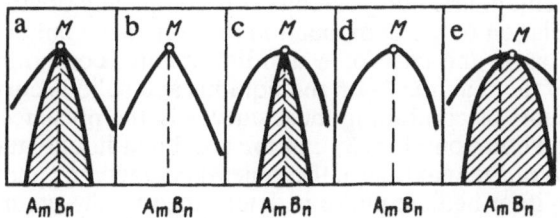

Fig. 21. Types of maxima of the liquidus and solidus curves in the A–B system in the presence of compound A_mB_n as a function of the latter's thermal stability in the solid and liquid states.

subsystems within which phase equilibrium may be established in one of the above-described forms of simple state diagrams. In this case we fix the position of the lines of the phase diagram in each particular system by means of the Gibbs free energy curves in the same way as for eutectic or peritectic systems. Figure 20 shows the state diagram of a system with stable congruently melting compounds A_mB_n, which on interacting with components A and B forms partial systems corresponding to eutectic and peritectic types. The character of the interaction in this case is chosen arbitrarily, while the construction of each partial system by means of the $G = f(x)$ curves is depicted in Figs. 13 and 15. The picture may become more complicated due to possible demixing of each partial system in the liquid phase, as well as on account of polymorphism of the components A or B, and of the compound A_mB_n.

The stability of the compound A_mB_n in the solid and liquid states is confirmed by the singular maximum at the melting point, corresponding to the stoichiometric composition where the branches of the liquidus and solidus curves of the limiting partial systems converge.

From Fig. 20 we may infer that the solid solutions of the components A and B formed on the basis of the A_mB_n compound are in essence different phases γ_1 and γ_2, rather than a single γ-phase as frequently asserted in a number of manuals on phase equilibria. Naturally, these solutions are formed on the basis of a single crystal lattice of the compound A_mB_n, so that they are isomorphs. Nevertheless, a number of examples are known when in eutectic-type systems the solid solutions based on the components have a crystal lattice of one type (for example the Eu–Ag system), while it is completely obvious that they should be thought of as different phases. In the present case, again using the principle of qualitative uniqueness of chemical compounds, one is led to conclude that the region of homogeneity splits, according to the stability of the compound, into two regions, corresponding to two different (though isomorphic) phases separated by the ordinate of the chemical compound.

If the compound A_mB_n is partially dissociated in the liquid phase, then the liquidus curve flattens somewhat at the maximum, the degree of dissociation being directly related to the radius of the liquidus curve. The compound A_mB_n then no longer divides the liquid solution into two different phases within a single homogeneous system, and as a result there will be only a single liquid phase in the A–B system. This should be accounted for by appropriate constructions with the help of geometrical thermodynamics. If the compound A_mB_n dissociates not only in the liquid but also in the solid state, the solidus curve will also acquire a flattened top, and, as in the liquid state, the radius of curvature will characterize the degree of dissociation of the compound. It should be borne in mind that in the intrinsic dissociation of the compound the maximum of the state diagram may be shifted with respect to the stoichiometric composition. It is completely obvious that on the basis of a partial dissociation of the compound in the solid state only one phase can form rather than two, as in the case of a nondissociating compound. Figure 21a–d shows the form of the maxima M on the liquidus and solidus curves, characterizing the different variants of combining these curves in the presence or absence of the dissociation of the compound A_mB_n, in the solid and liquid states. But, independently of the form of the maximum, the general character of phase equilibrium in a system having an intermediate phase is determined by the nature of its interaction with the components in the liquid and solid states and, accordingly, by the form of the resulting three-phase equilibrium.

If a chemical compound is formed in a peritectic reaction, the general principle of establishing phase relations in the vicinity of the peritectic horizontal pMb does

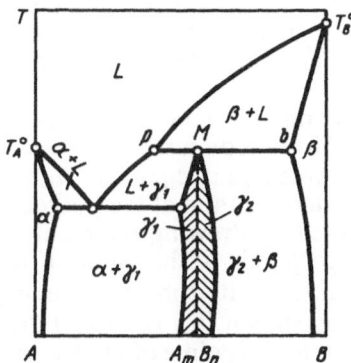

Fig. 22. State diagram of a system with incongruent melting of the chemical compound A_mB_n.

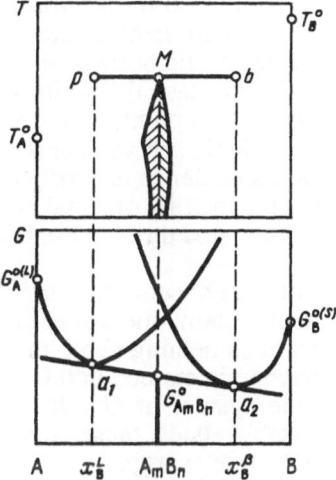

Fig. 23. Isobaric three-phase peritectic equilibrium in a two-
component system with incongruent melting of the compound
A_mB_n, derived by the geometrical thermodynamic method.

not differ from the simple peritectic case considered previously (cf. Fig. 15). But,
due to the formation of a solid solution based on component A and the corresponding two-phase equilibrium $\alpha \rightleftharpoons L$, it must be combined with the two-phase equilibria
which stabilize on cooling below the peritectic horizontal with the participation of
phases on the basis of the compound A_mB_n. As a result, various three-phase equilibria are possible. Figure 22 illustrates one possible state diagram with incongruent
melting of the compound A_mB_n. In trying to derive such a diagram by means of
Gibbs free energy curves, one should remember that either one or two phases may
form on the basis of the compound, so that in the geometrical constructions the

number of $G = f(x)$ curves should correspond to the number of phases. In other words, in a thermodynamic justification of the type of system shown in Fig. 22, one must take into account the principle of qualitative uniqueness of the relevant chemical compounds.

We note that the formation of a stable compound A_mB_n of stoichiometric composition in a peritectic transformation justifies the assumption that below the peritectic horizontal it plays the role of a component, with all the consequences that this entails. In particular, at the temperature of the peritectic horizontal the equilibrium of phases may be established by drawing the common tangent a_1a_2 through the point $G°_{A_mB_n}$ (which corresponds to the value of the Gibbs free energy of the compound A_mB_n) to the curves of its concentration dependence for the liquid (L) and the solid (β) phases, as shown in Fig. 23. Furthermore, by projecting the compositions corresponding to the points of contact onto the temperature T_p in the $T–x$ diagram, we obtain the fixed positions of the phases which at this temperature are jointly in equilibrium with the compound A_mB_n.

The variant depicted in Fig. 23 of forming a definite compound directly in a peritectic process is for a particular relation between the melting temperatures of the components and for three-phase equilibrium. Naturally, this relation may assume some other value, but the discussion will, in principle, remain the same. The compound A_mB_n may be joined to the peritectic horizontal at any point and may eventually coincide with the point p. In this case the melting of the compound A_mB_n is perceived as congruent at the transition point, and the three-phase equilibrium is treated as equilibrium between the liquid and solid compound A_mB_n and the β-phase [69]. However, such a physical coincidence is hardly ever attained, and its actual probability is quite small, so that the melting of a real compound will be either congruent or incongruent, and the associated three-phase horizontal will be eutectic or peritectic, respectively.

By using the construction of the concentration-dependent Gibbs free energies of the phases in a two-component system, one can predict extrapolations of monovariant equilibrium lines and learn something about their respective positions. The crux of the geometrical thermodynamic method in this context [49] is that, in drawing the common tangents to the system of $G = f(x)$ curves characterizing the metastable and corresponding stable states, these curves should be situated in regions of lower values of the Gibbs free energy.

Concluding our considerations of the general principles of constructing possible types of phase-equilibrium diagrams of two-component systems, we note that the examples presented above far from exhaust the variety of cases of combining one-, two-, and three-phase equilibria and, consequently, the possible types of state diagrams. In particular, we have not touched upon systems distinguished by partial mutual solubilities of their components, yet having a broad region of solid solutions characterized by a common point on the liquidus and solidus curves, i.e., including elements in the state diagram represented in Fig. 11. We do not consider systems with eutectic or peritectic three-phase equilibria, and many other cases. This is because the topic of our subsequent analysis of the thermodynamics of solutions is principally the two-phase equilibrium between solid and liquid solutions. In addition, a complete list of the possible forms of equilibrium is not possible even in principle, since this problem has not been solved completely at the present time.

As regards the role of pressure in the process of forming new combinations of monovariant equilibrium lines on the $T–x$ diagram, we call attention to the case dis-

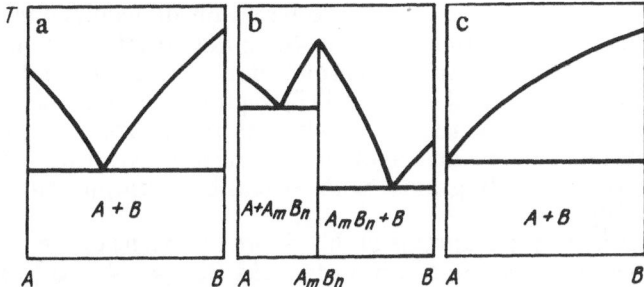

Fig. 24. State diagrams with vanishingly small regions of solubility in the solid phase.

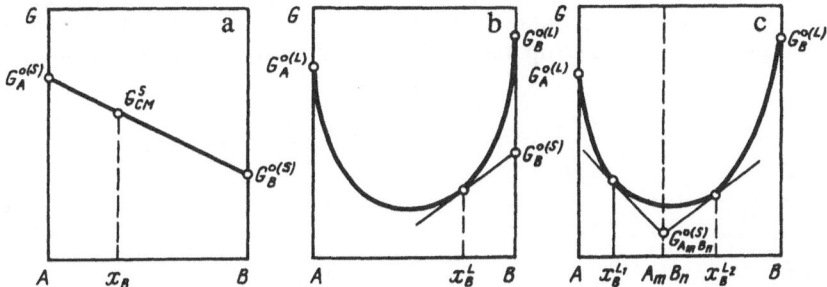

Fig. 25. Construction of diagrams of two-phase equilibria in systems with vanishingly small regions of solubility in the solid phase: a) a mixture of two solid components; b) equilibrium of the solid component of A and a liquid solution; c) equilibrium of the chemical compound with a liquid solution.

covered by one of the authors and developed in [70, 71], of isobaric phase interactions corresponding to the pressure at the triple point of one of the components.

On application of an external pressure other, basically new, types of state diagrams are possible for simple binary systems.

In considering possible combinations of regions of two-phase equilibrium in a two-component system, it was noted in [72] that, in addition to horizontal, lateral tangents are also possible, which may lead to fundamentally new types of state diagrams.

In treating the basic types of state diagrams of a two-component system, our starting point in all cases has been the fact of the formation of solid solutions on the basis of the components and the specific chemical compounds. Among these thermodynamics does not, in principle, rule out the possibility of the existence of a single-component phase in a multicomponent heterogeneous system. This assertion is based on the ideas of Gibbs concerning the reality of possible components, considered in the previous chapter. It was pointed out especially firmly by Putilov [16], in spite of the fact that an attempt to disprove absolute immiscibility by thermodynamic means was made by an authority of the stature of Planck [2].

Thus the proof of the existence of absolute immiscibility really involves a determination of which of the possible components actually exists, and the approach to the solution of the problem should be based on molecular-kinetic rather than thermodynamic considerations [73–78]. Such deliberations, connected with the molecular-kinetic treatment of the thermal motion of atoms and diffusion, have enabled one of the authors and Novikov [17] to prove the fundamental impossibility of the existence of the available components in heterophase multicomponent systems. From this follows an important conclusion concerning the fundamental impossibility of state diagrams with so-called "simple eutectics" and its modifications (Fig. 24).

The validation of the diagrams of this figure by means of the geometrical thermodynamic method, given in a series of works (cf. [41–43, 49]), should be considered conditional, or approximate (Fig. 25). More precisely, it should be formulated as follows: If a one-component phase exists in a heterogeneous system forming two or more components (in the present case nondissociating chemical compounds are also considered as components), then the derivations presented in [41–43, 49, etc.] are rigorous. If one assumes insignificantly small changes in the Gibbs free energy of a component with a very low solubility, then in the derivations one need consider only the temperature dependence of G for that component.

Chapter 4

FUNDAMENTALS OF THE
THERMODYNAMIC THEORY OF
SOLUTIONS OF CONDENSED SYSTEMS

1. GENERALIZED CONCEPT OF SOLUTION.
THERMODYNAMIC CLASSIFICATION OF SOLUTIONS

A solution is a system of variable composition in which one substance is distributed comparatively uniformly throughout one or more substances. We say "comparatively uniformly," since a completely uniform distribution is rare. A solution may generally have any state of aggregation, solid, liquid, or gaseous, so that one distinguishes gaseous, liquid, and solid solutions. Liquid solutions are divided into two groups: nonelectrolytic and electrolytic solutions. In gaseous systems it is more common to speak of gaseous mixtures rather than solutions. Solid solutions are less usual than liquid solutions, though thermodynamically they are analogous.

The process of dissolution can by no means be represented by a simple distribution of molecules or ions, since in the majority of cases it is associated with the appearance of various interactions among like and unlike particles. These specify the nature of the solution and the degree of departure from ideal behavior.

We stress that it is incorrect to compare the state of the dissolving substance with the state of its molecules in the gas, even in very dilute liquid solutions. A clear attestation of this is the fact that the heat of solution of a solid substance is usually very close to the heat of melting but considerably different from the heat of sublimation.

The simplest component parts of a solution which can be separated in a pure form and with which one may obtain solutions with any allowed composition by mixing are called the components of a solution.

From a thermodynamic point of view all components of a solution are equivalent and, therefore, dividing them into solvents and solutes is a convention, especially for systems representing a continuous series of solutions. Usually solvents are taken to be the components which are present in considerably larger quantities than the other components, or components which in a pure form under the given conditions have the same state of aggregation as the solution (whereas other components in pure form would have a different state of aggregation). In the following we shall designate solvents by the index 1, while the various solutes will be denoted 2, 3, ...

The most widely used method of describing concentrations in the thermodynamics of solutions is based on mole fraction representations [cf. Eqs. (1.1) and (1.2)].

In theory we distinguish two basic classes of solutions, ideal and nonideal (or real). Ideal solutions are subdivided into infinitely dilute solutions in which the mole fraction of each solute approaches zero, and perfect solutions, which remain ideal for all concentrations. The thermodynamic classification of solutions is based on the character of the equations for the chemical potentials of the solution components.

Thus, for an ideal solution the chemical potential of each component is defined by the expression

$$\mu_i^{id}(T, p, x) = \mu_i^{\circ}(T, p) + RT \ln x_i, \tag{4.1}$$

where $\mu_i^{\circ}(T, p)$ is the chemical potential of the component i at a standard temperature and pressure. The term $RT \ln x_i$ corresponds to the change in chemical potential due to mixing (in the course of forming the ideal solution).

We note that, in spite of the similarity in writing μ in Eqs. (2.53) and (4.1), its dependence on the applied pressure in these two cases is totally different. By analogy with (4.1) the chemical potentials of components in a nonideal solution may be written, according to [79], in the form

$$\mu_i(T, p, x) = \mu_i^{\circ}(T, p) + RT \ln x_i \gamma_i, \tag{4.2}$$

where γ_i is the activity coefficient.

The standard value of the chemical potential $\mu_i^{\circ}(T, p)$ obtains when the activity $(x_i\gamma_i = a_i)$ is unity.

There are two methods for normalizing the standard state:

1. The symmetric method for solutions in which both components are on an equal footing, in which case the state of the pure component serves as the standard state for each component. This is just the method used in computing the mixing functions on the basis of experimental data.

2. The asymmetric method, when as standard states of the components one chooses the states in the limit of dilute solutions. The essence of this method consists of the following:

a) for substance 1 (the solvent) we choose as standard its state in pure form;

b) for solvents we choose as standard their states in the dilute limit ($x_1 \to 1$, $x_i \to 0$), i.e.,

$$\mu_i^{\circ*}(T, p) = \lim_{x_i \to 0} (\mu_i - RT \ln x_i) \quad (i = 2, 3, \ldots, k).$$

The choice of standard states implies the normalization of the corresponding activity coefficients.

In the first case

$$\lim_{x_i \to 1} \gamma_i = 1, \quad (i = 1, 2, \ldots, k). \tag{4.3}$$

For the second case

$$\lim_{x_1 \to 1} \gamma_1^* = 1; \quad \lim_{x_i \to 0} \gamma_i^* = 1, \quad (i = 2, 3, \ldots, k). \tag{4.4}$$

By comparing (4.1) and (4.2) it is easy to show that the general condition for the ideality of a solution is that for all concentrations

$$\gamma_i (T, p, x_1, \ldots, x_{k-1}) = 1 \quad (i = 1, 2, \ldots, k).$$

The advantage of using the activity coefficients is that they enable one to preserve the formal similarity of the equations for nonideal and ideal solutions. For purely arithmetical reasons the activity coefficients are not as convenient a measure of departure from ideality for a solvent as for solutes. For solvents it is therefore more convenient to employ instead another correction factor, called the osmotic coefficient by Bjerrum and Guggenheim [18], and introduced in the following way:

$$\mu_1 = \mu_1^\bullet (T, p) + \Phi RT \ln x_1, \tag{4.5}$$

where $\Phi \to 1$ as $x_1 \to 1$ and $x_2, x_3 \ldots \to 0$. Comparing this expression with Eq. (4.2), we find

$$\Phi - 1 = \ln \gamma_1 / \ln x_1. \tag{4.6}$$

In using the osmotic coefficient the formal similarity with the equation for an ideal solution is partially lost, but in return the coefficient Φ is far more sensitive to small deviations from ideality than the activity coefficient.

2. THERMODYNAMIC FUNCTIONS OF MIXING. BASIC PROPERTIES AND THE LAW OF IDEAL SOLUTIONS

Before proceeding to the exposition of the basic properties of ideal solutions, we define functions of mixing. Let y be the molar value of some additive thermodynamic function (U, H, F, G, S, V, and so on). The molar function of mixing is denoted by y^M. By definition

$$y^M = y(x_1, \ldots, x_{k-1}, p, T) - \sum_i x_i y_i^\circ (p, T) \tag{4.7}$$

or

$$y^M = y(x_{\overline{1}}, \ldots, x_{k-1}, V, T) - \sum_i x_{\overline{i}} y_i^\circ (V, T), \tag{4.8}$$

where y_i° is the molar value of the thermodynamic function for the ith pure component.

It follows from the definition that the functions of mixing represent the changes in the thermodynamic functions of the solution during its formation from the pure components. With the aid of these functions one can describe the thermodynamic properties of solutions in a broad range of concentrations.

According to Eqs. (4.7) and (4.8) the functions of mixing vanish for pure components.

The partial molar function of mixing is defined by the difference

$$\bar{y}_i^M = \bar{y}_i - y_i^\circ, \tag{4.9}$$

where, according to (1.7),

$$\bar{y}_i = (\partial y/\partial n_i)_{p,\,T,\,n_{j\neq i}}.$$

Since by (1.9) any additive molar quantity is related to the partial molar quantities by the relation

$$y = \sum_{i=1}^{k} x_i \bar{y}_i, \tag{4.10}$$

we may represent the function of mixing, on account of (4.9), in the form

$$y^M = \sum_{i=1}^{k} x_i \bar{y}_i - \sum_{i=1}^{k} x_i y_i^\circ = \sum_{i=1}^{k} x_i (\bar{y}_i - y_i^\circ) = \sum_{i=1}^{k} x_i \bar{y}_i^M. \tag{4.11}$$

Therefore, the relations connecting the functions of mixing y^M and $\bar{y}_i^M$ are completely analogous to those between the additive thermodynamic functions and their partial molar values. On the basis of these relations it is easy to show that for a binary system we have

$$\mu_1^M = g^M - x_2 (\partial g^M/\partial x_2)_{p,\,T}, \tag{4.12}$$

$$\mu_2^M = g^M + x_1 (\partial g^M/\partial x_2)_{p,\,T}. \tag{4.13}$$

Similarly,

$$\bar{h}_1^M = h^M - x_2 (\partial h^M/\partial x_2)_{p,\,T}, \quad \bar{h}_2^M = h^M + x_1 (\partial h^M/\partial x_2)_{p,\,T}, \tag{4.14}$$

$$\bar{s}_1^M = s^M - x_2 (\partial s^M/\partial x_2)_{p,\,T}, \quad \bar{s}_2^M = s^M + x_1 (\partial s^M/\partial x_2)_{p,\,T}. \tag{4.15}$$

By utilizing the basic initial defining expression for an ideal solution (4.1), one may derive all other properties possessed by ideal solutions.

On the basis of (4.1) and (4.9), we may write

$$\mu_i^{M(id)} = \mu_i^{id} - \mu_i^\circ = RT \ln x_i. \tag{4.16}$$

In view of Eq. (4.11) this enables us to represent the Gibbs free energy of mixing in the following form:

$$g^{M(id)} = \sum_{i=1}^{k} x_i \mu_i^{M(id)} = RT \sum_{i=1}^{k} x_i \ln x_i. \tag{4.17}$$

Such a simple form for the free energy of mixing holds only for ideal solutions.

The enthalpy of mixing corresponding to this equation vanishes, as can be seen by using the relation between the Gibbs free energy and the enthalpy:

$$h^M = - T^2 [\partial (g^M/T)/\partial T]_{p, x_1, \ldots, x_k}. \tag{4.18}$$

Substituting from (4.17), we find

$$h^{M(id)} = -T^2 \sum_{i=1}^{k} x_i [\partial (\mu_i^{M(id)}/T)/\partial T]_{p, x_j} = 0. \tag{4.19}$$

In the formation of an ideal solution the volume of the system also remains unchanged, i.e., the volume of mixing of an ideal solution is equal to zero. This is seen to follow from the relation between the volume and the Gibbs free energy in (4.17):

$$v^{M(id)} = (\partial g^{M(id)}/\partial p)_{T, x_1, \ldots, x_k} = \sum_{i=1}^{k} x_i (\partial \mu^{M(id)}/\partial p)_{T, x_j} = 0. \tag{4.20}$$

From expressions (4.19) and (4.20) it also follows that the internal energy of mixing, equal to

$$u^M = h^M - p v^M, \tag{4.21}$$

should also vanish for an ideal solution.

Finally, differentiating (4.17) with respect to temperature, we get the entropy of mixing on forming an ideal solution:

$$s^{M(id)} = - (\partial g^{M(id)}/\partial T)_{p, x_1, \ldots, x_k} = -R \sum_{i=1}^{k} x_i \ln x_i. \tag{4.22}$$

All the properties of ideal solutions enumerated above are satisfied simultaneously. Thus a mixture formed without thermal effects and without volume change does not yet satisfy the definition of ideality. The additional requirement is that its entropy of mixing should satisfy relation (4.22).

On the basis of expressions (1.114), (2.16), (1.7), and (2.23), as well as the equation for the chemical potentials of the components of an ideal solution, it is easy to verify that in an ideal mixture the partial molar enthalpies $\bar{h}_i^{id}$ and partial molar volumes $\bar{v}_i^{id}$ of the components depend only on T and p:

$$\bar{h}_i^{id} = -T^2 \left[\frac{\partial \left(\mu_i^{id}/T \right)}{\partial T} \right]_{p, x_j} = -T^2 \left[\frac{\partial \left(\mu_i^{\circ} (T, p)/T \right)}{\partial T} \right]_p = h_i^{\circ},$$

$$\bar{v}_i^{id} = (\partial \mu_i^{id}/\partial p)_{T, x} = [\partial \mu_i^{\circ} (T, p)/\partial p]_T = v_i^{\circ}.$$

For the mean molar enthalpy and mean molar value of an ideal solution we may write, on the basis of the foregoing relations and (4.10),

$$h^{id} = \sum_{i=1}^{k} x_i \bar{h}_i^{id} = \sum_{i=1}^{k} x_i h_i^{\circ}, \tag{4.23}$$

$$v^{id} = \sum_{i=1}^{k} x_i \bar{v}_i^{id} = \sum_{i=1}^{k} x_i v_i^{\circ}. \tag{4.24}$$

It follows that $\bar{h}_i^{id}$ and $\bar{v}_i^{id}$ are equal in this case to the corresponding molar enthalpy h_i° and molar volume v_i° of the pure component i. The values of h and v depend on the composition of the solution since, in general, $y_i^{\circ} \neq y_j^{\circ}$.

For ideal perfect solutions formulas (4.23) and (4.24) are valid in the whole range of concentrations from $x_i = 1$ to $x_i = 0$.

We note that for solutions which become ideal only in the limit of low dilution, the preceding equations are satisfied only for x_1 near unity. In the limit of a dilute ideal solution we have for the solvent

$$\bar{h}_1^{*id} = h_1^{\circ}; \;\; \bar{v}_1^{*id} = v_1^{\circ},$$

while for the solutes we have, in general,

$$\bar{h}_i^{* \, (id)} \neq h_i^{\circ}; \;\; \bar{v}_i^{* \, (id)} \neq v_i^{\circ},$$

where the asterisk indicates that the given quantity is taken in the dilute solution limit.

This difference between ideal (perfect) and dilute ideal solutions is due to the fact that the enthalpy and volume of mixing which vanish for an ideal mixture will not necessarily equal zero for a dilute ideal solution. Let us consider the dilute limit for a binary solution. The enthalpy of its components due to mixing is equal to

$$H = n_1 h_1^{\circ} + n_2 h_2^{\circ},$$

whereas the enthalpy of a solution is defined by the equation

$$H = n_1 h_1^{\circ} + n_2 \bar{h}_2^{* \, (id)}.$$

From here it is simple to show that

$$h^M = x_2 \left(\bar{h}_2^{* \, (id)} - h_2^{\circ} \right).$$

Therefore, the formation of a limiting dilute solution from pure components at constant T and p is accompanied by thermal effects whose magnitude is determined by the preceding equation. Similarly, for the volume of mixing, we have

$$v^M = x_2 \left(\bar{v}_2^{* \, (id)} - v_2^{\circ} \right),$$

where the formation of a dilute ideal solution is, in general, accompanied by either compression or expansion. But, in spite of this fact, the mean molar enthalpy and mean molar volume depend linearly on the mole fraction in some concentration interval (corresponding to the formation of an ideal mixture).

Since for the asymmetrical method of normalizing in a binary dilute ideal mixture g is given by

$$g = x_1 \mu_1^{\circ} + x_2 \mu_2^{*} + RT \left(x_1 \ln x_1 + x_2 \ln x_2 \right),$$

we have

$$s = \left(x_1 s_1^{\circ} + x_2 \bar{s}_2^{* \, (id)} \right) - R \left(x_1 \ln x_1 + x_2 \ln x_2 \right).$$

This implies that the entropy of mixing of two partial components in forming the dilute ideal mixture does not equal the so-called "ideal" entropy of mixing, and differs from it by the quantity $x_2 \left(\bar{s}_2^{*(id)} - s_2^{\circ} \right)$. The "ideal" entropy of mixing pertains to the mixing process of component 1 with a hypothetical substance whose molar entropy is $\bar{s}_2^{*}$.

From the laws of thermodynamics it is difficult to determine the conditions for which solutions are ideal. Experimentally it is established that all sufficiently dilute solutions in which the mole fractions of all solutes are sufficiently close to zero may be considered ideal. Statistical thermodynamics leads to the same conclusion [80]. The concentration at which significant departures from ideality set in for a particular solution depends very much on the nature of the constituent substances. Solutions for which no deviations occur throughout the whole concentration range are extremely rare, and are met with only when the solvent and the solute possess similar chemical structures and have analogous thermodynamic characteristics (such as boiling temperature, energy of vaporization, vapor pressure at a given temperature, and so on) in their pure states.

Ideal solutions serve as convenient standards of comparison with real solutions. The ideal behavior of solutions is a limiting law; the more the solution components are similar to one another, the better the law is obeyed.

Using the gas form, Eq. (2.57), for the chemical potential of the solution component i, we obtain from the condition of phase equilibrium:

$$\mu_i^{\bar{-}} \equiv \mu_i^{\circ} (T, \ p) + RT \ln x_i = \mu_i^{\circ \, (G)} (T) + RT \ln \varphi_i, \tag{4.25}$$

where φ_i is the fugacity of component i in the saturated vapor above the solution. From this we find that

$$\varphi_i = k_i x_i. \tag{4.26}$$

Here

$$k_i^- \equiv \exp\left\{[\mu_i^\circ(T,\ p) - \mu_i^{\circ\,(G)}(T)]/RT\right\}. \tag{4.27}$$

If we assume that the vapor phase is a mixture of ideal gases, then relation (4.26) becomes

$$p_i^{id} = k_i x_i, \tag{4.28}$$

where p_i is the partial pressure of the saturated vapor of component i above the solution.

Formulas (4.26) and (4.28) are expressions of Henry's law for ideal solutions: the partial vapor pressure of any substance of the solution is proportional to its mole fraction. The quantity k_i is Henry's coefficient. In an ideal (perfect) solution $k_i = p_i^\circ$, where p_i° is the saturated vapor pressure of substance i.

It follows from (4.27) that the quantity k_i is a function only of p and T and is independent of the concentration. One may also show that at ordinary pressures k_i is practically unaffected by pressure. The dependence of k_i on temperature may be determined, according to (4.27), by the formula

$$\frac{d \ln k_i}{dT} = \frac{1}{R} \frac{d\left[\mu_i^{\circ\,(L)}(T,\ p)/T\right]}{dT} - \frac{1}{R} \frac{d\left[\mu_i^{\circ\,(G)}(T)/T\right]}{dT}$$

$$= \frac{-h_i^{\circ\,(L)} + h_i^{\circ\,(G)}}{RT^2} = \frac{\Delta h_{i,\,V}^\circ}{RT^2}, \tag{4.29}$$

where $\Delta h_{i,\,V}^\circ$ is the molar heat of vaporization of component i from the ideal solution in equilibrium with its vapor.

A narrower definition of an ideal solution could be the statement that all solution components obey Henry's law [cf. formula (4.26) or (4.28)].

Starting from Henry's law and using the Gibbs–Duhem relation, we are led to Raoult's law:

$$\varphi_1 = \varphi_1^\circ x_1.$$

For an ideally behaving gaseous mixture above a solution, Raoult's law becomes

$$p_1^{id} = p_1^\circ x_1, \tag{4.30}$$

or, in a slightly different form,

$$(p_1^\circ - p_1)/p_1^\circ = \sum_i^{k-1} x_i. \tag{4.31}$$

This equation shows that the relative lowering of the partial pressure of the solvent vapor is equal to the sum of mole fractions of the solutes.

The total vapor pressure of a solution is determined by Dalton's law. Hence, by using (4.30), we obtain for a two-component solution:

$$p = p_1^\circ (1 - x_2) + p_2^\circ x_2 = p_1^\circ + (p_2^\circ - p_1^\circ) x_2, \tag{4.32}$$

so that the total vapor pressure of an ideal binary solution is also a linear function of the mole fraction.

The application of conditions (4.30) or (4.31) to several different concentrations indicates that the chemical potentials of the components have the form (4.1) and, consequently, that the solution is ideal. Therefore, by studying the vapor one may evaluate the ideality of a given solution.

The compositions of an ideal solution and its saturated vapor are in general different, i.e., $x^{(L)} \neq x^{(G)}$. The condition $x^{(L)} = x^{(G)}$ for all concentrations is obeyed only when the saturated vapor pressures of both pure components are equal. The dependence of $x^{(G)}$ on $x^{(L)}$ may change sharply with temperature.

The equilibrium constant of a reaction proceeding in a solvent which does not participate is given by Eq. (2.93). If we substitute into the law of mass action (2.93) from Henry's law (4.28), which is obeyed by each component in an ideal solution, we obtain

$$\frac{\left(x_1^{'v_1'} x_2^{'v_2'} \ldots\right)}{\left(x_1^{v_1} x_2^{v_2} \ldots\right)} = K_p^{id}(T) \frac{k_1^{v_1} k_2^{v_2} \ldots}{k_1^{'v_1'} k_2^{'v_2'} \ldots} = K_x^{id}(T, p), \tag{4.33}$$

where $K_x^{id}(T, p)$ is a constant for a given solvent, temperature, and pressure.

For strong dilutions the equilibrium constant (4.33) may be expressed in terms of the molar concentration or the molality [24]. To determine the temperature dependence of $K_x^{id}(T, p)$, we have, on the basis of van't Hoff's isobaric equation [10] and relations (4.33) and (4.29),

$$\left[\frac{\partial \ln K_x^{id}(T, p)}{\partial T}\right]_p = \frac{\Delta H^{\circ (G)}}{RT^2} - \sum_i \frac{v_i \Delta h_{iV}^{\circ}}{RT^2} = \frac{\Delta H^{\circ (\alpha)}}{RT^2} \quad (\alpha = L, S), \tag{4.34}$$

where ΔH° is the heat of reaction at constant pressure in the ideal solution.

Proceeding the same way for the pressure dependence of $K_x^{id}(T, p)$ on the basis of (4.33), (4.27), and (2.23), we find

$$\left[\frac{\partial \ln K^{id}(T, p)}{\partial p}\right]_T = -\sum_i \frac{v_i v_i^\circ}{RT} = -\frac{\Delta V^{\circ (\alpha)}}{RT} \quad (\alpha = L, S), \tag{4.35}$$

where ΔV° is the change in volume of the reaction proceeding in the ideal solution.

Let us consider two phases, α and β, each of which is an ideal solution. The general condition of chemical equilibrium with respect to component i will be

$$\mu_i^\alpha = \mu_i^\beta. \tag{4.36}$$

Furthermore, in accordance with (4.25) and in view of (4.26), we obtain

$$x_i^\beta / x_i^\alpha = k_i^\alpha / k_i^\beta = K_i^- (p, T). \tag{4.37}$$

where $K_i(p, T)$ is the distribution coefficient of the solution component i among the two phases α and β.

Formula (4.37) is an expression of the Nernst distribution law for ideal solutions, whereby the relative mole fractions of the components in two ideal solutions in equilibrium with each other depend only on temperature and pressure, and not on the composition of the solutions.

Comparing (4.29) and (4.37), we obtain the dependence of K_i on temperature:

$$\left[\frac{\partial \ln K_i\,(p,\,T)}{\partial T}\right]_p = \left(\frac{\partial \ln k_i^\alpha}{\partial T}\right)_p - \left(\frac{\partial \ln k_i^\beta}{\partial T}\right)_p = \frac{\Delta h_i^{\circ\,(\alpha\to\beta)}}{RT^2}, \tag{4.38}$$

where $\Delta h_i^{\circ(\alpha \to \beta)}$ is the standard heat of transition of one mole of component i frόm the α into the β phase.

3. THE LAW OF NONIDEAL SOLUTIONS. EXCESS THERMODYNAMIC FUNCTIONS. CLASSIFICATION OF DEVIATIONS FROM IDEALITY

As already pointed out, the activity coefficient of a component in a nonideal solution is a function of temperature, pressure, and composition of the solution. On the basis of the Gibbs–Helmholtz equation (1.114), as well as (2.16) and (1.7), we have

$$\left[\frac{\partial\,(\mu_i/T)}{\partial T}\right]_{p,\,x_j} = -\frac{h_i}{T^2}. \tag{4.39}$$

It is easy to verify from this that the activity coefficient of the ith component in a nonideal solution depends on the temperature:

$$(\partial \ln \gamma_i/\partial T)_{p,\,x_i} = \frac{1}{R}\,\{\partial/\partial T\,[(\mu_i - \mu_i^\circ)/T]\}_{p,\,x_i}$$
$$= -\,(h_i - h_i^\circ)/RT^2, \tag{4.40}$$

where h_i° is the partial molar enthalpy of the ith component in the standard state. Similarly, on the basis of relations (2.23) and (4.2), we obtain

$$(\partial \ln \gamma_i/\partial p)_T = (\bar v_i - \bar v_i^\circ)/RT, \tag{4.41}$$

where $\bar v_i^\circ$ is the partial molar volume of the ith component in the standard state.

If Eq. (4.2) is substituted into the Gibbs–Duhem equation (2.25), and if we use the operator D to designate the change in the composition at constant temperature and pressure, then for one mole of the mixture we obtain

$$\sum_{i=1}^k x_i D\,(\mu_i)_{p,\,T} = RT\left\{\sum_{i=1}^k x_i D\,(\ln x_i) + \sum_{i=1}^k x_i D\,(\ln \gamma_i)\right\} = 0. \tag{4.42}$$

But the first expression on the right-hand side of (4.42) vanishes identically. Hence the second sum must also equal zero:

$$\sum_{i=1}^{k} x_i D \left(\ln \gamma_i\right)_{p,\,T} = 0. \tag{4.43}$$

The state of a binary solution may be defined by the variables T, p, x_2, which enable one to write Eq. (4.43) in the form

$$(1 - x_2) \left(\partial \ln \gamma_1 / \partial x_2\right)_{p,\,T} + x_2 \left(\partial \ln \gamma_2 / \partial x_2\right)_{p,\,T} = 0. \tag{4.44}$$

Then

$$\ln \gamma_1 = - \int_0^{x_2} (x_2 / x_1) \, d \ln \gamma_2. \tag{4.45}$$

Thus, to calculate, for example, the activity coefficient of the first component of a binary mixture in some region of the composition of the solution starting with $x_2 = 0$, it is sufficient to know the activity coefficient of the second component. This is very important in cases where any one of the components of the solution has a measurable vapor pressure.

The integration of Eq. (4.45) is carried out by the usual graphical method. The calculation of the activity coefficient of the first component is difficult near $x_2 = 1$, since the integrand tends to infinity near this point. In order to reduce the computational error, it is desirable to have available data for the activity coefficient of the second component for compositions which are extremely dilute relative to the first component, when the accuracy of experimentally determined quantities entering into (4.45) is considerably lower. Therefore, in the remaining concentration region adjoining the second pure component, one employs various interpolation methods for the activity coefficients. One of these consists of expanding the logarithm of the activity coefficient in a power series in the concentration:

$$\ln \gamma_2 = a_0 x_1^2 + a_1 x_1^3 + \cdots .$$

Near $x_2 = 1$ the series converges rapidly, and we need only retain the first two terms. Then the quantities a_0 and a_1 are easily determined by plotting the function

$$\ln \gamma_2 / x_1^2 = a_0 + a_1 x_1.$$

The quantities of a_0 and a_1 obtained this way are used in the integration of Eq. (4.45) for constructing the remaining part of the curve of $\ln \gamma_2$ as a function of x_2 / x_1.

Introducing the osmotic coefficient of the solvent determined by Eq. (4.5), we may write for the change of composition at constant temperature and pressure, in accordance with expression (4.43):

$$-x_1 \ln \left(1/x_1\right) D\Phi + (1 - \Phi) \sum_{i=2}^{k-1} dx_i + \sum_{i=2}^{k-1} x_i D \ln \gamma_i = 0. \tag{4.46}$$

This equation was introduced by Bjerrum. In particular, for a binary solution Eq. (4.46) may be written in the form

$$(1 - x_2) \ln (1 - x_2) D\Phi + (1 - \Phi) dx_2 + x_2 D \ln \gamma_2 = 0 \qquad (4.47)$$

or in integral form

$$-\ln \gamma_2 = \int_0^{x_2} [(1 - x_2)/x_2] \ln (1 - x_2) D\Phi + \int_0^{x_2} (1 - \Phi) d \ln x_2. \qquad (4.48)$$

We see, therefore, that one may calculate the activity coefficient of a solute in a given nonideal binary solution, provided one knows the osmotic coefficient of the solvent for all solutions which are more dilute than the actual solution in question at the same temperature and pressure.

Starting from (4.2) we obtain for the vapor pressure of a solution component of a nonideal solution, instead of the ideal result (4.28),

$$p_i = k_i x_i \gamma_i. \qquad (4.49)$$

For symmetric normalization $k_i = p_i^\circ$. With the aid of Eqs. (4.49), (4.29), and (4.40) it is easy to show that

$$\left(\frac{\partial \ln p_i}{\partial T} \right)_{p, \, x_j} = \left(\frac{\partial \ln k_i}{\partial T} \right)_{p, \, x_j} + \left(\frac{\partial \ln \gamma_i}{\partial T} \right)_{p, \, x_j} = \frac{\Delta h_{i, \, v} (T)}{RT^2}, \qquad (4.50)$$

where $\Delta h_{i, \, v}(T)$ is the molar heat of vaporization of the ith substance from the given solution.

In the same way we obtain, on the basis of formulas (4.49), (4.27), (4.23), and (4.41) with the symmetric method of normalization,

$$\left(\frac{\partial \ln p_i}{\partial p} \right)_{T, \, x_j} = \frac{v_i^{\circ \, (L)}}{RT} + \frac{\bar{v}_i^L - v_i^{\circ \, (L)}}{RT} = \frac{\bar{v}_i^L}{RT}. \qquad (4.51)$$

The numerical values of k_i and γ_i depend on the choice of the method of normalization, but the product $k_i \gamma_i$ is equal to the ratio p_i/x_i and does not depend on the choice of the standard state.

The formulas for the vapor pressure of the solvents are completely analogous to those derived for the solutes. In our case they are given by the following:

$$p_1 = p_1^\circ x_1 \gamma_1, \qquad (4.52)$$

$$(\partial \ln p_1/\partial T)_{p, \, x_i} = \Delta h_{1, \, v}/RT^2, \qquad (4.53)$$

$$(\partial \ln p_1/\partial p)_{T, \, x_i} = \bar{v}_1^\alpha/RT \quad (\alpha = L, \ S). \qquad (4.54)$$

With the asymmetric normalization method we may write, for a nonideal solution,

$$p_1 = p_1^\circ x_1 \gamma_1^*, \qquad (4.55)$$

$$p_2 = k_2 x_2 \gamma_2^*. \tag{4.56}$$

In the case under consideration the limiting law for the solvent will be Raoult's law (4.30), while for the solute it will be Henry's law (4.28).

Comparing expressions (4.55) and (4.30), or (4.56) and (4.28), one finds that with asymmetric normalization the activity coefficient of any component is expressed in terms of the ratio of vapor and ideal pressures:

$$\gamma_1^* = p_1/p_1^{id} = p_1/p_1^{\circ} x_1; \quad \gamma_2^* = p_2/p_2^{id} = p_2/k_2 x_2. \tag{4.57}$$

With the symmetric normalization of (4.3) it follows from Eq. (4.49) for the vapor pressure that

$$\begin{aligned} x_1 &= 1 \quad p_1^{\circ} = k_1, \\ x_2 &= 1 \quad p_2^{\circ} = k_2, \end{aligned} \tag{4.58}$$

i.e.,

$$p_1 = p_1^{\circ} x_1 \gamma_1; \quad p_2 = p_2^{\circ} x_2 \gamma_2. \tag{4.59}$$

Hence

$$\gamma_1 = p_1/p_1^{id} = p_1/(p_1^{\circ} x_1); \quad \gamma_2 = p_2/p_2^{id} = p_2/(p_2^{\circ} x_2). \tag{4.60}$$

Thus the advantage of symmetric normalization is that all the components may be treated by one and the same method. The activity coefficients (4.60) have a simple meaning. They are the ratio of the observed partial pressure to that which would be observed if the solution were ideal. Therefore, within the symmetric normalization the activity coefficients serve as a direct measure of the deviation of the solution from ideality. Substituting the values of the activity coefficients from (4.59) into Eq. (4.44), we obtain

$$(1 - x_2)(\partial \ln p_1/\partial x_2)_{T,\, p} + x_2 (\partial \ln p_2/\partial x_2)_{T,\, p} = 0. \tag{4.61}$$

In accordance with expressions (4.54) and (4.51) the partial pressures of the components depend on the total pressure, but for usual pressures the possible changes of p_i with pressure are negligibly small in view of the small values of $\bar{v}_i$.

Equation (4.61) reduces in practice to the Duhem–Margules equation

$$(1 - x_2)(\partial \ln p_1/\partial x_2)_T + x_2(\partial \ln p_2/\partial x_2)_T = 0, \tag{4.62}$$

in which only the temperature is held constant in the derivatives. This equation, relating the partial pressures to one another, is valid for any deviation of the solution from ideality. In deriving it we assumed only that the gaseous phase is an ideal gaseous mixture, and that the partial molar volumes of the components in the solution are negligibly small.

We assume that p_1 and p_2 have been determined experimentally. Then Eq. (4.62) leads to a simple criterion with which one may estimate the accuracy of experimental results. Thus Eq. (4.62) may be rewritten in the form

$$\frac{\partial p_1/\partial x_1}{p_1/x_1} = \frac{\partial p_2/\partial x_2}{p_2/x_2}. \tag{4.63}$$

For a graphical interpretation of this result one may note that $\partial p_2/\partial x_2$ is the slope of the tangent to the curve $p_2 = f(x_2)$ at the point x, while p_2/x_2 is the slope of the line joining x with the origin of the coordinates. Similar inferences are arrived at by considering the curve $p_1 = f(x_1)$.

It follows from Eqs. (4.30) and (4.57) that

$$\gamma_1 = \gamma_1^* = p_1/(p_1^0 x_1) \tag{4.64}$$

and

$$\gamma_2/\gamma_2^* = k_2/p_2^0. \tag{4.65}$$

But since $\overset{*}{\gamma_2} \to 1$ when $x_i \to 0$, the preceding equation means that

$$\lim_{x_2 \to 0} \gamma_2 = k_2/p_2^0. \tag{4.66}$$

Therefore expression (4.65) may be expressed in the form

$$\gamma_2/\gamma_2^* = \lim_{x_2 \to 0} \gamma_2 \tag{4.67}$$

or

$$\gamma_2^*/\gamma_2 = \lim_{x_2 \to 0} \gamma_2^{-1}. \tag{4.68}$$

These equations enable one to go from one method of normalization to another. If, for example, we define γ_2 as a function of state, then formula (4.67) enables one to compute $\overset{*}{\gamma_2}$.

The equilibrium condition for a chemical reaction occurring in a solvent which does not participate in it is given by Eq. (2.90). Substituting the values of μ_i from (4.2) into this equation, it is easy to verify that the law of mass action in a nonideal mixture gives

$$\ln \prod_i (x_i \gamma_i)^{\nu_i} = - \sum_i \nu_i \mu_i^0 (T,\ p)/RT = \ln K_a (T,\ p) \tag{4.69}$$

or

$$\prod_i (x_i \gamma_i)^{\nu_i} = (x_1' \gamma_1')^{\nu_1'} (x_2' \gamma_2')^{\nu_2'} \ldots / (x_1 \gamma_1)^{\nu_1} (x_2 \gamma_2)^{\nu_2} \ldots = K_a (T,\ p), \tag{4.70}$$

where the equilibrium constant $K_a(T, p)$ is constant for a given solvent, temperature, and pressure. Equation (4.70) may equally be written in the form

$$(x_1')^{\nu_1'} (x_2')^{\nu_2'} \ldots / (x_1)^{\nu_1} (x_2)^{\nu_2} \ldots = K_a (T,\ p) (\gamma_1)^{\nu_1} (\gamma_2)^{\nu_2} \ldots / (\gamma_1')^{\nu_1'} (\gamma_2')^{\nu_2'} \ldots = K_x', \tag{4.71}$$

where the quantity K_x' depends on the concentration.

The dependence of $K_a(T, p)$ on temperature and pressure for a nonideal solution, according to (4.69), (4.39), and (2.23), is given by

$$\left[\frac{\partial \ln K_a(T,\ p)}{\partial T}\right]_p = \frac{\Delta H^{\circ\ (\alpha)}}{RT^2}; \quad \left[\frac{\partial \ln K_a(T,\ p)}{\partial p}\right]_T = -\frac{\Delta V^{\circ\ (\alpha)}}{RT}. \quad (4.72)$$

In accordance with expression (4.71), we have

$$\left(\frac{\partial \ln K_x'}{\partial T}\right)_p = \left(\frac{\partial \ln K_a(T,\ p)}{\partial T}\right)_p - \sum_i \nu_i \left(\frac{\partial \ln \gamma_i}{\partial T}\right)_p. \quad (4.73)$$

Combining this with (4.40) and (4.72), we get

$$(\partial \ln K_x'/\partial T)_p = \Delta H^\alpha/RT^2, \quad (4.74)$$

where ΔH^α is the heat of reaction in the given solution. Similarly, it is easily verified that

$$(\partial \ln K_x'/\partial p)_T = -\Delta V^\alpha/RT, \quad (4.75)$$

where ΔV^α is the change in volume during the reaction in the given solution.

On the basis of relations (2.90) and (4.2), we find that for the equilibrium of two phases

$$\begin{aligned} \mu_i^\alpha - \mu_i^\beta &= \mu_i^{\circ\ (\alpha)} - \mu_i^{\circ\ (\beta)} + RT \ln[(x_i\gamma_i)^\alpha/(x_i\gamma_i)^\beta] \\ &= RT \ln K_i(T,\ p) + RT \ln[(x_i\gamma_i)^\alpha/(x_i\gamma_i)^\beta] = 0. \end{aligned} \quad (4.76)$$

In Eq. (4.76) we introduced the notation

$$\mu_i^{\circ\ (\alpha)} - \mu_i^{\circ\ (\beta)} = RT \ln K_i(T,\ p). \quad (4.77)$$

Hence

$$K_i(T,\ p) = \frac{x_i^\beta \gamma_i^\beta}{x_i^\alpha \gamma_i^\alpha}, \quad (4.78)$$

where the distribution coefficient $K_i(T, p)$ of a given solvent vapor is a function of only temperature and pressure. Equation (4.78) represents the most general form of the Nernst distribution law [81].

If one of the phases is a limiting dilute ideal solution while the other is not, then Eq. (4.78) enables one to calculate the activity coefficient γ_i in the nonideal phase.

By analogy with the preceding, we have

$$[\partial \ln K_i(T,\ p)/\partial T]_p = (h_i^{\circ\ (\beta)} - h_i^{\circ\ (\alpha)})/RT^2 = \Delta h_i^{\circ\ (\alpha \to \beta)}/RT^2, \quad (4.79)$$

$$[\partial \ln K_i(T,\ p)/\partial p]_T = -\Delta v_i^{\circ\ (\alpha \to \beta)}/RT, \quad (4.80)$$

where $\Delta h_i^{\circ(\alpha \to \beta)}$ is the standard change in enthalpy, while $\Delta v_i^{\circ(\alpha \to \beta)}$ is the standard change in volume in transferring one mole of component i from the solution α

to solution β. The quantities $\Delta h_i^{\circ(\alpha \to \beta)}$ and $\Delta v_i^{\circ(\alpha \to \beta)}$ are functions of only T and p and do not depend on the composition of the phases.

Furthermore, it follows from relations (4.78), (4.40), and (4.79) that

$$\left[\frac{\partial \ln \left(x_i^\beta / x_i^\alpha \right)}{\partial T} \right]_p = (\bar{h}_i^\beta - \bar{h}_i^\alpha)/RT^2 = \Delta \bar{h}_i^{(\alpha \to \beta)}/RT^2. \tag{4.81}$$

Knowledge of the activity and the activation coefficient of each component is not sufficient to account for deviation from ideality of a real solution. For this reason the thermodynamics of solutions makes extensive use of the so-called excess thermodynamic functions.

Earlier, in the example of ideal solutions, we introduced the idea of thermodynamic functions of mixing. The definition of these functions is easily extended to nonideal solutions. The difference between a thermodynamic function of mixing y^M for a real solution and its value for an ideal solution $y^{M(id)}$ at the same T and p is called the excess thermodynamic function. The excess thermodynamic functions are denoted by y^E. Hence, by definition, we have

$$y^E = y^M - y^{M(id)}. \tag{4.82}$$

The quantity y^E is the excess (positive or negative) of a given thermodynamic property of the solution relative to which the property of the ideal solution comprising the same components is compared. For a pure component, obviously $y^E = 0$.

The Gibbs free energy for the mixing of the components leading to the formation of the solution is found not from (4.17) but from the following equation:

$$g^M = \sum_{i=1}^{k} x_i \mu_i^M = RT \sum_{i=1}^{k} x_i \ln x_i \gamma_i. \tag{4.83}$$

For the derivation of this equation, as for the other excess thermodynamic functions, one uses the symmetric method of normalization. The mutual relations between the excess thermodynamic functions have been considered in [82].

Comparing expressions (4.17) and (4.83), we obtain

$$g^E = g^M - g^{M(id)} = RT \sum_{i=1}^{k} x_i \ln \gamma_i, \tag{4.84}$$

where g^E is the excess Gibbs free energy. On the basis of (4.84) and using (4.12) and (4.13), we may write for a two-component solution

$$\begin{aligned} \mu_1^E &= RT \ln \gamma_1 = g^E - x_2 \left(\partial g^E / \partial x_2 \right)_{T,p}, \\ \mu_2^E &= RT \ln \gamma_2 = g^E + x_1 \left(\partial g^E / \partial x_2 \right)_{T,p}. \end{aligned} \tag{4.85}$$

Analogous expressions hold for the other excess thermodynamic functions:

$$\bar{s}_1^E = s^E - x_2 \left(\partial s^E / \partial x_2 \right)_{p,T}, \quad \bar{s}_2^E = s^E + x_1 \left(\partial s^E / \partial x_2 \right)_{p,T}; \tag{4.86}$$

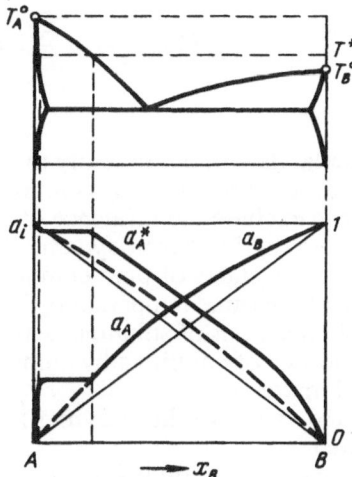

Fig. 26. Diagram of phase equilibria and the activities of the components of a binary alloy with a heterogeneous region: a_A and a_B are the activities of components A and B relative to the pure liquids of substances A and B at temperature T^*; a^* is the activity of component A relative to the pure solid of substance A.

$$\bar{h}_1^E = h^E - x_2 (\partial h^{E'}/\partial x_2)_{p,T}, \quad \bar{h}_2^E = h^E + x_1 (\partial h^E/\partial x_2)_{p,T}. \quad (4.87)$$

The excess thermodynamic functions can be easily computed from general thermodynamic relations. On the basis of (1.90) the excess entropy is determined by the derivative of Eq. (4.84) with respect to temperature:

$$s^E = -RT \sum_{i=1}^{k} x_j (\partial \ln \gamma_i/\partial T)_{p,x_i} - R \sum_{i=1}^{k} x_i \ln \gamma_i . \quad (4.88)$$

The excess enthalpy is found by substituting (4.84) and (4.88) into (1.113):

$$h^E = -RT^2 \sum_{i=1}^{k} x_j [\partial \ln \gamma_i/\partial T)_{p,x_i}. \quad (4.89)$$

To calculate the excess volume, we differentiate (4.84) with respect to pressure:

$$v^E = RT \sum_{i=1}^{k} x_j (\partial \ln \gamma_i/\partial p)_{T,x_i}. \quad (4.90)$$

The excess internal energy is determined by substituting (4.89) and (4.90) into the relation $u^E = h^E - pv^E$:

$$u^E = -RT \left[T \sum_{i=1}^{k} x_i \, (\partial \ln \gamma_i / \partial T)_{p, \, x_i} + p \sum_{i=1}^{k} x_i \, (\partial \ln \gamma_i / \partial p)_{T, \, x_j} \right]. \qquad (4.91)$$

In processing experimental data one frequently meets a situation in which the condensed solutions do not form a one-phase system for the concentration regimes studied. However, in order to exclude the influence of the various states of aggregation of the pure components on the thermodynamic functions of mixtures, it is usual to adopt as standard for the states of pure components the same state of aggregation as for the solution at the same temperature. This enables one to compare with ease the thermodynamic functions of solutions with one another, and leads to a general analysis of the properties of solutions by comparing them with the corresponding quantities for ideal mixtures.

The experimentally found values of the activities are converted in a new standard state in the following manner.

Suppose we determine experimentally the activity of component A in a liquid solution at temperature T^* (which is below the melting temperature of A) relative to the pure solid substance A. Then the new standard state is taken to be the state occupied by the supercooled liquid component A at the experimental temperature T^*. Considering the process of transition of component A between the three phases, the supercooled liquid component A, the liquid solution A–B, and the pure solid substance A, we obtain the following relation for the value of the activity A rescaled to the new standard state [83]:

$$a_A = Aa_A^*. \qquad (4.91a)$$

Here

$$A = \exp \, (1/RT) \left\{ -\Delta h_{m, \, A}^{\circ} \, (1 - T^*/T_{m, \, A}^{\circ}) \right.$$

$$\left. - \int_{T^*}^{T_{m, \, A}^{\circ}} \Delta c_{p, \, A}^{\circ} \, dT + T^* \int_{T^*}^{T} (\Delta c_{p, \, A}^{\circ}/T) dT \right\},$$

a_A is the activity of component A in the solution relative to the supercooled liquid, a_A^* is the activity of component A in the solution relative to the pure solid substance, and $\Delta c_{p, \, A}^{\circ}$ is the difference in the heat capacity of A between the solid and liquid states, calculated for one mole. The activity of the second component B for all concentrations of the solution is found in the usual way by means of Eq. (4.45).

If the activity of the lower melting component B is determined experimentally (Fig. 26), then the calculation proceeds in the following stages: 1) integration of the Gibbs–Duhem equation yields the activity a_A^*; 2) the activity of substance A is recalculated according to (4.91a) to the new standard state of the supercooled liquid of component A; 3) the functions a_A and a_B are extrapolated to the heterogeneous region and to the solid solutions of component B in the substance A, provided this can be done with sufficient accuracy.

The excess thermodynamic functions are intimately related to experimentally measured quantities such as the vapor pressure above the solution, the heat of mixing at constant pressure, etc. Comparison with the excess thermodynamic functions of real solutions provides information on the intermolecular interaction of their constituents. In this sense one may judge the nature of the solutions and classify them according to the character of their deviation from ideality. Two limiting cases may be discerned:

1. Regular solutions, for which

$$h^E \gg T \,|\, s^E \,|, \tag{4.92}$$

and, consequently,

$$g^E \approx h^E. \tag{4.93}$$

In this case the deviation from ideality is specified mainly by the heat of mixing.

2. Athermal solutions, for which

$$h^E \ll T \,|\, s^E \,|, \tag{4.94}$$

and, consequently,

$$g^E \approx - T s^E. \tag{4.95}$$

In this case the deviation from ideality is, to a considerable extent, specified by entropy changes.

4. THE CONCEPT OF A REGULAR SOLUTION. LATTICE MODEL OF SOLUTIONS. STRICTLY REGULAR SOLUTION AND ITS MODIFICATIONS

As was shown in Chapters 2 and 3, a rigorous thermodynamic analysis of phase transitions may be carried out by utilizing two basically interrelated approaches. Nevertheless, neither approach, though completely rigorous, is able to treat concrete systems, inasmuch as they provide only a qualitative picture of phase relations. In going to numerical solutions, one has to make model assumptions concerning the character of intermolecular interactions in solutions to obtain explicit expressions for the thermodynamic functions and to establish relations between the state parameters of any particular system.

The adoption of definite models enables one to compute not only certain properties of a particular system and compare them with experiment, but also to make important inferences concerning the divergence between the model and the real system, which also provides invaluable information about the physicochemical features of the phases forming this system.

Starting with the work of Biron [84] and van Laar [85, 86], an important role in the thermodynamic description of phase transformation has been played by the regular solution model and its subsequent modifications. This model is based on the assumption that the energy of a system is independent of the character of the atomic distribution in it. It takes into account only the difference between the inter-action energies of like and unlike nearest-neighbor atoms.

The actual term "regular solution" was introduced for the first time by Hilde-brand [87–89], who defined it as a solution whose excess volume and excess en-tropy of mixing vanish, and whose concentration-dependent entropy of mixing is determined the same way as for an ideal solution. Van Laar's method was devel-oped further in the works of Carlson and Colburn [90], Scatchard [91], Hildebrand [88, 89, 92, 93], Herzfeld and Heitler [94], and others.

Scatchard [91] and Hildebrand and Wood [95] obtained an expression for the excess internal energy of one mole of a liquid solution, allowing one to express the activity coefficient of a regular mixture in the form

$$RT \ln \gamma_1 = v_1^{\circ} \Phi_2^2 (\delta_1 - \delta_2)^2, \quad RT \ln \gamma_2 = v_2^{\circ} \Phi_1^2 (\delta_1 - \delta_2)^2, \qquad (4.96)$$

where Φ_i and δ_i are, respectively, the volume fraction and the solubility parameter introduced by Hildebrand. We rewrite the equations in (4.96) in a form convenient for calculation, assuming that $v_1^{\circ} = v_2^{\circ}$:

$$RT \ln \gamma_1 = \omega x_2^2, \qquad (4.97)$$

$$RT \ln \gamma_2 = \omega x_1^2, \qquad (4.98)$$

where ω is a parameter (the exchange energy) which is calculated from experimental data.

Let us note a property of a regular solution implied by its definition. The con-dition $s^E = 0$ characterizing a regular solution implies that

$$\ln \gamma_1 = f(1/T), \quad \ln \gamma_2 = f(1/T). \qquad (4.99)$$

Indeed, s^E vanishes for all values of the number of moles of the components, i.e.,

$$\partial s^E / \partial n_1 = 0; \quad \partial s^E / \partial n_2 = 0. \qquad (4.100)$$

In accordance with expression (1.90), we may write, on the basis of (4.100),

$$\partial^2 g^E / \partial n_1 \partial T = 0; \quad \partial^2 g^E / \partial n_2 \partial T = 0. \qquad (4.101)$$

One easily verifies from this, using (4.84) and the Gibbs–Duhem relation, that

$$\partial RT \ln \gamma_1 / \partial T = 0; \qquad (4.102)$$

$$\partial RT \ln \gamma_2 / \partial T = 0, \qquad (4.103)$$

which is equivalent to the expressions in (4.99). Conversely, the satisfaction of (4.99) implies that $s^E = 0$. Hence, in regular solutions the logarithm of the activity coefficient is inversely proportional to the absolute temperature.

It follows from (4.84), (4.97), and (4.98) that the excess thermodynamic potential in forming a regular solution, and consequently the enthalpy, is determined by the simple expression

$$g^E \approx h^E = \omega x_1 x_2. \tag{4.104}$$

Since the excess entropy of a regular solution is taken to be zero, it follows from (4.104), on account of (1.90), that the parameter ω should not depend on temperature.

The behavior of some real solutions approximately satisfies the parabolic equation (4.104). However, the parabolicity of the functions h^E and g^E is not yet proof of the regularity of a solution.

The theory of regular solutions suffers from internal inconsistencies. As already remarked, the entropy of a regular solution should be equal to the entropy of an ideal solution and the deviation from ideality is specified only by the finite value of the enthalpy. At the basis of this conclusion is the assumption that the entropy change in forming the mixture is independent of the thermal effect, which is strictly applicable only to ideal mixtures, where h^M is zero. The finite values of h^M, whether positive or negative, affect the value of the entropy of mixing.

The quantity s^E plays an essential role for many properties of mixtures [96], even in cases when it is an insignificant part of $s^{M(id)}$.

Using the Gibbs–Duhem relation (4.43), it is simple to verify that for concentrations at which the activity coefficient of the solvent is determined by an equation of the form (4.97), the activity coefficient of the solute equals

$$\ln \gamma_2 = \alpha x_1^2 + J, \tag{4.105}$$

where J is a constant of integration and α is a constant parameter.

It is obvious that the system of equations

$$\ln \gamma_1 = \alpha x_2^2, \quad \ln \gamma_2 = \alpha x_1^2 + J \tag{4.106}$$

is not identical to the system (4.97) and (4.98). From a comparison of the latter and (4.106) it follows that if equations of the form (4.97) and (4.98) hold for the whole concentration range from $x_1 = 0$ to $x_1 = 1$, then the integration constant in (4.106) must vanish. However, this conclusion is not in accord with experimental data for many real systems [97].

In the opinion of Lesnik [98], the class of regular solutions is not as broad as was supposed in introducing the concept of regular solution, though they are met with more frequently than one would expect on general considerations.

The theory of strictly regular solutions refers to a number of simple lattice theories of solutions. It is quite well founded on statistical thermodynamics [80, 99, 100], whereas its modifications (subregular, quasiregular, and other models) are empirical, since they introduce corrections to the temperature, concentration, etc., without sufficient physical justification. An exposition of such theories appears in several monographs [101–103].

The basic parameter in the theory of strictly regular solutions, namely the exchange energy $\bar{\omega}_{12}$ of the reaction forming a single bond 1–2, is defined by the relation

$$\bar{\omega}_{12} = u_{12} - 1/2 \, (u_{11} + u_{22}) \tag{4.107}$$

and characterizes the difference between the interaction energies of like and unlike molecules [104].

The quantity $2\bar{\omega}_{12}$ describes the change in the potential energy of the system in replacing the pairs 1–1 and 2–2 by two pairs of 1–2:

$$(1 - 1) + (2 - 2) = 2 \, (1 - 2), \tag{4.108}$$

i.e., this change pertains to so-called quasichemical reaction processes.

An approximate method of estimating the configuration integral by combinatorics on the basis of an "independent pair" assumption (cf. such methods in [101–103]) leads to the result

$$\bar{X}_{12}^2 / [(N_1 - \bar{X}_{12}) \, (N_2 - \bar{X}_{12})] = \exp \, (-2\bar{\omega}_{12}/kT), \tag{4.109}$$

where N_1 and N_2 are the number of molecules of the first and second kind, and $\bar{X}_{12}$ is the mean number of 1–2 pairs in the solution.

The preceding relation is termed the quasichemical formula or the equation of quasichemical equilibrium because of its formal similarity to the expression (4.108) for the equilibrium constant of a quasichemical reaction. The equation of quasichemical equilibrium plays an important role in the theory of strictly regular solutions. It establishes a relation between the mean number of 1–2 pairs in the solution and the exchange energy.

It follows from (4.109) that when $\bar{\omega}_{12} = 0$, then clearly $\bar{X}_{12} = N_1 N_2 / (N_1 + N_2)$, i.e., any distribution of molecules in the solution is then equally probable, or, in other words, the molecules are distributed randomly. When $\bar{\omega}_{12} \neq 0$, we have departures from randomness, i.e., there is an ordering effect. If $\bar{\omega}_{12} < 0$, then $\bar{X}_{12} > N_1 N_2 / (N_1 + N_2)$, and if $\bar{\omega}_{12} > 0$, then $\bar{X}_{12} < N_1 N_2 / (N_1 + N_2)$. Therefore, $\bar{\omega}_{12}$ characterizes the structure of the solution.

For convenience in calculating thermodynamic functions, one introduces two parameters η and β into the theory of strictly regular solutions, defined by

$$\eta = \exp \, [\bar{\omega}_{12}/kT], \quad \bar{X}_{12} = 2N_1 N_2 / [(N_1 + N_2) \, (\beta + 1)]. \tag{4.110}$$

The molar free energy of mixing of a strongly regular solution may be calculated from the relation [102]

$$\begin{aligned}
f^M \approx g^M &= N \, (x_1 \bar{\mu}_1^M + x_2 \bar{\mu}_2^M) = RT \Big\{ (1 - x_2) \ln \, (1 - x_2) \\
&+ x_2 \ln x_2 + \left(\frac{z}{2} \right) [(1 - x_2) \ln \, (\beta + 1 - 2x_2)/(1 - x_2) \, (1 + \beta) \\
&+ x_2 \ln \, (\beta - 1 + 2x_2)/x_2 \, (1 + \beta)] \Big\}.
\end{aligned} \tag{4.111}$$

where z is the coordination number. The excess free energy is equal to

$$f^E \approx g^E = RT \left(\frac{z}{2}\right) \left[(1 - x_2) \ln \frac{\beta + 1 - 2x_2}{(1 - x_2)(\beta + 1)} \right.$$
$$\left. + x_2 \ln \frac{\beta - 1 + 2x_2}{x_2(1 + \beta)} \right].$$
(4.112)

The activity coefficient of the components, via (4.112), will be

$$\gamma_1 = \left[\frac{\beta + 1 - 2x_2}{(1 - x_2)(1 + \beta)}\right]^{z/2},$$
(4.113)

$$\gamma_2 = \left[\frac{\beta - 1 + 2x_2}{(1 + \beta) x_2}\right]^{z/2}.$$
(4.114)

In view of the Gibbs–Helmholtz relation the change of enthalpy in forming the solution from the pure components may be estimated as follows:

$$h^M \simeq u^M = z\overline{\omega}_{12}\overline{X}_{12} = [2z/(1 + \beta)] x_2 (1 - x_2) N\overline{\omega}_{12}.$$
(4.115)

If the exchange energy $\overline{\omega}_{12} = 0$, then by (4.110), $\eta = 1$ and $\beta = 1$. Substituting $\beta = 1$ into formulas (4.113) and (4.114), we get $\gamma_1 = 1$ and $\gamma_2 = 1$, i.e., the solution here is ideal.

From $s^E = (h^M - g^E)/T$ and (4.112) and (4.115), we may find an expression for the entropy of a strictly regular solution.

It is simple to verify that for a given x_2 the functions g^E/RT and h^M/RT, determined by (4.112) and (4.115), depend only on the dimensionless parameter $\overline{\omega}_{12}/kT$, called the reduced exchange energy of a single 1–2 bond. Formula (4.111) for the change in the free energy g^M, and its consequences (4.113) and (4.114) for the activity coefficients, are considered as "first approximations" of the theory of strictly regular solutions to the description of the real behavior of solutions.

Apart from this "first approximation" one may consider a rough "zeroth approximation," which is obtained when, for small values of $\overline{\omega}_{12}$, the parameter β is expanded in a power series in terms of $\overline{\omega}_{12}/kT$ up to first order in $\overline{\omega}_{12}/kT$.

In the zeroth approximation we shall have

$$g^M = RT \left[(1 - x_2) \ln (1 - x_2) + x_2 \ln x_2\right] + x_2 (1 - x_2) Nz\overline{\omega}_{12},$$
(4.116)

$$g^E = h^M = x_2 (1 - x_2) Nz\overline{\omega}_{12},$$
(4.117)

$$RT \ln \gamma_1 = zN\overline{\omega}_{12}x_2^2,$$
(4.118)

$$RT \ln \gamma_2 = zN\overline{\omega}_{12}x_1^2,$$
(4.119)

$$s^E = 0.$$
(4.120)

It follows from formulas (4.113)–(4.115) and (4.117)–(4.119) that when $\overline{\omega}_{12}$ is positive (the formation of a 1–2 pair is energetically unfavorable), an endothermic effect is observed on mixing, and there is a positive deviation from the ideal behavior of the solution; when $\overline{\omega}_{12}$ is negative, the mixing is exothermic and the deviation from ideal behavior is negative. If $\overline{\omega}_{12} > 0$, then the system exhibits a tendency to demix.

All the equations considered above may be derived on condition that the exchange energy depends on the temperature [105]. In practical calculations this dependence is assumed linear. Nevertheless, in Rastogi's work [106], which assumed the constancy of the exchange coefficient broadening [107], it was shown that the dependence $\bar{\omega}_{12} = f(T)$ is exponential, in agreement with experimental data [108, 109]. The theory of conformal solutions [110] also leads to an exponential dependence of $\bar{\omega}_{12}$ on temperature [111].

The assumption of a linear dependence of $\bar{\omega}_{12}$ on T does not contradict the theoretically obtained conclusion of exponential dependence, since in a narrow temperature interval the exponential curve can always be approximated by a straight line. It should be understood, however, that the assumption of the linear temperature dependence of the exchange energy is only an approximation, and that the exponential dependence of $\bar{\omega}_{12}$ on T is more accurate.

In the work of Esin [112] it was shown that the equations of the first approximation of strictly regular solutions reduce to those of the zeroth approximation not only for small absolute values of $\bar{\omega}_{12}$ but also for any large $|\bar{\omega}_{12}|$, provided $\bar{\omega}_{12} < 0$ (a strongly disordered system) and the concentration of one of the components is not large. In these cases the corresponding parameters have a somewhat different physical meaning than in the equations of regular solutions. In addition, the relative exchange energies, in contrast to regular solutions, are not identical in the formulas for the activity coefficients and for the heat of mixing.

The "ordering effect" does not usually determine entirely the magnitude of the excess entropy of mixing s^E, which may be expressed as a sum of thermal (isochoric), volume, and configurational parts. In turn, the thermal contribution to the entropy may be split into vibrational, electronic, and sometimes paramagnetic components. At high temperatures the thermal part of the excess entropy may be assumed independent of temperature, which is equivalent to a linear temperature dependence of the exchange energy.

In view of the foregoing, we obtain in the zeroth approximation of quasiregular solutions the following expressions for the solution properties [113]:

$$h^E = Nz\bar{\omega}_0 x_1 x_2, \tag{4.121}$$

$$s^E = Nz\bar{\omega}_{01} x_1 x_2, \tag{4.122}$$

$$g^E = Nz\left(\bar{\omega}_0 - \bar{\omega}_{01}T\right) x_1 x_2, \tag{4.123}$$

where $\bar{\omega}_0$ and $\bar{\omega}_{01}$ are parameters independent of temperature.

In the first approximation of the quasichemical model, taking into account the temperature dependence of the exchange energy leads to the formulas

$$h^E = Nz\bar{\omega}_0 x_1 x_2 \left[1 - 2\left(\bar{\omega}_0 - \bar{\omega}_{01}T\right)/kT\right], \tag{4.124}$$

$$g^E = Nz\left(\bar{\omega}_0 - \bar{\omega}_{01}T\right) x_1 x_2 \left\{1 - \left[\left(\bar{\omega}_0 - \bar{\omega}_{01}T\right)/kT\right] x_1 x_2\right\}. \tag{4.125}$$

The deviation of a solution from regular behavior is manifested in a marked asymmetry of the concentration dependence of thermodynamic properties. This asymmetry is due to the fact that the energy of mixing on forming the solution is not completely determined by giving the values of the most probable number of pairs

$\bar{X}_{12}$ of unlike atoms and of the corresponding energy of interparticle interactions u_{ij}. This energy is not strictly constant, but depends on the temperature and the composition of the solution.

The simplest relation may be derived on assuming that the exchange energy $\bar{\omega}_{12}$ changes linearly with composition. Although this assumption has no rigorous justification, it is widely used, and in a number of cases the theoretical conclusions based on it are in agreement with experimental data.

Hardy [114] has attempted to take into account the dependence of the exchange energy on composition by means of a model of subregular solutions. In this model one assumes that the exchange energy is a linear function of composition, while the entropy of mixing coincides with its ideal value. The formula of the zeroth approximation of the quasichemical theory for g^E becomes

$$g^E = h^E = zNx_2(1 - x_2)(\bar{\omega}_0 - \bar{\omega}_1 x_2). \tag{4.126}$$

From (4.85) and (4.126) it is easy to verify that in the subregular solution model approximation the activity coefficients of the components are determined by the expressions

$$RT \ln \gamma_1 = Nzx_2^2[(\bar{\omega}_0 - \bar{\omega}_1) + 2x_2\bar{\omega}_1], \tag{4.127}$$

$$RT \ln \gamma_2 = Nz(1 - x_2)^2(\bar{\omega}_0 + 2x_2\bar{\omega}_1). \tag{4.128}$$

The application of the model of subregular solutions to the thermodynamic analysis of phase diagrams in a binary system has been considered in [115].

In accordance with the results of [113, 116, 117] relation (4.126) describes satisfactorily the isotherms of the thermodynamic properties of a series of silicate melts and sulfide and metallic solutions, which are characterized not only by asymmetric curves of g^E and h^E but also by alternating deviations from Raoult's law. Nevertheless, (4.126) does not reflect the temperature dependence of g^E.

The assumption of the linear dependence of the energy of mixing on temperature and composition leads to the equations [113, 118, 119]

$$g^E = Nzx_2(1 - x_2)[\bar{\omega}_0 - \bar{\omega}_{01}T + x_2(\bar{\omega}_1 - \bar{\omega}_{11}T)], \tag{4.129}$$

$$s^E = Nzx_2(1 - x_2)\bar{\omega}_{01} + Nzx_2^2(1 - x_2)\bar{\omega}_{11}, \tag{4.130}$$

$$RT \ln \gamma_1 = Nzx_2^2[\bar{\omega}_0 - \bar{\omega}_1 - (\bar{\omega}_{01} - \bar{\omega}_{11}T)] + 2Nzx_2^3(\bar{\omega}_1 - \bar{\omega}_{11}T), \tag{4.131}$$

$$RT \ln \gamma_2 = Nz(1 - x_2)^2[\bar{\omega}_0 - \bar{\omega}_{01}T + 2x_2(\bar{\omega}_1 - \bar{\omega}_{11}T)]. \tag{4.132}$$

Since the models of strongly regular and subregular solutions have limited applicability, attempts have been made to modify them. The goal of these attempts is the broadening of the range of application of the models to include calculations of phase equilibria, as well as the thermodynamic analysis of experimentally constructed state diagrams and thermodynamic properties. One of the more successful attempts was carried out by Sharkey and co-workers [120], who assumed a more complicated model of a subregular solution in which both binary and ternary groups of atoms were taken into account.

The enthalpy of mixing obtained in this work has the form

$$h^M \approx \alpha_1 x_1^2 x_2 + \alpha_2 x_1 x_2^2 - \alpha_3 x_1^2 x_2^2, \tag{4.133}$$

where

$$\alpha_3 = 2\bar{\omega}_{12}^2/zRT, \tag{4.134}$$

and α_1, α_2 are parameters that take into account structural and exchange effects on mixing, as well as the configurational enthalpy of mixing. The parameter α_3 should depend, in principle, on the composition, since the radial distribution of the atoms changes with composition, but in the calculation it was assumed to be a constant.

From the analysis of (4.133) it is easy to verify that when $\alpha_1 = \alpha_2$ and $\alpha_3 = 0$, we recover the regular solution approximation; for $\alpha_1 \neq \alpha_2$ and $\alpha_3 = 0$, we get back the subregular model, and for $\alpha_1 = \alpha_2$ and $\alpha_3 < 0$, the quasichemical approximation.

The result obtained in [120] will be modified if the linear nature of the temperature dependence of the exchange energy is taken into account. Let us transform (4.133) into a more convenient form, which coincides with the form written in the Margules approximation:

$$h^M \approx x_1 x_2 \left(\omega_0 + \omega_1 x_2 + \omega_2 x_2^2 \right). \tag{4.135}$$

A similar power series expansion may be carried out for the excess molar entropy of mixing:

$$s^E = x_1 x_2 \sum_{n=0} \omega_{n,1} x^n. \tag{4.136}$$

Retaining only the first three terms, we obtain

$$s^E = x_1 x_2 \left(\omega_{01} + \omega_{11} x_2 + \omega_{21} x_2^2 \right). \tag{4.137}$$

Combining this with (4.135) and (4.137), one may write for the excess free energy

$$g^E = x_1 x_2 \left[\omega_0 - \omega_{01} T + (\omega_1 - \omega_{11} T x_2) + (\omega_2 - \omega_{21} T) x_2^2 \right]. \tag{4.138}$$

Using Eq. (4.85), we obtain the activity coefficients of the components in this approximation:

$$RT \ln \gamma_1 = x_2^2 \left[\omega_0 - \omega_1 - (\omega_{01} - \omega_{11}) T \right] \\ + x_2^3 \left[2 \left(\omega_1 - \omega_{11} T \right) + (\omega_2 - \omega_{21} T)(3 x_2 - 2) \right], \tag{4.139}$$

$$RT \ln \gamma_2 = x_1^2 \left[\omega_0 - \omega_{01} T + 2 \left(\omega_1 - \omega_{11} T \right) x_2 \right. \\ \left. + 3 \left(\omega_2 - \omega_{21} T \right) x_2^2 \right]. \tag{4.140}$$

The most inadequate part of the theory of strictly regular solutions is the assumption that the change of volume on mixing is zero. The consequences ensuing from this assumption within the framework of the quasichemical method of calculating thermodynamic functions have been thoroughly analyzed in the work of Krichevskii [121]. The formulas he obtained differ from those of the zeroth approximation of the theory of strictly regular solutions by the extra terms $(p - p_i{}^o)v$, where v is the volume of the mixture.

Subsequently, starting from intermolecular forces, other methods of calculating thermodynamic functions of solutions have been developed, which could account for the influence of "exchange" effects. Prigogine et al. [122–124] have put forward a "free volume" theory of solutions, based on an application of the free volume theory of liquids developed by Lennard-Jones and Devonshire [125, 126]. Another interesting approach to this problem was presented by Longuet-Higgins [110]. In his conformal theory of solutions no model of the liquid state is employed, but rigorous restrictions are imposed on the intermolecular potential. The molecules of a "conformal" solution must have similar shapes, sizes, and intermolecular interactions. In the theory of conformal solutions the activity coefficients of a binary mixture are determined by the expression

$$RT \ln \gamma_i = x_j^2 E_s d_{12},\qquad(4.141)$$

which is very similar to the zeroth approximation result of the theory of strictly regular solutions. The parameters E_s and d_{12} in Eq. (4.141) are calculated from experimental data on the heat of evaporation and the critical parameters, respectively. A comparative analysis of the conformal and regular models of solutions has been given in a review [127].

In conclusion we note that in a real solution the shape and size of the component molecules may differ considerably, which is often the source of the deviation of the properties of a real solution from ideality.

In the original works of Flory [128–130], Huggins [131, 132], and others, special attention was devoted to studying the simplest case, in which the excess energy of mixing was equal to zero, so that the excess entropy uniquely expressed the deviation from ideal behavior (athermal solutions).

In recent years great progress has been made in the analytical treatment of the thermodynamic properties of solutions, thanks to the appearance of the new correlation equations of Wilson, Renon–Prauznitz, UNIQUAL, and others [133]. These equations are based, much more than the previous ones, on a molecular picture, though none of them may be said to provide a systematic molecular model of a solution. Therefore, the parameters of these equations can also be found by essentially empirical methods.

5. FORMAL DESCRIPTION OF THE THERMODYNAMIC PROPERTIES OF REAL SYSTEMS

Since the explicit dependence of the thermodynamic functions on the state parameters cannot be obtained with sufficient accuracy, one frequently resorts to a formal description. We shall now consider one widely applicable interpolation formula.

Any integral excess thermodynamic property y^E, or a partial excess property of the ith component $\overline{y}_i{}^E$ of a binary system A–B, may be represented in a simple power series of the type proposed by Margules [134]:

$$y^E = x_1 x_2 \left(q_0 + q_1 x_2 + q_2 x_2^2 + \cdots\right), \tag{4.142}$$

$$\bar{y}_1^E = x_2^2 \left(a_0 + a_1 x_2 + a_2 x_2^2 + \cdots\right), \tag{4.143}$$

$$\bar{y}_2^E = x_1^2 \left(b_0 + b_1 x_2 + b_2 x_2^2 + \cdots\right), \tag{4.144}$$

where y may be the Gibbs free energy, the enthalpy, the entropy, etc.

The above coefficients are not independent. The connection between them is determined by the relations

$$
\begin{aligned}
a_n &= (n+1)(q_n - q_{n+1}), \quad b_n = (n+1)q_n, \\
a_n &= b_n - b_{n+1}(n+1)/(n+2).
\end{aligned}
\tag{4.145}
$$

If one assumes that all coefficients except q_0, a_0, b_0 vanish, one recovers the equations of a regular solution, (4.97) and (4.98). If we retain only the first two terms of the expansion, we are led to the subregular solution model.

The use of the Margules relations is inconvenient, since they contain different coefficients in all three equations. Furthermore, replacing x_2 by x_1 in Eqs. (4.142)–(4.144) leads to a new set of coefficients. Another drawback is that within the concentration range from $x = 0$ to $x = 1$ the Margules series are far from orthogonal. For example, the degree of nonorthogonality $P = 0.9931$ if one includes the sixth terms in the series. For nonorthogonal polynomials the values of the first coefficients are strongly dependent on the order of the series, i.e., as the order is increased all previous values of the coefficients have to be recalculated. It follows from this that one cannot assign a definite physical meaning to the numerical values of the coefficients of nonorthogonal polynomials. This makes it impossible to compare properties of different systems through a comparison of the numerical values of the corresponding coefficients.

Another disadvantage of the Margules equations is that for small values of x and large n the magnitude of x^n becomes negligibly small. This requires the coefficient a_n to be rather large, to ensure that the nth term $a_n x^n$ is significant. In this context we note that for $P > 0.99$ the Margules coefficients must be calculated to a considerably greater accuracy than the original data [135]. In [136] it was shown that in a computer calculation of the Margules equation with 19 terms even the second iteration was inadequate.

The above drawbacks may be avoided by representing the concentration dependence of the thermodynamic properties in terms of orthogonal polynomials [137].

Although various orthogonal functions may be chosen to approximate the composition dependence of thermodynamic properties in binary systems, the number of coefficients needed to achieve agreement between theory and experiment to a required accuracy depends very much on the choice of polynomials. Since by assumption the quasichemical theory represents the composition dependence of excess thermodynamic properties algebraically rather than by trigonometric or transcendental equations, algebraic polynomials are preferred. The orthogonal functions should be chosen so as to lead to the fulfillment of Raoult's law, and to guarantee the finite value and smooth character of the function $\omega = f(x)$ in the limit of dilute solutions [137]. In addition, since many thermodynamic properties must sat-

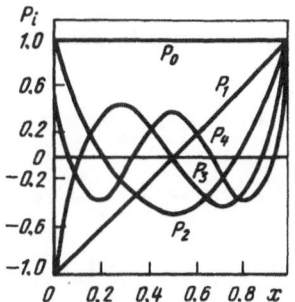

Fig. 27. Behavior of the first five orthogonal Legendre polynomials in the concentration interval $0 \leq x \leq 1$.

isfy the Gibbs–Duhem relation, the chosen orthogonal polynomials should be simply differentiable and integrable. It is also desirable that by retaining the first or first two terms of the orthogonal function series we obtain expressions corresponding to the regular or subregular solutions, respectively.

The conditions enumerated above are fulfilled by the Legendre polynomials which are strictly orthogonal in the interval of concentrations $0 \leq x \leq 1$. The first five Legendre functions in this region of concentrations have the form

$$P_0\,(x) = 1,$$

$$P_1\,(x) = 2x - 1,$$

$$P_2\,(x) = 6x^2 - 6x + 1, \qquad\qquad (4.146)$$

$$P_3\,(x) = 20x^3 - 30x^2 + 12x - 1,$$

$$P_4\,(x) = 70x^4 - 140x^3 + 90x^2 - 20x + 1.$$

The general recurrence relation for the polynomial $P_n(x)$ is written in the following form [133]:

$$P_n\,(x) = \frac{(2n-1)\,(2x-1)}{n}\,P_{n-1} - \frac{n-1}{p}\,P_{n-2}\,(x). \qquad (4.147)$$

Figure 27 illustrates graphically the first five Legendre functions, where we see that for $0 \leq x \leq 1$ none of the polynomials exceed ± 1. In contrast to the functions x, x^2, x^3, ..., which reach their maxima at $x = 1$ and vanish at $x = 0$, the Legendre polynomials reach their extremal values at various values of x. Furthermore, starting with $P_3(x)$, each succeeding Legendre polynomial has one more extremum than the preceding one. Therefore, the disparity between the corresponding experimental and theoretical curves does not increase near $x = 1$ or $x = 0$, in contrast to the simple power series case. Figure 27 also shows clearly the lack of interdependence of the coefficients of the Legendre polynomials, since each function $P_n(x)$ is characterized by a different form of functional dependence. The activity coefficients of the components and the excess free energy of mixing of a solution will be represented by the following general expressions in terms of the orthogonal Legendre polynomials:

$$g^E = x_1 x_2 \sum_{n=0}^{n'} q_n P_n (x_2),$$

$$RT \ln \gamma_1 = x_2^2 \sum_{n=0}^{n'} a_n P_n (\bar{x_2}), \qquad (4.148)$$

$$RT \ln \gamma_2 = x_1^2 \sum_{n=0}^{n'} b_n P_n (x_2).$$

In these equations the coefficients q_n, a_n, and b_n are determined by the relations

$$a_n = (n+1) q_n + (2n+1) \sum_{k=1}^{k=n'-n} q_{n+k} (-1)^k, \qquad (4.149)$$

$$b_n = (n+1) q_n + (2n+1) \sum_{k=1}^{k=n'-n} q_{n+k}, \qquad (4.150)$$

$$q_n = a_n/(n+1) + (2n+1) \sum_{k=1}^{k=n'-n} \frac{a_{n+k}}{(n+k)(n+k+1)}, \qquad (4.151)$$

$$q_n = b_n/(n+1) + (2n+1) \sum_{k=1}^{k=n'-n} \frac{(-1)^k b_{n+k}}{(n+k)(n+k+1)}, \qquad (4.152)$$

$$b_n = a_n + (2n+1) \sum_{k=1}^{k=n'-n} \frac{k(2n+k+1) a_{n+k}}{(n+k)(n+k+1)}. \qquad (4.153)$$

Retaining only the first three terms and taking into account (4.146), (4.149), and (4.150), relations (4.149) may be written in the form

$$q^E = x_1 x_2 [q_0 + q_1 (2x_2 - 1) + q_2 (6x_2^2 - 6x_2 + 1)], \qquad (4.154)$$

$$RT \ln \gamma_1 = x_2^2 [q_0 + q_1 (4x_2 - 3) + q_2 (18x_2^2 - 24x_2 + 7)], \qquad (4.155)$$

$$RT \ln \gamma_2 = x_1^2 [q_0 + q_1 (4x_2 - 1) + q_2 (18x_2^2 - 12x_2 + 1)]. \qquad (4.156)$$

The expressions for the activity coefficients as presented in (4.155) and (4.156) are no longer orthogonal functions. Therefore, in order to have strictly orthogonal polynomials in all three equations, one should keep three sets of different coefficients.

The representation of thermodynamic functions in terms of orthogonal Legendre polynomials is especially suitable for systems with strong chemical interactions, characterized by the complex nature of the dependences of thermodynamic properties. However, as shown by the examples of the binary systems Sn–Hg, Sn–Cd, and Sn–Zn [137], these polynomials are also useful in a comparative description of the behavior of thermodynamic properties of simpler systems, where accord between experimental and calculated curves is attained by taking into account five or fewer coefficients, which in this case can be given a definite physical meaning in terms of a molecular-statistical description.

Williams and Brouwer [138, 139] proposed a trigonometric Fourier sine series representation for the excess integral functions of a solution

$$y^E = \sum_{n=1}^{n'} q_n \sin(n\pi x), \tag{4.157}$$

which is orthogonal in the concentration interval $0 \le x \le 1$. The basic objection to this representation is that approximating thermodynamic properties by trigonometric function in some concentration regions does not satisfy certain thermodynamic requirements [137]. The expressions for the excess partial molar quantities

$$\bar{y}_1^E = \sum_{n=1}^{n'} q_n [\sin(n\pi x_2) - n\pi x_2 \cos(n\pi x_2)],$$

obtained on the basis of Eq. (4.159), are also not orthogonal functions.

One may also represent the thermodynamic properties of a solution in terms of a Fourier cosine series [137]:

$$\bar{y}_1^E = x_2^2 \left[a_0 + \sum_{n=1}^{n'} a_n \cos(n\pi x_2) \right], \tag{4.158}$$

$$\bar{y}_2^E = x_1^2 \left\{ a_0 + \frac{1}{x_1^2} \left[C - a_0 - \sum_{n=1}^{n'} a_n (\sin(n\pi x_2)/n\pi \right. \right.$$
$$\left. \left. + x_1 x_2 \cos(n\pi x_2)) \right] \right\}, \tag{4.159}$$

$$y^E = x_2 \left[C - a_0 x_2 - \sum_{n=1}^{n'} a_n \sin(n\pi x_2)/n\pi \right]. \tag{4.160}$$

where C is a constant of integration.

The functions in (4.159) and (4.160) are also not orthogonal. In addition, to achieve the same degree of agreement between experimental data and calculation, the trigonometric series for thermodynamic properties requires more terms than the simple power series or the Legendre polynomials.

The excess thermodynamic properties of solutions may be successfully described by the well-studied Chebyshev polynomials. The expressions analogous to (4.148) will, in this case, be of the form

$$y^E = x_1 x_2 \sum_{n=0}^{n'} q_n T_n(x), \quad \bar{y}_1^E = x_2^2 \sum_{n=0}^{n'} a_n T_n(x),$$

$$\bar{y}_2^E = x_1^2 \sum_{n=0}^{n} b_n T_n(x),\qquad(4.161)$$

where the first three Chebyshev polynomials transformed to the interval $0 \le x \le 1$ are defined by

$$T_0(x) = 1, \ T_1(x) = 2x - 1, \ T_2(x) = 8x^2 - 8x + 1.\qquad(4.162)$$

The general recursion relation for the polynomials T_n is

$$T_n(x) = 2(2x - 1)T_{n-1}(x) - T_{n-2}(x).\qquad(4.163)$$

The coefficients in (4.161) are related to one another as follows:

$$a_n = (n+1)q_n + \sum_{k=1}^{k=n'-n} (-1)^k (2n + 2\delta k) q_{n+k}\qquad(4.164)$$

$$(\delta = 1/2 \quad\text{if}\quad n = 0, \ \delta = 1 \quad\text{if}\quad n > 0),$$

$$b_n = (n+1)q_n + \sum_{k=1}^{k=n'-n} (2n + 2\delta k) q_{n+k}.\qquad(4.165)$$

Multiplying the Chebyshev polynomials by $(4x - 4x^2)^{-1/4}$ one may transform them into orthogonal forms in the interval $0 \le x \le 1$, but this procedure leads to severe mathematical complications and does not provide any advantages over the use of Legendre polynomials.

To represent the concentration dependence of thermodynamic properties one may also use the Redlich–Kister equations [139]:

$$y^E = x_1 x_2 \sum_{j=1}^{k} q_j (1 - 2x_2)^{j-1},\qquad(4.166)$$

$$\bar{y}_1^E = x_2^2 \sum_{j=1}^{k} q_j \{(2j - 1 - 2jx_2](1 - 2x_2)^{j-2}\},\qquad(4.167)$$

$$\bar{y}_2^E = x_1^2 \sum_{j=1}^{k} q_j [(1 - 2jx_2)(1 - 2x_2)^{j-2}].\qquad(4.168)$$

For Eq. (4.166) with seven terms the largest value of the nonorthogonality parameter P is 0.9669, considerably better than the value 0.9931 in the Margules equation.

The Laguerre and Hermite polynomials are orthogonal in the intervals $0 \le x \le \infty$ and $-\infty \le x \le \infty$, respectively, and are therefore unsuitable for approximating the thermodynamic functions of binary systems.

Williams [138] proposed a method of grouping polynomials into $Z_n(x)$ functions and represented the excess properties, though not ω, in a series in $Z_n(x)$.

The advantage of such an approximation is that it agrees with the quadratic Darken dependence for dilute solutions. Its only disadvantages are the approximate orthogonality of the Z functions, and the absence of recurrence formulas for terms beyond $n = 6$. Other orthogonal functions not mentioned here are not usually convenient for simple representations of thermodynamic properties.

Krupkowski [140] proposed taking into account the concentration asymmetry of the excess free energy of mixing by introducing a temperature-independent coefficient of asymmetry k by the equation

$$g^E = \frac{Ra}{(k-1)\,T^{r-1}}\, x_2\,(1-x_2)^{k-1} \tag{4.169}$$

or, if the asymmetry is reversed, by the formula

$$g^E = \frac{Ra}{(k-1)\,T^{r-1}}\, x_1\,(1-x_1)^{k-1}. \tag{4.170}$$

The activity coefficients of the components will be, accordingly,

$$\ln \gamma_1 = \omega'\,(T)\,\{(1-x_2)^k - [k/(k-1)]\,(1-x_2)^{k-1} + 1/(k-1)\}, \tag{4.171}$$

$$\ln \gamma_2 = \omega'\,(T)\,(1-x_2)^k \tag{4.172}$$

or

$$\ln \gamma_1 = \omega'\,(T)\,(1-x_1)^k, \tag{4.173}$$

$$\ln \gamma_2 = \omega'\,(T)\,\{(1-x_1)^k - [k/(k-1)]\,(1-x_1)^{k-1} + 1/(k-1)\}, \tag{4.174}$$

where $\omega'(T) = a/T^r$, a and r are constants characterizing a particular solution, and $\omega'(T)$ is the reduced value of the exchange energy, equal to $\omega(T)/RT$. It is easily verified for $k = 2$ that the Krupkowski formula reduces to the symmetric relations of regular solutions.

Relations (4.171)–(4.174) satisfy the Gibbs–Duhem equation. The choice of one or other of the equations depends on the nature of the symmetry possessed by the system [141]. The parameters ω and k are found from experimental data. In practical calculations it is usually assumed that $\omega(T)$ depends linearly on temperature and, consequently,

$$\omega'\,(T) = \omega\,(T)/RT = a/T \pm b. \tag{4.175}$$

When the deviation from Raoult's law is positive, relation (4.179) is applied to the component with the smaller atomic radius [142]. From experiments it is found that, as a rule, k is between 1 and 2 for positive, and larger than 2 [143] for negative deviations from Raoult's law. Such an analysis for the functions $f_1 = x_2^k$ and $f_2 = x_2^k - [k/(k-1)]x_2^{k-1} + 1/(k-1)$ and for the values 0, $1 \le k \le 3$ was given in [144]. It was also established that the larger the asymmetry of such thermodynamic properties as the enthalpy of mixing or the excess free energy, the nearer the asymmetry coefficient is to unity.

Although the method of Krupkowski allows one to take into account slight asymmetries of the concentration dependence of thermodynamic properties, it is not convenient for systems with alternating deviations from Raoult's law [116]. The basic inadequacy of this method consists of the fact that in a particular approximation the expression for the activity coefficients of the components is a product of two functions:

$$\gamma_i (x, \ T) = f(T) f(x),$$

one of which describes the dependence on temperature, the other on concentration. The consequence of such a formal representation of thermodynamic functions is that, for T = const, the integral and partial molar quantities are strictly proportional to one another, while the excess quantities do not depend on temperature [145].

In analyzing phase transformation data one uses most frequently some sort of linear temperature approximation for the parameter ω. In this approximation the previously considered polynomials are written formally as before, but it is assumed that each of the parameters associated with them depends linearly on temperature.

In accordance with the foregoing, the expressions for the activity coefficients of a binary solution may be represented in the various polynomial approximations, as follows:

for the Margules series:

$$
\begin{aligned}
RT \ln \gamma_1 &= x_2^2 [q_0 - q_1 - (q_{01} - q_{11}) T] + x_2^3 [2 (q_1^- - q_{11}T) \\
&\quad + (q_2 - q_{21}T) (3x_2 - 2) + \cdots], \\
RT \ln \gamma_2 &= x_1^2 [q_0 - q_{01}T + 2 (q_1 - q_{11}T) x_2 + 3 (q_2 - q_{21}T) x_2^2 + \cdots];
\end{aligned}
\tag{4.176}
$$

for the Legendre polynomial approximation:

$$
\begin{aligned}
RT \ln \gamma_1 &= x_2^2 [q_0 - q_{01}T + (q_1 - q_{11}T) (4x_2 - 3) \\
&\quad + (q_2 - q_{21}T) (18x_2^2 - 24x_2 + 7) + \cdots], \\
RT \ln \gamma_2 &= x_1^2 [q_0 - q_{01}T + (q_1 - q_{11}T) (4x_2 - 1) \\
&\quad + (q_2 - q_{21}T) (18x_2^2 - 12x_2 + 1) + \cdots];
\end{aligned}
\tag{4.177}
$$

for the Redlich–Kister polynomial approximation:

$$
\begin{aligned}
RT \ln \gamma_1 &= x_2^2 [q_0 - q_{01}T + (q_1 - q_{11}T) (3 - 4x_2) \\
&\quad + (q_2 - q_{21}T) (5 - 6x_2) (1 - 2x_2) + \cdots], \\
RT \ln \gamma_2 &= x_1^2 [q_0 - q_{01}T + (q_1 - q_{11}T) (1 - 4x_2) \\
&\quad + (q_2 - q_{21}T) (1 - 6x_2) (1 - 2x_2) + \cdots];
\end{aligned}
\tag{4.178}
$$

for the Fourier sine series approximation:

$$RT \ln \gamma_1 = \sum_{k=1} q_k [\sin (k\pi x_2) - k\pi x_2 \cos (k\pi x_2)]$$

$$-T \sum_{k=1} q'_k [\sin (k\pi x_2) - k\pi x_2 \cos (k\pi x_2)],$$

$$RT \ln \gamma_2 = \sum_{k=1} q_k [\sin (k\pi x_2) + k\pi x_1 \cos (k\pi x_2)]$$

$$-T \sum_{k=1} q'_k [\sin (k\pi x_2) + k\pi x_1 \cos (k\pi x_2)]. \tag{4.179}$$

Brebrick [146] proposed a model in the linear-temperature approximation which provides equations relating the activity coefficients to the excess partial molar properties of solutions:

$$\mu_1^E = RT \ln \gamma_1 = \bar{h}_1^E - T\bar{s}_1^E, \tag{4.180}$$

$$\mu_2^E = RT \ln \gamma_2 = \bar{h}_2^E - T\bar{s}_2^E. \tag{4.181}$$

The quantities $\bar{h}_i^E$ and $\bar{s}_i^E$ do not depend on the temperature but have an arbitrary dependence on the composition. By using hyperbolic functions to represent the concentration dependence of the excess partial molar properties, the model gives the following expressions for the activity coefficients [146]:

$$RT \ln \gamma_1 = ax_2^2 \{1 + \operatorname{sh} C (x_2 - x_0) - C (1 - x_2) \operatorname{ch} C (x_2 - x_0)\}$$
$$-Tbx_2^2 \{1 + \operatorname{sh} D (x_2 - x_0) - D (1 - x_2) \operatorname{ch} D (x_2 - x_0)\},$$

$$RT \ln \gamma_2 = a (1 - x_2)^2 \{1 + \operatorname{sh} C (x_2 - x_0) + Cx_2 \operatorname{ch} C (x_2 - x_0)\}$$
$$-Tb (1 - x_2)^2 \{1 + \operatorname{sh} D (x_2 - x_0) + Dx_2 \operatorname{ch} D (x_2 - x_0)\}, \tag{4.182}$$

where C and D are parameters which are independent of temperature and composition; $x_2 = x_0$ is the coordinate of the point relative to which calculations are made and can assume any value between 0 and 1.

Temperature dependences for ω which are more complicated than linear have been assumed in [139, 147]. The method proposed in [139] was based on a Taylor series expansion of the function $g^E(T, x)$ in terms of temperature. Retaining the first three terms of the expansion and using the well-known thermodynamic relation

$$g^E = h^E - Ts^E$$

together with (1.90) and (1.97), we find

$$g^E (T, x) = h^E - Ts^E - c_p^E \left(T - \theta - T \ln \frac{T}{\theta}\right), \tag{4.183}$$

where θ is the temperature at which the expansion is carried out, and the values of h^E, s^E, and c_p^E are taken at $T = \theta$.

The relation for $g^E(T, x)$ may also be represented in terms of one of the above polynomials. For example, its Redlich–Kister expansion is

$$g^E(T,\ x) = x_1 x_2 \sum_{j=1}^{k} \omega_j (T) (1 - 2x_2)^{j-1}. \tag{4.184}$$

Comparing (4.183) and (4.184), we may write

$$g^E(T,\ x) = x_1 x_2 \sum_{j=1}^{k} \left[h_j^E - T s_j^E - c_j^E \left(T - \theta - T \ln \frac{T}{\theta} \right) \right] (1 - 2x_2)^{j-1}. \tag{4.185}$$

Hiskes and Tiller [147] obtained an expression for the activity coefficients in the form of a double Taylor series in temperature and pressure, representing the quantity $x_2(\partial\mu_2/\partial x_2)_{T,p}$ as follows:

$$x_2 (\partial\mu_2/\partial x_2)_{T,\ p} = \sum_{n=0}^{N} \sum_{m=0}^{M} \beta_{mn} (T - T_0)^m (x_2 - x_0)^n, \tag{4.186}$$

where x_0, T_0 are the coordinates of the point around which the expansion is carried out, and β_{mn} are the expansion coefficients. Writing $(x_2 - x_0)^n$ in Eq. (4.186) as a binomial series

$$(x_2 - x_0)^n = \sum_{q=0}^{n} c_n^q (-1)^q x_0^q x_2^{n-q} \tag{4.187}$$

and dividing both sides of Eq. (4.186) by x_2, we obtain

$$(\partial\mu_2/\partial x_2)_{T,\ p} = (1/x_2) \sum_{n=0}^{N} \sum_{m=0}^{M} \beta_{mn} (T - T_0)^m (-1)^n x_0^n$$
$$+ \sum_{n=0}^{N} \sum_{m=0}^{M} \beta_{mn} (T - T_0)^m \sum_{q=0}^{n-1} c_n^q (-1)^q x_0^q x_2^{n-q-1}. \tag{4.188}$$

Integrating Eq. (4.186) with respect to x_2, we get

$$\mu_2 (x,\ T) = \ln x_2 \sum_{n=0}^{N} \sum_{m=0}^{M} \beta_{mn} (T - T_0)^m (-1)^n x_0^n$$
$$+ \sum_{n=1}^{N} \sum_{m=0}^{M} \beta_{mn} (T - T_0)^m \left\{ \sum_{q=0}^{n-1} c_n^q (-1)^q x_0^q x_2^{n-q}/(n - q) \right\} + J_2, \tag{4.189}$$

where J_2 is the constant of integration.
In accordance with Raoult's law

$$\lim_{x_2 \to 1} \mu_2 (x,\ T) = \mu_2^{\circ}(x,\ T) + RT \ln x_2^-. \tag{4.190}$$

From a comparison of expressions (4.189) and (4.190) it follows that

$$RT = \sum_{n=0}^{N} \sum_{m=0}^{M} \beta_{mn} (T - T_0)^m (-1)^n x_0^n, \tag{4.191}$$

$$J_2 = \mu_2^\circ (x,\ T) - \sum_{n=1}^{N} \sum_{m=0}^{M} \beta_{mn} (T - T_0)^m \sum_{q=0}^{n-1} c_n^q (-1)^q x_0^q/(n - q). \tag{4.192}$$

Using (4.191) and (4.192), expression (4.189) may be written in the form

$$\mu_2 (x,\ T) = \mu_2^\circ (T) + RT \ln x_2 + \sum_{n=1}^{N} \sum_{m=0}^{M} \beta_{mn} (T - T_0)^m$$

$$\times \left[\sum_{q=0}^{n-1} c_n^q (-1)^q x_0^q (x_2^{n-q} - 1)/(n - q) \right]. \tag{4.193}$$

Using the Gibbs–Duhem relation and Eq. (4.186), we obtain

$$(1 - x_2) (\partial \mu_1/\partial x_2)_{T,\ p} = \sum_{n=0}^{N} \sum_{m=0}^{M} (-1)^n \beta_{mn} (T - T_0)^m (x_2 - x_0)^n. \tag{4.194}$$

Performing the same transformations, we obtain a similar expression for $\mu_1(x, T)$:

$$\mu_1 (x,\ T) = \mu_1^\circ (T) + RT \ln (1 - x_2) + \sum_{n=1}^{N} \sum_{m=0}^{M} \beta_{mn} (-1)^n (T - T_0)^m$$

$$\times \left\{ \sum_{q=0}^{n-1} (-1)^q c_n^q (1 - x_0)^q [(1 - x_2)^{n-q} - 1]/(n - q) \right\}, \tag{4.195}$$

where

$$RT = \sum_{n=0}^{N} \sum_{m=0}^{M} \beta_{mn} (T - T_0)^m (1 - x_0)^n. \tag{4.196}$$

On account of formulas (4.193) and (4.195) the expression for the excess free energy of mixing may be written in the form

$$g^E = \sum_{m=0}^{M} \sum_{n=1}^{N} \beta_{mn} (T - T_0)^m \left\{ (-1)^n (1 - x_2) \sum_{q=0}^{n-1} (-1)^q c_n^q (1 - x_0)^q \right.$$

$$\times [(1 - x_2)^{n-q} - 1]/(n - q) + x_2 \sum_{q=0}^{n-1} (-1)^q c_n^q x_0^q (x_2^{n-q} - 1)/(n - q) \bigg\}. \tag{4.197}$$

Applying relation (1.9) to excess quantities, it is easy to verify that

$$s^E = \sum_{m=1}^{M} \sum_{n=1}^{N} \beta_{mn} (T - T_0)^{m-1} m \left\{ (-1)^n (1 - x_2) \sum_{q=0}^{n-1} (-1)^q c_n^q \right.$$

$$\times (1 - x_0)^q [(1 - x_2)^{n-q} - 1]/(n - q) + x_2 \sum_{q=0}^{n-1} (-1)^q c_n^q x_0^q$$

$$\left. \times (x_2^{n-q} - 1)/(n - q) \right\}. \tag{4.198}$$

By applying (1.113) to excess quantities and using relations (4.197) and (4.198), we obtain

$$h^E = \sum_{m=0}^{M-1} \sum_{n=1}^{N} [\beta_{mn} + (m + 1)\beta_{m+1, n}T](T - T_0)^m \left\{ (-1)^n (1 - x_2) \right.$$

$$\times \sum_{q=0}^{n-1} (-1)^q c_n^q (1 - x_0)^q [(1 - x_2)^{n-q} - 1]/(n - q)$$

$$+ x_2 \sum_{q=0}^{n-1} (-1)^q c_n^q x_0^q (x_2^{n-q} - 1)/(n - q) + \sum_{n=1}^{N} \beta_{mn} (T - T_0)^m$$

$$\times \left[(-1)^n (1 - x_2) \sum_{q=0}^{n-1} (-1)^q c_n^q (1 - x_0)^q [(1 - x_2)^{n-q} - 1]/(n - q) \right.$$

$$\left. \left. + x_2 \sum_{q=0}^{n-1} (-1)^q c_n^q x_0^q (x_2^{n-q} - 1)/(n - q) \right] \right\}. \tag{4.199}$$

For several simple cases one may obtain from (4.193) and (4.195) expressions for the chemical potentials of the components which are in agreement with results of known models of solutions [147]. This implies a direct connection between interpolation formulas and theoretical equations, and suggests that the higher-order terms correspond to deviations from the validity of theoretical representations.

6. ASSOCIATED SOLUTION MODEL

It has already been remarked that in regular solutions the heat of mixing is an order of magnitude smaller than the thermal energy RT (~2500 J/mole at ordinary temperatures). However, there is a large class of real solutions in which the energy of interaction between molecules is of the same order or even much larger than the thermal energy and, as a result, relatively stable molecular configurations of various sorts may be formed in the solution. The formation of these configurations leads to a change in the rotational and vibrational spectra of the constituent molecules.

In general we classify the molecules in a solution into two groups:

1) free molecules or monomers (monomolecules), not altered by the presence of neighboring molecules;

2) associated complexes, which are comparatively loose associations of like or unlike molecules whose rotational and vibrational states are altered as a result of their interactions; for brevity we shall term these "aggregates."

In 1908 Dolezalek [150] put forward the idea that all nonideal solutions should follow Raoult's law, but that the observed departures from ideal behavior are in some way related to unaccounted solvations which lead to the formation of various complexes. In essence, Dolezalek's work dealt with aggregates of various sorts which complicate the molecular composition of a solution.

Solutions in which one has specific directional interactions besides the van der Waals forces are called associated solutions. There are two approaches to a theoretical treatment of their equilibrium properties [148]. The more fundamental of the currently applied methods is based on the theory of associated solutions, in which the solution is considered to be a mixture of monomers and various kinds of aggregates whose equilibrium is determined by the law of mass action. The other approach is based on a lattice model of a solution formed by monomers of molecules interacting with strong directional forces.

The two methods apply in different regions [149]. The lattice model, worked out in its most general form by Barker, is valid for studying systems with directional interactions which do not lead to the formation of definitive compounds (in particular, the theory can describe systems with aggregates that are closed rings or branched chains of various sizes and structures).

In studying systems in which nonadditive interactions play an essential role (the case most frequently encountered), one must have recourse to the theory of associative equilibrium. Its starting point is a definitive model for the reaction of association, for example, a reaction of the type

$$mA + nB \rightleftarrows A_mB_n$$

or a reaction of successive associations:

$$A_{i-1} + A_1 \rightleftarrows A_i,$$

leading to the formation of aggregates of various sizes (where $i = 1, 2, ..., \infty$). The fundamental equation of the theory is the law of mass action, and the basic parameters are the equilibrium constant of the reaction of association, and the enthalpy of association.

In an associated solution, which by definition consists of monomers and aggregates, one does not consider those interactions between the various molecular forms which would be strong enough to lead to association. It follows from this that the interactions between aggregates and monomers are of the van der Waals kind, and, therefore, the properties of associated mixtures should not be different from those of simple solutions of nonpolar molecules. In other words, the major part of the deviation from ideality in associated solutions is due to the interactions which lead to the very formation of the aggregates.

Depending on the degree of approximation, a solution of aggregates and monomers may be ideal, regular, athermal, etc. [101, 150–153].

The ideal associated solution (ideal mixture of monomers and aggregates for all concentrations) is the simplest model, in which one neglects all differences in the energy of the van der Waals interaction between molecular complexes which would result from differences in the kind and size of these complexes.

Stecki [154] generalized the concept of ideal associated solutions by assuming that the difference in the van der Waals interactions between particles may be taken into account by means of the activity coefficients of the molecular complexes, within the approximation of a regular solution model. Consequently, such a model is a regular mixture of monomers and aggregates and is termed a regular associated solution.

This theory is widely used at present to analyze phase transitions in binary and quasibinary systems, including semiconducting compounds where the presence of aggregates in the liquid phase is most likely [155].

6.1. The Chemical Potential of Components in an Associated Solution

Following Prigogine [24], we consider an associated solution with components A and B. Generally one may note that the binary solution A–B includes molecular species of the type

$$A_r, \; B_t, \; A_k B_l \;\; (r, \, t, \, k, \, l = 1, \, 2, \, 3, \, ...).$$

If the total numbers of moles of A and B in the solution are n_A and n_B, while the numbers of moles of the actually existing various molecular forms in the solution are n_{A_r}, n_{B_t}, $n_{A_k B_l}$, then calculating in terms of monomer units we have

$$n_A = \sum_r rn_{A_r} + \sum_k \sum_l kn_{A_k B_l}, \tag{4.200}$$

$$n_B = \sum_t tn_{B_t} + \sum_k \sum_l ln_{A_k B_l}. \tag{4.201}$$

The chemical potentials of the molecular forms present in the solution are determined by the expression

$$\mu_\alpha = (\partial G/\partial n_\alpha)_{p, \, T, \, n_{\beta \neq \alpha}} \;\; (\alpha = A_r; \; B_t; \; A_k B_l). \tag{4.202}$$

The experimentally measurable chemical potentials of the components A and B are

$$\mu_A = (\partial G/\partial n_A)_{p, \, T, \, n_B}; \;\; \mu_B = (\partial G/\partial n_B)_{p, \, T, \, n_A}. \tag{4.203}$$

The aggregates in the solution are in equilibrium among themselves and with the monomers A_1 and B_1. The possible reactions among these parts can be represented by the equations

$$A_r = rA_1; \; B_t = tB_1; \; A_k B_l = kA_1 + lB_1. \tag{4.204}$$

In equilibrium the following conditions should hold:

$$\mu_{A_r} = r\mu_{A_1}; \; \mu_{B_t} = t\mu_{B_1}; \; \mu_{A_k B_l} = k\mu_{A_1} + l\mu_{B_1}. \tag{4.205}$$

We write the total differential of G at constant T and p for the associated system and treat the latter as a mixture of various aggregates. Then

$$dG = \sum_r \mu_{A_r} \, dn_r + \sum_t \mu_{B_t} \, dn_t + \sum_k \sum_l \mu_{A_k B_l} \, dn_{A_k B_l}. \qquad (4.206)$$

Using the conditions of (4.205), we obtain, on the basis of (4.206),

$$dG = \mu_{A_1} \sum_r r \, dn_{A_r} + \mu_{B_1} \sum_t t \, dn_{B_t} + \mu_{A_1} \sum_k \sum_l k \, dn_{A_k B_l}$$
$$+ \mu_{B_1} \sum_k \sum_l l \, dn_{A_k B_l}. \qquad (4.207)$$

According to Eqs. (4.200) and (4.201),

$$dn_A = \sum_r r \, dn_{A_r} + \sum_k \sum_l k \, dn_{A_k B_l},$$
$$dn_B = \sum_t t \, dn_{B_t} + \sum_k \sum_l l \, dn_{A_k B_l}, \qquad (4.208)$$

from which one may write

$$dG = \mu_{A_1} \, dn_A + \mu_{B_1} \, dn_B. \qquad (4.209)$$

For any binary system at constant T and p we have

$$dG = \mu_A \, dn_A + \mu_B \, dn_B. \qquad (4.210)$$

Equations (4.209) and (4.210) must be identical for all dn_A and dn_B, which is possible only when

$$\mu_A = \mu_{A_1}, \quad \mu_B = \mu_{B_1}. \qquad (4.211)$$

Therefore, the experimentally measured chemical potentials of the components μ_A and μ_B are equal to the chemical potentials of the monomolecules. The relations in (4.211) are general and are valid for any system in which an association reaction is taking place, irrespective of the type of this reaction or the degree of departure of the associated mixture from ideality. They apply not only to liquid systems but also to associated gases.

On the basis of (4.211) and the law of mass action, together with well-known thermodynamic relations, one may obtain formulas for calculating the thermodynamic functions of mixing of associated solutions.

6.2. Activities of Components of Regular Associated Solutions

We shall show now how one calculates the activity coefficients of a regular associated solution. Keeping in mind condition (4.211), we may write the Gibbs–

Duhem equation at constant T and p for a binary mixture in which molecular association is taking place, as follows:

$$n_A \, d\mu_{A_1} + n_B \, d\mu_{B_1} = 0. \tag{4.212}$$

Substituting expressions (4.200) and (4.201) into (4.212), we obtain

$$\sum_r r n_{A_r} d\mu_{A_1} + \sum_k \sum_l k n_{A_k B_l} d\mu_{A_1} + \sum_t t n_{B_t} d\mu_{B_1}$$
$$+ \sum_k \sum_l l n_{A_k B_l} d\mu_{B_1} = 0. \tag{4.213}$$

From here, because of relation (4.205), we find

$$\sum_r n_{A_r} d\mu_{A_r} + \sum_t n_{B_t} d\mu_{B_t} + \sum_k \sum_l n_{A_k B_l} d\mu_{A_k B_l} = 0. \tag{4.214}$$

Substituting into Eq. (4.214) the expressions for the chemical potentials of the molecular complexes in expanded form, it is easily verified that

$$\sum_r x_{A_r} d\ln \gamma_{A_r} + \sum_t x_{B_t} d\ln \gamma_{B_t} + \sum_k \sum_l x_{A_k B_l} d\ln \gamma_{A_k B_l} = 0, \tag{4.215}$$

where γ_{A_r}, γ_{B_t}, and $\gamma_{A_k B_l}$ are the activity coefficients of the various molecular entities present in the solution. It is clear from this that Eq. (4.215) is identical in form to the Gibbs–Duhem equation for a mixture consisting of the substances A_r, B_t, and $A_k B_l$.

It is known that the activity coefficient of the ith component of a regular n-component solution is determined by the equation [155]

$$RT \ln \gamma_i = \sum_{j=1}^{n} \omega_{ij} x_j^2 + \sum_{k=1}^{n} \sum_{j=1}^{n} x_k x_j (\omega_{ij} + \bar{\omega}_{ik} - \omega_{kj}), \tag{4.216}$$

where ω_{ij} is the exchange energy characterizing the pair interaction of particles in the solution.

Relation (4.216) satisfies Eq. (4.215) and is used to determine the activity coefficients of molecular entities in a regular associated solution.

Chapter 5

THERMODYNAMIC ANALYSIS OF PHASE DIAGRAMS WITHIN VARIOUS APPROXIMATE THEORIES OF SOLUTIONS

1. ANALYSIS OF PHASE EQUILIBRIA CURVES IN TWO-COMPONENT SYSTEMS BASED ON THE THEORY OF IDEAL SOLUTIONS

1.1. Schröder's Equation. The Entropy of Melting and the Shape of the Liquidus Curve. Calculation of the Eutectic Point

Schröder's logarithmic equation is a mathematical expression of the liquidus curve in a state diagram having a eutectic point.

Before giving the explicit form of the equation of the liquidus within the ideal solution approximation, it is appropriate to consider some general thermodynamic results which follow from the van der Waals–Storonkin differential equation [12, 13].

We consider a two-component two-phase system at constant pressure in which the components form a continuous series of ideal liquid solutions and have negligibly small solubilities in the solid phase. A schematic form of the state diagram of such a system is presented in Fig. 24a. We choose T and x_2 as the independent parameters which fix the state of the system.

Let us now write the van der Waals differential equation for equilibrium between the ideal liquid solution and the solid component 1. Since in this case $x_1^S = 1$ and $x_2^S = 0$, the van der Waals equation becomes

$$\left(\frac{\partial T}{\partial x_2}\right)_p^L = \frac{x_2^L \left(\partial^2 g/\partial x_2^2\right)_{p,\,T}^L}{s_1^{\circ\,(S)} - s^L + x_2^L \left(\partial s/\partial x_2\right)_{p,\,T}^L}, \tag{5.1}$$

where x_2^L is the molar fraction of component B in the saturated solution. Bearing in mind that the liquid phase is ideal, we may use the relations

$$s^L = s_1^{\circ\,(L)} x_1 + s_2^{\circ\,(L)} x_2, \tag{5.2}$$

$$(\partial^2 g / \partial x_2^2)_{p,T}^L = (\partial \mu_2 / \partial x_2)_{p,T}^L / (1 - x_2^L), \tag{5.3}$$

to readily transform (5.1) into the form

$$[\Delta h_{1,m}^{\circ}(T)/R]\, d\,(1/T) = -d\ln x_1^L, \tag{5.4}$$

where $\Delta h_{1,m}^{\circ}(T)$ is the heat of melting of the first component at temperature T.

Integrating Eq. (5.4) from the melting temperature of the first component to some temperature T, we obtain

$$RT \ln x_1^L = -\Delta h_{1,m}^{\circ}(1 - T/T_{1,m}^{\circ}) + \int\limits_{T}^{T_{1,m}^{\circ}} \left[\int\limits_{T}^{T_{1,m}^{\circ}} \Delta c_{p,1}^{\circ}\, d\ln T \right] dT, \tag{5.5}$$

where $\Delta c_{p,1}^{\circ}$ is the difference in the heat capacity of the first component between the solid and liquid states.

The explicit form of Eq. (5.5) will depend on the method of deriving it. At least two other forms of the liquidus curve of a binary simple eutectic system are known, obtained by the method of cycles and by integrating the Planck function G/T.

Malinovsky [156] showed that all three forms are equivalent and that the order of integration on the right-hand side of Eq. (5.5) must be preserved.

Thus, Eq. (5.5) determines the curve of the initial crystallization of the first component (the left branch of the liquidus in Fig. 24a), provided the solution is ideal and no solid solutions are formed. The difference of heat capacities between the solid and liquid states $\Delta c_{p,1}^{\circ}$ does not, in general, depend on the temperature. If $\Delta c_{p,1} = 0$, then

$$-\ln x_1^L = (\Delta h_{1,m}^{\circ}/R)\,(1/T - 1/T_{1,m}^{\circ}). \tag{5.6}$$

An analogous relation holds for the curve of initial crystallization of the second component (the right branch of the liquidus in Fig. 24a):

$$-\ln x_2^L = (\Delta h_{2,m}^{\circ}/R)\,(1/T - 1/T_{2,m}^{\circ}). \tag{5.7}$$

Equations such as (5.6) and (5.7) were presented in this form by Schröder [157, 158] in his determination of the solubilities of low-solubility substances (limiting dilute solutions).

Schröder's equations are widely used in calculations of the heats of meltings of particular substances based on phase transition data, in eutectic-type systems when the ideality of solutions is confirmed by some indirect method.

If one plots the logarithm of the molar fraction of a solute as a function of $1/T$ then, if (5.6) or (5.7) are satisfied, the curve will be a straight line whose slope gives us the value of the heat of melting of the solvent, while the point on the graph corresponding to the molar fraction $x_i = 1$ gives the melting temperature of the ith component.

When $\Delta h_{i,m}°(T)$ depends on temperature it is useful to construct the graph of $\ln x_i^{(L)}$ as a function of $\ln (T/T_{i,m}°)$. In order to illustrate this situation we shall rewrite (5.4) in the form

$$(\partial \ln x_i^L/\partial \ln T)_p = \Delta h_{i, m}°(T)/RT . \tag{5.8}$$

Differentiating this expression with respect to $\ln T$, we obtain

$$[\partial^2 \ln x_i^L/(\partial \ln T)^2]_p = \Delta c_{p, i}°/R - \Delta h_{i, m}°(T)/RT . \tag{5.9}$$

If the dependence of $\ln x_i^L$ on $\ln T$ is linear, then the second derivative in (5.9) is zero at the melting point $\Delta h_{i,m}° = T_{i,m}°\Delta s_{i,m}°$. Consequently, we are justified in approximating

$$\begin{aligned}\Delta s_{i, m}° &\approx \Delta c_{p, i}°, \\ \Delta h_{i, m}°(T) &= \Delta h_{i, m}° - \Delta c_{p, i}° (T_{i, m}° - T) = \Delta s_{i, m}° T.\end{aligned} \tag{5.10}$$

Substituting (5.10) into (5.8) and integrating from $T_{i,m}°$ to T, we obtain

$$\ln x_i^L \approx (\Delta s_{i, m}°/R) \ln (T/T_{i, m}°) . \tag{5.11}$$

In accordance with this expression the graph of $\ln x_i^{(L)}$ vs. $\ln (T/T_{i,m}°)$ will also be a straight line whose slope fixes the entropy of melting of the solvent $\Delta s_{i,m}°$.

It must, of course, be remembered that equations such as (5.6) and (5.7) apply only to ideal solutions, and even for these they are an approximation.

Using relation (5.6) we may also show that the shape of the crystallization curve near the ordinates of pure components enables one to recognize, in the absence of calorimetric data, the compound with the lower entropy of melting.

Let us rewrite (5.6) in the form

$$T = \Delta h_{i, m}°/(\Delta s_{i, m}° - R \ln x_i^L) . \tag{5.12}$$

A double differentiation of this equation with respect to x_i leads to the expression

$$(\partial^2 T/\partial x_i^2)_p = \frac{\Delta h_{i, m}° R}{(\Delta s_{i, m}° - R \ln x_i^L)^2} \left\{\frac{2R}{\Delta s_{i, m}° - R \ln x_i^L} - 1\right\} \frac{1}{x_i^2} . \tag{5.13}$$

The sign of this derivative determines the curvature of the liquidus line and itself depends on the sign of the expression

$$2R - \Delta s_{i, m}° + R \ln x_i^L . \tag{5.14}$$

It is obvious that near the beginning of the curve, i.e., as $x_i^{(L)} \to 1$,

$$(\partial^2 T/\partial x_i^2)_p < 0 \quad \text{if} \quad \Delta s_{i, m}° > 2R, \tag{5.15}$$

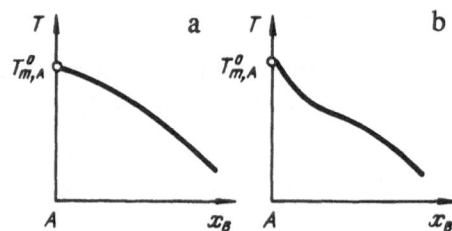

Fig. 28. Possible forms of liquidus curves near the ordinates of the pure components: a) $\Delta s_{1,m}° > 2R$; b) $\Delta s_{1,m}° < 2R$.

$$(\partial^2 T/\partial x_i^2)_p > 0 \quad \text{if} \quad \Delta s_{i,\,m}° < 2R. \qquad (5.16)$$

Both cases are illustrated schematically in Fig. 28. The first of these corresponds to an entropy of melting of the solvent at the melting point which is larger than $2R$, while the second one is for an entropy smaller than $2R$.

It is of interest to analyze Eq. (5.6) for the case where $x_2^L \to 0$. Then $\ln(1 - x_2^L)$ may be simply approximated by $-x_2^L$, so that (5.6) is transformed into

$$T - T_{1,\,m}° = -\left(RT_{1,\,m}°T/\Delta h_{1,\,m}°\right) x_2^L \qquad (5.17)$$

or approximately

$$T - T_{1,\,m}° = -\left(RT_{1,\,m}^2/\Delta h_{1,\,m}°\right) x_2^L < 0. \qquad (5.18)$$

Relation (5.18) expresses the well-known van't Hoff's law of the lowering of the freezing temperature.

Tyrkiel [159] attempted, by means of thermodynamics, to answer the question whether the temperature of initial solidification of a two-component system differs from the temperature of crystallization of the pure component. Starting from the condition of thermodynamic equilibrium, he showed that the source of such disparity is the influence of the second component on the vapor pressure and, consequently, on the chemical potential of the first component. In his opinion this interpretation becomes a general rule applying to both ideal and nonideal solutions, to systems whose components are practically immiscible in the solid state, as well as to systems having significant solubilities in the solid state.

1.2. Systems with a Continuous Series of Solid Solutions. Equations of the Solidus and Liquidus Curves

The theory of ideal solutions applied to liquid and solid solutions allows one to describe only the "cigar"-shaped state diagrams shown schematically in Fig. 9.

Let components 1 and 2 form an ideal solution in the liquid and solid states. Let T and x_2 be the independent variables.

It is easily verified, by using (5.2), (5.3), and (4.1), that the van der Waals differential equation for this case has the form

$$\left(\frac{\partial T}{\partial x_2}\right)_p^L = -\frac{RT^2}{\left(\Delta h_{1,\ m}^\circ x_1^{(S)} + \Delta h_{2,\ m}^\circ x_2^{(S)}\right)} \frac{x_2^{(S)} - x_2^{(L)}}{x_1^{(L)} x_2^{(L)}}, \tag{5.19}$$

$$\left(\frac{\partial T}{\partial x_2}\right)_p^S = -\frac{RT^2}{\left(\Delta h_{1,\ m}^\circ x_1^{(L)} + \Delta h_{2,\ m}^\circ x_2^{(L)}\right)} \frac{x_2^{(L)} - x_2^{(S)}}{x_1^{(S)} x_2^{(S)}}. \tag{5.20}$$

We transform these into a form more convenient to integrate:

$$\left(x_1^L x_1^S \Delta h_{1,\ m}^\circ - x_1^L x_1^S \Delta h_{2,\ m}^\circ + x_1^L \Delta h_{2,\ m}^\circ\right) dT$$
$$= -\left(x_2^S - x_2^L\right) RT^2\, d\ln x_2^L, \tag{5.21}$$

$$\left(x_1^L x_1^S \Delta h_{1,\ m}^\circ - x_1^L x_1^S \Delta h_{2,\ m}^\circ\right.$$
$$\left. + x_1^S \Delta h_{2,\ m}^\circ\right) dT = -\left(x_2^L - x_2^S\right) RT^2\, d\ln x_2^S. \tag{5.22}$$

Subtracting (5.21) from (5.22), we obtain

$$\left(x_1^S - x_1^L\right) \Delta h_{2,\ m}^\circ\, dT = -\left(x_2^L - x_2^S\right) RT^2 \left(d\ln x_2^S - d\ln x_2^L\right). \tag{5.23}$$

Noting that

$$x_1^S - x_1^L = x_2^L - x_2^S,$$

eq. (5.23) may be reduced to the form

$$\frac{d\ln\left(x_2^S / x_2^L\right)}{d\left(1/T\right)} = \frac{\Delta h_{2,\ m}^\circ (T)}{R}. \tag{5.24}$$

Similarly, we obtain

$$\frac{d\ln\left(x_1^S / x_1^L\right)}{d\left(1/T\right)} = \frac{\Delta h_{1,\ m}^\circ (T)}{R}. \tag{5.25}$$

The quantity $\Delta h_{i,m}^\circ(T)$ in expressions (5.24) and (5.25) is understood to be the heat of melting of the ith component at a given temperature T. Integrating Eq. (5.24) from $T_{2,m}^\circ$ to T, assuming that $\Delta h_{2,m}^\circ$ does not depend on temperature, we obtain

$$x_2^S / x_2^L = \exp - \left[\left(\Delta h_{2,\ m}^\circ / R\right)\left(1/T_{2,\ m}^\circ - 1/T\right)\right]. \tag{5.26}$$

We obtain an analogous equation on the basis of (5.25):

$$x_1^S / x_1^L = \exp - \left[\left(\Delta h_{1,\ m}^\circ / R\right)\left(1/T_{1,\ m}^\circ - 1/T\right)\right]. \tag{5.27}$$

This may be rewritten in the form

$$\left(1 - x_2^S\right)/\left(1 - x_2^L\right) = \exp - \left[\left(\Delta h_{1,\ m}^\circ / R\right)\left(1/T_{1,\ m}^\circ - 1/T\right)\right]. \tag{5.28}$$

Introducing the designation

$$\lambda_1 = (\Delta h^\circ_{1,\,m}/R)\,(1/T^\circ_{1,\,m} - 1/T), \tag{5.29}$$

$$-\lambda_2 = (\Delta h^\circ_{2,\,m}/R)\,(1/T^\circ_{2,\,m} - 1/T), \tag{5.30}$$

and solving simultaneously Eqs. (5.26) and (5.28), we obtain for the liquidus line

$$x^L_2 = (e^{-\lambda_1} - 1)/(e^{-\lambda_1} - e^{\lambda_2}) \tag{5.31}$$

and for the solidus line

$$x^S_2 = (e^{-\lambda_1} - 1)/(e^{-(\lambda_1+\lambda_2)} - 1). \tag{5.32}$$

These relations were obtained by van Laar and may be applied, in particular, to the calculation of the liquidus and solidus curves in ideal systems which form a continuous series of liquid and solid solutions.

An especially simple case arises when the melting temperature and the heat of melting of both components are almost identical. Under such conditions we obtain, after solving Eqs. (5.26) and (5.28),

$$x^L_2 = x^S_2 \quad \text{and} \quad -\lambda_1 = \lambda_2.$$

It is simple to verify on the basis of expressions (5.29) and (5.30) that the preceding equation holds if $T = T_m^\circ$, i.e., in this case the liquidus and solidus curves coincide and degenerate into a single horizontal straight line. Kurnakov [44] has shown that this sort of state diagram applies, for example, to systems whose components are left and right optical isomers.

Equation (5.26) may be derived more simply. The chemical potential of component i in an ideal solution is given by Eq. (4.1). In equilibrium we shall then have

$$\left.\begin{aligned}
\mu^{\circ\,(L)}_1\,(T,\,p) - \mu^{\circ\,(S)}_1\,(T,\,p) &= RT\,\ln x^S_1/x^L_1, \\
\mu^{\circ\,(L)}_2\,(T,\,p) - \mu^{\circ\,(S)}_2\,(T,\,p) &= RT\,\ln x^S_2/x^L_2.
\end{aligned}\right\} \tag{5.33}$$

The left-hand side of (5.33) represents the change of the Gibbs free energy per mole of a melting component. At the particular chosen temperature T this change may be written, in accordance with (1.113) and (2.37), as

$$\mu^{\circ\,(L)}_i - \mu^{\circ\,(S)}_i = \Delta g^\circ_{i,\,m} = \Delta h^\circ_{i,\,m}(T) - T\Delta s^\circ_{i,\,m}(T). \tag{5.33a}$$

At the melting point of component i the quantity $\Delta g_{i,m}^\circ$ vanishes. Hence, noting that

$$\Delta s^\circ_{i,\,m} = \Delta h^\circ_{i,\,m}/T^\circ_{i,\,m}$$

and assuming that the quantities $\Delta h_{i,m}^\circ$ and $\Delta s_{i,m}^\circ$ change only slightly with the temperature, we may write approximately

$$\mu^{\circ\,(L)}_i - \mu^{\circ\,(S)}_i \approx \Delta h^\circ_{i,\,m}\,(1 - T/T^\circ_{i,\,m}). \tag{5.33b}$$

Substitution of this expression into (5.33) yields formulas (5.26) and (5.28).

1.3. Differential Equation of the Liquidus in a System of Congruently Melting Compounds. Integration of the Equation for Ideal Solutions

A practically important case is the equilibrium of two phases in a system with congruently melting chemical compounds of the type A_mB_n, distinguished by a negligibly small region of homogeneity. Prigogine and Defay [24] have treated this case by means of the chemical affinity function. The possibility of a concrete solution to the problem on the basis of the model of regular solutions has been considered by the present authors [160], who have described a method of deriving the integral form of the liquidus curve in a system with congruently melting A_mB_n-type compounds.

We shall show how to derive the differential equation of the liquidus curve in such a system on the basis of the principle of equilibrium displacements, in the form given by van der Waals and Storonkin, and shall supply its solution in the approximation of the ideal solution model.

Let us consider a two-component two-phase system with the compound A_mB_n at constant pressure. The state diagram of such a system is shown in Fig. 20. Let T and x_2 (the molar fraction of the B component) be the independent variables fixing the state of the system. We may write the van der Waals differential equation in the present case, as follows:

$$\left(\frac{\partial T}{\partial x_2}\right)^L_p = -\frac{(x_2^S - x_2^L)\,(\partial^2 g/\partial x_2^2)^L_{p,\,T}}{s^S - s^L - (x_2^S - x_2^L)\,(\partial s/\partial x_2)^L_{p,\,T}}. \tag{5.34}$$

Since the differential molar heat of formation of the solid phase from the liquid is determined by the expression (cf. [13])

$$\bar{q}^{(L \to S)} = [s^S - s^L - (x_2^S - x_2^L)\,(\partial s/\partial x_2)^L_{p,\,T}]\,T, \tag{5.34a}$$

and since

$$x_2^S x_1^L - x_2^L x_1^S = x_2^S - x_2^L, \tag{5.34b}$$

we may use (5.3) to transform Eq. (5.34) into

$$\left(\frac{\partial T}{\partial x_2}\right)^L_p = -T\,(1 - x_2^S)\left(\frac{x_2^S}{1 - x_2^S} - \frac{x_2^L}{1 - x_2^L}\right)\left(\frac{\partial \mu_2}{\partial x_2}\right)^L_{p,\,T}/\bar{q}^{(L \to S)}. \tag{5.35}$$

In view of the fact that we have here, in equilibrium with the liquid solution, a solid phase which consists of a congruently melting compound with a narrow range of homogeneity, we may assume that its composition is practically constant and, for a given temperature interval, coincides with the composition of the compound. Consequently,

$$x_2^S \approx n/(m+n), \tag{5.36}$$

where m and n are the stoichiometric coefficients. We rewrite (5.35) with the help of (5.36):

$$\left(\frac{\partial T}{\partial x_2}\right)_p^L = -mT\left(\frac{n}{m} - \frac{x_2^L}{1-x_2^{(L)}}\right)\left(\frac{\partial \mu_2}{\partial x_2}\right)_{p,T}^L \Big/ [(m+n)\,\bar{q}^{(L \to S)}]. \tag{5.37}$$

Since one mole of the phase responsible for the composition of the compound corresponds to the formation of $[1/(m+n)]$ moles of the analytic compound $A_m B_n$, we may consider the quantity $(m+n)\bar{q}^{(L \to S)}$ as the differential molar heat of crystallization of this compound from the saturated solution, i.e.,

$$(m+n)\,\bar{q}^{(L \to S)} = \bar{q}_{A_m B_n}^{(L \to S)}. \tag{5.38}$$

Using this in Eq. (5.37), we may write the latter in the form

$$\left(\frac{\partial T}{\partial x_2}\right)_p^L = -\left[mT\left(\frac{n}{m} - \frac{x_2^L}{1-x_2^L}\right)\left(\frac{\partial \mu_2}{\partial x_2}\right)_{p,T}^L\right]\Big/ \bar{q}_{A_m B_n}^{(L \to S)}. \tag{5.39}$$

This equation expresses in differential form the functional dependence of the temperature of the liquidus of a congruently melting compound on the composition, i.e., it is an analytical expression for the liquidus curve, corresponding to the initial crystallization of the compound $A_m B_n$. In addition, Eq. (5.39) is identical to the expression obtained by Prigogine and Defay [24]. We draw the important conclusion that two quite different approaches to the analysis of heterogeneous equilibrium provide identical results. This is additional evidence of the fruitfulness of using the principle of equilibrium displacements as formulated by van der Waals and Storonkin [12, 13]. It also indicates that the criticism of Semenchenko [39] concerning the lack of independence of the van der Waals equation is unfounded.

The analysis of Eq. (5.39) shows, above all, that if the liquid has the same composition as the compound which is in equilibrium with it, i.e., if

$$x_2^L/(1 - x_2^L) = n/m,$$

and if the derivative of the chemical potential $(d\mu_2/dx_2)_{p,T}^L$ is not infinite (in other words, the degree of dissociation of the compound in the liquid does not vanish), then $(\partial T/\partial x_2)_p^L = 0$, i.e., on the liquidus curve the temperature goes through an extremum.

If we restrict our treatment to stable states, for which by (2.141) we have

$$(\partial \mu_2/\partial x_2)_{p,T} > 0,$$

then it is easily verified that the extremum is in fact a maximum. Indeed, in all known cases the magnitude of the preceding molar heat of crystallization satisfies $\bar{q}_{A_m B_n}^{(L \to S)} < 0$, and in accordance with Eq. (5.39)

$$(\partial T/\partial x_2)_p^L > 0 \quad \text{for} \quad x_2^L/(1 - x_2^L) < n/m,$$

$$(\partial T/\partial x_2)_p^L < 0 \quad \text{for} \quad x_2^L/(1 - x_2^L) > n/m.$$

To integrate Eq. (5.39), one must choose a model which allows one to write the quantities entering into it in explicit form. The process of crystallization of a compound with a narrow range of homogeneity at some temperature T may be represented in the form

$$m\,(A)_{\text{sat}}^L + nB_{\text{sat}}^L \rightleftarrows (A_m B_n)^S. \tag{5.40}$$

The change of the Gibbs free energy in this process is determined from the expression

$$-\Delta G = m\mu_A^L + n\mu_B^L - \mu_{A_m B_n}^{\circ\,(S)} \tag{5.41}$$

or, expanding the values of the chemical potentials, from

$$-\Delta G = -\Delta G^{id} + mRT \ln \gamma_1 + nRT \ln \gamma_2. \tag{5.42}$$

Using the Gibbs–Helmholtz equation as well as relation (5.42), we may generally write the following expression for the thermal effect $\bar{q}_{A_m B_n}^{(L \to S)}$ in (5.39):

$$\bar{q}_{A_m B_n}^{(L \to S)} = \bar{q}_{A_m B_n}^{id\,(L \to S)} + RT^2\,[m\,(\partial \ln \gamma_1/\partial T)_{p,\,x} + n\,(\partial \ln \gamma_2/\partial T)_{p,\,x}]. \tag{5.43}$$

In the general case,

$$(\partial\mu_2/\partial x_2)_{p,\,T}^L = RT/x_2^L + RT\,(\partial \ln \gamma_2/\partial x_2)_{p,\,T}^L. \tag{5.44}$$

We solve the problem within an ideal solution model. The phenomenological differential molar heat of crystallization of the compound $A_m B_n$ from the saturated solution may be represented by two terms: the heat of segregation and the heat of crystallization of the compound at the given temperature:

$$\bar{q}_{A_m B_n}^{(L \to S)} = -\Delta h_{m,\,A_m B_n}^{\circ}\,(T) + q^E, \tag{5.44a}$$

where q^E is the heat of segregation.

For an ideal solution q^E is zero, so that the quantity $\bar{q}_{A_m B_n}^{(L \to S)}$ is equal to the heat of crystallization of the compound at the given temperature.

Furthermore, in accordance with expression (5.44) we have, in an ideal solution,

$$(\partial\mu_2/\partial x_2)_p^L = RT/x_2^L. \tag{5.45}$$

Substituting this relation into Eq. (5.39), we obtain

$$(\Delta h_{m,\,A_m B_n}^{\bullet}/RT^2)\,dT = (n/x_2^L - m/(1 - x_2^L))\,dx_2^L. \tag{5.46}$$

If we now take the heat of melting of the compound to be a constant in the temperature interval considered, and if we integrate Eq. (5.39) from the point which corresponds to the composition of the congruently melting compound, at which $x_2^L = n/(m + n)$ and $T = T_{m,A_mB_n}$, to the point with coordinates x_2^L and T, we obtain

$$T = \frac{\Delta h^\circ_{m,A_mB_n}}{\Delta s^\circ_{m,A_mB_n} - R \ln \left\{ [(m+n)^{m+n}/m^m n^n] (1 - x_2^L)^m (x_2^L)^n \right\}}. \tag{5.47}$$

This equation may be written in a different form, more convenient for graphical representation:

$$\frac{\Delta h^\circ_{m,A_mB_n}}{R} \left(\frac{1}{T} - \frac{1}{T^\circ_{m,A_mB_n}} \right) = - \ln \left[\frac{(m+n)^{m+n}}{m^m n^n} (1 - x_2^L)^m (x_2^L)^n \right]. \tag{5.48}$$

The graph of the function $\ln [(x_2^L)^n (1 - x_2^L)^m] = f(1/T_{liq})$ is a straight line whose slope determines the heat of melting of the compound A_mB_n.

Gorbunov [161] has treated, on the basis of the van der Waals differential equation, the local features of the shape of the curves characterizing equilibrium in binary systems with ideal phases under isobaric conditions. Assuming that the liquid solution is ideal (in which case the solid solution cannot be an ideal phase) and bearing in mind (5.34b), the van der Waals differential equation may be reduced to the form

$$(\Delta s^{(L \to S)}/RT)\, dT + (x_2^S/x_2^L - x_1^S/x_1^L)\, dx_1^L = 0, \tag{5.49}$$

where $\Delta s^{(L \to S)} < 0$ is the differential molar entropy of crystallization of the ideal phase.

At the points $x_1^S = x_1^L = 0$ and $x_1^S = x_1^L = 1$ the ratios x_i^S/x_i^L become indeterminate (0/0).

Using l'Hospital's rule we may find the limiting value of the function $(x_2^S/x_2^L - x_1^S/x_1^L)$ at the points $x_1^S = x_1^L = 0$ and $x_1^S = x_1^L = 1$:

$$\lim_{\substack{x_1^S \to 0 \\ x_1^L \to 0}} \left(\frac{x_2^S}{x_2^L} - \frac{x_1^S}{x_1^L} \right)_T = \left[\left(\frac{\partial x_1^L}{\partial x_1^S} \right)_{T, x_1=0} - 1 \right] \Big/ \left(\frac{\partial x_1^L}{\partial x_1^S} \right)_{T, x_1=0}, \tag{5.50}$$

$$\lim_{\substack{x_1^S \to 1 \\ x_1^L \to 1}} \left(\frac{x_2^S}{x_2^L} - \frac{x_1^S}{x_1^L} \right)_T = \left[1 - \left(\frac{\partial x_1^L}{\partial x_1^S} \right)_{T, x_1=1} \right] \Big/ \left(\frac{\partial x_1^L}{\partial x_1^S} \right)_{T, x_1=1}. \tag{5.51}$$

At the points $x_1 = 0$ and $x_1 = 1$, Eq. (5.49) may be transformed by (5.50) and (5.51) into the following form:

$$-[\Delta h_{2,\,m}^{\circ}/R\,(T_{2,\,m}^{\circ})^{2}]\,(\partial T/\partial x_{1}^{S})_{p,\,x_{1}=0} = (\partial x_{1}^{L}/\partial x_{1}^{S})_{p,\,x_{1}=0} - 1, \qquad (5.52)$$

$$-[\Delta h_{1,\,m}^{\circ}/R\,(T_{1,\,m}^{\circ})^{2}]\,(\partial T/\partial x_{1}^{S})_{p,\,x_{1}=1} = 1 - (\partial x_{1}^{L}/\partial x_{1}^{S})_{p,\,x_{1}=1}. \qquad (5.53)$$

In view of the obvious inequalities $\Delta h_{m,i}^{\circ} > 0$ and $R > 0$, Eqs. (5.52) and (5.53) provide the correlation between the signs of the temperature derivatives and of the quantity $(dx_{1}^{L}/dx_{1}^{S})_{p,x_{i}}$ at points which correspond to the pure components of the two-phase binary system: solid–ideal melt. Thus,

1) if $(\partial T/\partial x_{1}^{S})_{p,x_{1}=0} \wedge 0$, then $1 \wedge (\partial x_{1}^{L}/\partial x_{1}^{S})_{p,x_{1}=0}$

2) if $(\partial T/\partial x_{1}^{S})_{p,x_{1}=1} \wedge 0$, then $(\partial x_{1}^{L}/\partial x_{1}^{S})_{p,x_{1}=1} \wedge 1$,

where the sign $\wedge$ expresses in each case the same relation (larger, equal, or smaller).

Relations (5.52) and (5.53) characterize the connection between the limiting behavior of the solidus line and the composition lines of the ideal–nonideal phases.

According to the third Gibbs–Konowalow law [1, 30], in binary systems the equilibrium quantities x_{i}^{S} and x_{i}^{L} always change in the same direction (under conditions of constant temperature and pressure). Therefore, the derivatives $(\partial T/\partial x_{1}^{S})_{p}$ may be replaced by $(\partial T/\partial x_{1}^{L})_{p}$. As a result of such a replacement we obtain a similar behavior for the composition lines of the coexisting phases and for the liquidus line.

2. INTERRELATION OF VARIOUS TYPES OF PHASE EQUILIBRIUM DIAGRAMS AND THE NATURE OF INTERMOLECULAR INTERACTIONS

Starting with well-known expressions for the thermodynamic potential, one may study various types of state diagrams both analytically and by the methods of geometrical thermodynamics.

There are three possible approaches to deriving the equations for the curves of monovariant equilibrium in a two-component two-phase system. The most widespread method is based on the equilibrium conditions of Gibbs, written in terms of the equality of chemical potentials:

$$\mu_{1}^{\alpha} = \mu_{1}^{\beta}, \quad \mu_{2}^{\alpha} = \mu_{2}^{\beta}. \qquad (5.54)$$

The second method relies on equations which express the phase equilibria in terms of the thermodynamic potentials of the phases:

$$
\begin{aligned}
G^{\alpha} - x_{2}^{\alpha}\,(\partial G/\partial x_{2})^{\alpha} &= G^{\beta} - x_{2}^{\beta}\,(\partial G/\partial x_{2})^{\beta}, \\
G^{\alpha} + (1 - x_{2}^{\alpha})\,(\partial G/\partial x_{2})^{\alpha} &= G^{\beta} + (1 - x_{2}^{\beta})\,(\partial G/\partial x_{2})^{\beta}.
\end{aligned}
\qquad (5.55)
$$

If $x_{2} = x^{\alpha} = x^{\beta}$, then we may write, on the basis of the first relation in (5.55),

$$G^{\alpha} - G^{\beta} = x_{2}\,[(\partial G/\partial x_{2})^{\alpha} - (\partial G/\partial x_{2})^{\beta}].$$

In view of expression (3.27) we are led to an equation which lies at the basis of the third approach:

$$G^\alpha = G^\beta. \tag{5.56}$$

The theory of ideal solutions applied to liquid and solid solutions allows one to describe only "cigar"-shaped state diagrams. Apart from the peritectic diagrams, the simplest diagrams will be obtained by assuming for the liquid state a model of ideal solutions, and, for the solid state, nonideal solutions. However, even for the liquid state, ideal solutions are rarely encountered, so that a more meaningful analysis is possible when a model of nonideal solutions is assumed for both the solid and the liquid states. In tackling this problem the theory of regular solutions has played an important part, and its application to a number of current problems has been very fruitful.

The application of the theory of regular solutions to coexisting phases provided the opportunity to present for the first time a unified description of all simple types of state diagrams.

The use of the theory of regular solutions in the theoretical study of phase equilibria was initiated by the work of Becker [162], who treated a special case of phase equilibrium, dissociation in a solid solution, occurring without a phase transition (i.e., when the crystal lattice of both components is identical). Becker's approximation was inadequate in that he assumed that the dissociation curve was symmetric, a situation rarely observed in real systems. Nevertheless, the essential point in Becker's equation is that use of it enables one to estimate the role of intermolecular interactions in the behavior of solutions by considering experimental data.

Pines [163, 164] generalized Becker's method by considering dissociation (demixing) in solutions accompanied by a phase transition. The results concerning the connection between the character of the intermolecular interaction and the form of the equilibrium diagram were further developed and refined in the work of Kamenetskaya [165–167].

Employing the second approach, we shall obtain, within the model of a regular solution, equations for the curves of demixing and of monovariant equilibrium corresponding to the initial crystallization of the components.

The molar Gibbs free energy for a two-component regular solution at $p = $ const may be written in the following manner:

$$g(T, x_2) = (1 - x_2)\,\mu_1^\circ(T) + x_2\mu_2^\circ(T)$$
$$+ RT\,[(1 - x_2)\ln(1 - x_2) + x_2 \ln x_2] + h^{(E)}(x), \tag{5.57}$$

where

$$\mu_i^\circ(T) = h_i^\circ - Ts_i^\circ. \tag{5.58}$$

Differentiating (5.57) with respect to x_2 leads to an expression of the following form:

$$(\partial g/\partial x_2)_T = RT \ln[x_2/(1 - x_2)] + (\partial h^E(x_2)/\partial x_2)_T$$
$$- \mu_1^\circ(T) + \mu_2^\circ(T), \tag{5.59}$$

$$(\partial^2 g/\partial x_2^2)_T = RT/[x_2 (1 - x_2)] + (\partial^2 h^E (x_2)/\partial x_2^2)_T. \qquad (5.60)$$

The equations of the binodal (demixing in the liquid phase) or the spinodal (dissociation in the solid phase) are easily obtained by equating (5.60) to zero.

As in [168], the enthalpy of mixing may be represented in the form of a Fourier sine series:

$$h^{(E)} (x_2) = \sum_{k=1}^{n} a_k \sin k\pi x_2. \qquad (5.61)$$

This way of writing the enthalpy satisfies the condition $h^E = 0$ at $x = 0$ and at $x = 1$, and admits an arbitrary dependence of the parameters of the intermolecular interaction on the concentration. Thus, on the basis of Eq. (5.61), one may carry out a more general analysis of the dissociation or demixing process than in the approximation of the strictly regular model, according to which the process is only allowed for positive values of ω, while the dissociation curve is symmetric.

Utilizing Eqs. (5.60) and (5.61) the equation for the spinoidal may be written in the form

$$RT = (1 - x_2) x_2 \pi^2 \sum_{k=1}^{n} k^2 a_k \sin k\pi x_2.$$

The critical point ($T = T_k$ and $x = x_k$) is the maximum of the spinodal. For $T < T_k$ the solid solution dissociates.

Figure 29 illustrates the dissociation curves corresponding to enthalpies of mixing in (5.61) with $k = 1, 2, 3$ and positive and negative values of a_k. One sees from this diagram that for regular solutions the character of the dissociation curve depends on the ratio of the parameter of intermolecular interaction and the value of RT. Generally speaking, one might say, for example, that actually all systems having positive heats of mixing ($h^E > 0$) should be immiscible in the solid phase. Experimentally this is not confirmed. The source of this discrepancy is the following. If h^E is not large enough, then T_k is so low that the rate of the process of dissociation in the solid state will be infinitely slow; thus, as a rule, a solid solution is formed which will actually be metastable. Due to the small rate of diffusion, a large portion of the diagram is probably inaccurate in the region of low temperatures.

For simplicity we shall consider the interrelation between the intermolecular interaction and the form of the phase diagram within the framework of the model of strictly regular solutions.

Substituting (5.57) and (5.59) into (5.55), using (5.58) and (4.104), and neglecting the temperature dependence of the heat of melting, we obtain formulas relating the equilibrium temperature to the concentrations of the coexisting phases in the strictly regular solution approximation:

$$T = \frac{(x_2^2 \omega)^L - (x_2^S)^2 \omega^S + \Delta h_{1, m}^\circ}{\Delta s_{1, m}^\circ - R \ln[(1 - x_2^{(L)})/(1 - x_2^S)]}, \qquad (5.62)$$

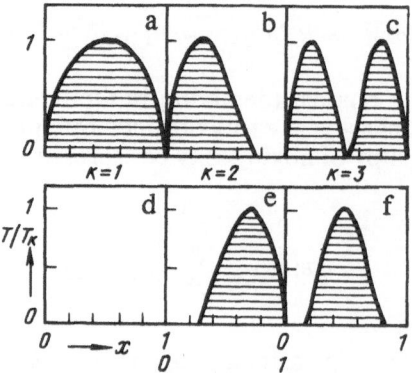

Fig. 29. Dissociation regions corresponding to values of the excess enthalpy $h^E = \sum_{k=1}^{n} a_k \sin k\pi x$ for $k = 1, 2$, and 3: upper curves a_k positive, lower curves a_k negative; values of RT_k/a_k: a) 2.47; b) 7.85; c) 13.54; d) –2.47; e) –7.85; f) –22.21.

$$T = \frac{\left(1 - x_2^L\right)^2 \omega^L - \left(1 - x_2^S\right)^2 \omega^S + \Delta h_{2,\,m}^\circ}{\Delta s_{2,\,m}^\circ - R \ln\left(x_2^L/x_2^S\right)}. \tag{5.63}$$

In the absence of demixing the appropriate stability conditions from (2.140) should apply in each phase:

$$\left(\frac{\partial^2 g}{\partial x_2^2}\right)_{T,p}^L = -2\omega^L + \frac{RT}{x_2^L\left(1 - x_2^L\right)} > 0,$$

$$\left(\frac{\partial^2 g}{\partial x_2^2}\right)_{T,p}^S = -2\omega^S + \frac{RT}{x_2^S\left(1 - x_2^S\right)} > 0. \tag{5.64}$$

If $\omega^L < 0$ and $\omega^S < 0$, these inequalities apply at any temperature. If $\omega^L > 0$ or $\omega^S > 0$, then the above inequalities become equalities at some temperature and concentration; on further lowering of the temperature the quantities $(\partial^2 g/\partial x_2^2)_{T,p}^L$ or $(\partial^2 g/\partial x_2^2)_{T,p}^S$ become negative in some concentration range, which means that the $g^L(x^L)$ or $g^S(x^S)$ curves acquire a maximum.

If one were able to find the explicit dependence of T on x^L and x^S, the calculation of the equilibrium curves and their dependence on the various constants referred to would become quite simple. Unfortunately, one cannot find the explicit solution of these equations, although in some special cases we were able to treat the diagrams analytically.

Case I. If in both phases the bonds between like and unlike atoms are assumed to be identical, i.e., $\omega^L = \omega^S = 0$, and if in the pure components a phase transformation occurs with a thermal effect (e.g., melting), then a "cigar"-type phase diagram results with a continuous series of solid solutions. It will be shown below that this "cigar" shape is also obtained for different values of ω^L and ω^S. In the present case one may solve Eqs. (5.62) and (5.63) explicitly and obtain the equations of the liquidus (5.31) and the solidus (5.32).

Cases II and III. Here one assumes that in the temperature regime within which a phase transformation occurs there is no dissociation or demixing. In this case the equilibrium curves will have a common maximum or minimum (see Fig. 11).

If one assumes $x^L = x^S = x_e$ in Eqs. (5.62) and (5.63), one obtains

$$
x_e = \frac{\Delta s^\circ_{1, m} - \sqrt{\begin{array}{c}(\Delta s^\circ_{1, m})^2 - (\Delta s^\circ_{1, m} - \Delta s^\circ_{2, m})\,[\Delta s^\circ_{1, m} - \\ - (\Delta h^\circ_{1, m}\Delta s^\circ_{2, m} - \Delta h^\circ_{2, m}\Delta s^\circ_{1, m})/(\omega^L - \omega^S)]\end{array}}}{\Delta s^\circ_{1, m} - \Delta s^\circ_{2, m}}, \qquad (5.65)
$$

$$
T_e = T^\circ_{1, m} + \frac{\omega^{(L)} - \omega^{(S)}}{\Delta s^\circ_{1, m}}\, x_e^2 = T^\circ_{2, m} + \frac{\omega^L - \omega^S}{\Delta s^\circ_{2, m}}\,(1 - x_e)^2. \qquad (5.66)
$$

If now $\omega^L > \omega^S$, then $T_e > T_{i,m}^\circ$, i.e., the temperature of equal concentrations is above the melting point of the pure components. One obtains a state diagram with a maximum. If $\omega^L < \omega^S$, then $T_e < T_{i,m}^\circ$, and we obtain a phase diagram with a minimum.

Thus, if the tendency for the clustering of unlike atoms is much more pronounced in a solid solution than in liquid, we shall obtain a state diagram with a maximum, while in the opposite case we have one with a minimum. We may also infer from Eqs. (5.65) and (5.66) that state diagrams with a maximum or a minimum are a direct indication of the nonideal behavior of a mixture, since for ideal mixtures this type of diagram is in principle impossible [169–172].

Basing his considerations on Eq. (5.65), Kostarev [173] has sharpened the conditions for the formation of state diagrams with continuous series of solid solutions. Using the condition for the existence of extrema on the liquidus and solidus curves, he came to the conclusion that when the extremum occurs at the edges of the concentration interval (0–1), or when it is completely absent, one obtains a "cigar"-type state diagram. Under the specified conditions his analysis leads to the following inequalities:

$$
\Delta s^{02}_{1, m} - (\Delta s^\circ_{1, m} - \Delta s^\circ_{2, m})
$$
$$
\times \left[\Delta s^\circ_{1, m} - \frac{\Delta s^\circ_{1, m}\Delta s^\circ_{2, m}\,(T^\circ_{1, m} - T^\circ_{2, m})}{\omega^L - \omega^S}\right] < 0; \qquad (5.67)
$$
$$
x_e \geqslant 1, \quad x_e \leqslant 0.
$$

Systematic solution of the inequality (5.67) leads to a necessary and sufficient condition for obtaining cigar-shaped state diagrams:

$$
\Delta s^\circ_{1, m}\,(T^\circ_{2, m} - T^\circ_{1, m}) \leqslant \omega^S - \omega^L \leqslant \Delta s^\circ_{2, m}\,(T^\circ_{2, m} - T^\circ_{1, m}). \qquad (5.68)
$$

The quantity $\Delta s_{i,m}^\circ$ may be calculated from the state diagram in terms of the slope of the curves at the points $x_2^L = x_2^S = 0$ and $x_2^L = x_2^S = 1$:

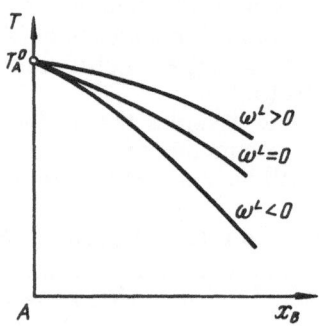

Fig. 30. Liquidus curves for various values of the energy of mixing.

$$\Delta s^\circ_{1,m} = RT^\circ_{2,m} \left\{ \ln \left[\left(\frac{\partial T}{\partial x_2} \right)^L \Big/ \left(\frac{\partial T}{\partial x_2} \right)^S \right]_{T=T^\circ_{2,m}} \right\} \Big/ (T^\circ_{1,m} - T^\circ_{2,m})$$

and

$$\Delta s^\circ_{2,m} = RT^\circ_{1,m} \left\{ \ln \left[\left(\frac{\partial T}{\partial x_2} \right)^L \Big/ \left(\frac{\partial T}{\partial x_2} \right)^S \right]_{T=T^\circ_{1,m}} \right\} \Big/ (T^\circ_{2,m} - T^\circ_{1,m}) .$$

The width of the cigar depends on $\Delta s_{i,m}{}^\circ$, the smaller the latter, the narrower the cigar.

Similarly, on the basis of (5.65) and the condition for the existence of extrema in the interval $0 < x_e < 1$, one may obtain inequalities [173, 174] which apply when the state diagrams have a minimum:

$$\omega^S - \omega^L > \Delta s^\circ_{2,m} (T^\circ_{2,m} - T^\circ_{1,m}) \tag{5.69}$$

or a maximum:

$$\omega^S - \omega^L < \Delta s^\circ_{1,m} (T^\circ_{2,m} - T^\circ_{1,m}). \tag{5.70}$$

Case IV. For very large energies of mixing in the solid phase, the mutual solubility of the components is practically negligible. In this case Eqs. (5.62) and (5.63) assume a very simple form and will describe the liquidus curves of eutectic-type systems.

a. The equation of the initial crystallization of the first components becomes

$$x_2 = 0, \quad T = \frac{\Delta h^\circ_{1,m} + (x_2^2 \omega)^L}{\Delta s^\circ_{1,m} - R \ln (1 - x_2^L)} . \tag{5.71}$$

b. The equation of the initial crystallization of the second component is

$$x_1 = 0, \quad T = \frac{\Delta h^\circ_{2,m} + (1 - x_2^L)^2 \omega^L}{\Delta s^\circ_{2,m} - R \ln x_2^L} . \tag{5.72}$$

If $\omega^L = 0$, Eqs. (5.71) and (5.72) coincide with the Schröder–Le Chatelier equations of the liquidus lines. Therefore, the terms $(x_2{}^2\omega)^L$ and $(1 - x_2{}^L)^2\omega^L$ characterize the deviation of the solution from ideal behavior. Figure 30 indicates schematically the influence of ω^L on the behavior of the solubility curves. At the points $x = 0$ or $x = 1$ all three curves have a common tangent whose slope determines the magnitude of $\Delta s_{i,m}{}^{\circ}$.

It follows from an analysis of various solubility curves, with differing signs of ω^L, that for $\omega^L = 0$ Raoult's law applies in a larger range of concentrations than for either $\omega^L < 0$ or $\omega^L > 0$, i.e., as one might expect, the solution is closest to ideality for $\omega^L = 0$.

We note that among the eutectic diagrams one may meet all three types of liquidus curves ($\omega^L = 0$, $\omega^L > 0$, and $\omega^L < 0$), although apparently for the majority of eutectics a positive value of ω^L is typical.

As pointed out by Kostarev [173], the analysis of eutectic- or peritectic-type state diagrams is more complicated than that of diagrams with a continuous series of solid solutions, although even for the former one may give a qualitative and even a semiquantitative description of the condition of realization of one type of equilibrium or another.

Thus, by using the rough approximation of Hildebrand [92], according to which a state diagram with a eutectic or a peritectic may be considered to result from the imposition of equilibrium with a continuous series of solid solutions on the dissociation curve, we have [173], in addition to (5.69), another sufficient condition for dissociation of the solid phase of a simple eutectic:

$$\omega^S > 2RT^{\circ}_{i,\,m}, \tag{5.73}$$

where $T_{i,m}{}^{\circ}$ is the melting temperature of the lower melting component.

Since for a simple peritectic the liquid phase, unlike the solid, does not dissociate, we have $\omega^S > \omega^L$, and the inequality (5.68) becomes the condition [173]

$$0 < \omega^S - \omega^L \leqslant \Delta s^{\circ}_{2,\,m}\,(T^{\circ}_{2,\,m} \overset{.}{-} T^{\circ}_{1,\,m}), \tag{5.74}$$

complementing the condition for the dissociation of the solid phase [82]:

$$\omega^S > R\,(T^{\circ}_{1,\,m} + T^{\circ}_{2,\,m}). \tag{5.75}$$

The theory of regular solutions does not enable one to supply a quantitative criterion for the formation of intermediate phases, and for the formation of the corresponding state diagrams in two-component systems. Nevertheless, one may treat within its framework the relative position of the liquidus curves in systems with congruently melting compounds.

We shall analyze the equation of the liquidus curve of a congruently melting compound A_mB_n, Eq. (5.39), within the approximation of a regular solution. For a regular solution we have

$$RT \ln \gamma_i = \omega\,(1 - x_i)^2.$$

Then Eq. (5.44) may be transformed into the form

$$(\partial\mu_2/\partial x_2)^L_{p,T} = RT/x_2^L - 2\omega^L(1 - x_2^L),$$

On substituting this expression into Eq. (5.39), we obtain

$$\left(\frac{\partial T}{\partial x_2}\right)^L_p = -mT\left(\frac{n}{m} - \frac{x_2^L}{1 - x_2^L}\right)\left[\frac{RT}{x_2^L} - 2\omega^L(1 - x_2^L)\right]\Bigg/ \bar{q}^{(L\to S)}_{A_mB_n}. \qquad (5.76)$$

This describes in differential form the equilibrium of the original crystals of A_mB_n with the regular liquid solution.

For a regular solution one may neglect the quantity q^E in Eq. (5.44a). Then the approximate form of relation (5.76) becomes, for the present case,

$$\left(\frac{\partial T}{\partial x_2}\right)^L_p = mT\left(\frac{n}{m} - \frac{x_2^L}{1 - x_2^L}\right)\left[\frac{RT}{x_2^L} - 2\omega^L(1 - x_2^L)\right]\Bigg/ \Delta h^\circ_{m,\,A_mB_n}. \qquad (5.77)$$

In analyzing this expression one must consider two cases, corresponding to positive and negative deviations from Raoult's law. For negative deviations the activity coefficients are smaller than unity and, consequently, $\omega^L < 0$. It then follows from Eq. (5.77) that with $\omega^L < 0$ the following inequality is satisfied for any value of x_2^L:

$$|\partial T/\partial x_2|^L > |\partial T/\partial x_2|^{id\,(L)}.$$

This implies that if the solution is characterized by a negative deviation from Raoult's law, then the maximum of the congruently melting compound is more peaked than for an ideal solution.

For positive deviations from Raoult's law the activity coefficients are larger than unity and $\omega^L > 0$. In this case, for all values of x_2^L, we have

$$|\partial T/\partial x_2|^L < |\partial T/\partial x_2|^{id\,(L)},$$

namely, the maximum is flatter than for an ideal solution. For very large positive values of ω^L a region of immiscibility may arise. The phase diagram for this case is illustrated in Fig. 31, which corresponds to a system with a lower critical temperature of mixing [24].

We now consider possible solutions of (5.62) and (5.63) by the geometric method of Rozeboom, which consists of studying free-energy functions of varying degrees of curvature for the coexisting phases.

Following Kamenetskaya [166], we first apply the Rozeboom method to study the free-energy surfaces of the coexisting phases, determined by the relations

$$g^L(T, x^L) = (1 - x_2^L)h_1^{\circ\,(L)} + x_2^L h_2^{\circ\,(L)}$$
$$- T[(1 - x_2^L)s_1^{\circ\,(L)} + x_2^L s_2^{\circ\,(L)}]$$
$$+ RT[x_2^L \ln x_2^L + (1 - x_2^L)\ln(1 - x_2^L)] + \omega^L x_2^L(1 - x_2^L), \qquad (5.78)$$

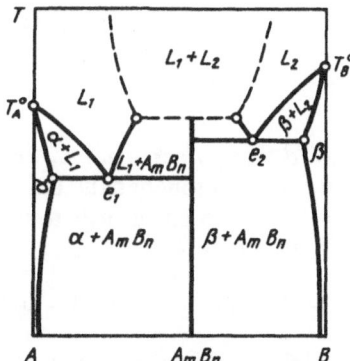

Fig. 31. State diagram with restricted solubility for the liquid phase in the presence of a lower critical point.

$$g^S(T, x_2^S) = (1 - x_2^S)\, h_1^{\circ\,(S)} + x_2^S h_2^{\circ\,(S)}$$
$$- T\,[(1 - x_2^S)\, s_1^{\circ\,(S)} + x_2^{(S)} s_2^{\circ\,(S)}]$$
$$+ RT\,[(1 - x_2^S)\ln(1 - x_2^S) + x_2^S \ln x_2^S] + \omega^S x_2^S (1 - x_2^S). \qquad (5.79)$$

These equations represent, in the space of the coordinates g, T, and x, certain trough-shaped surfaces g^L and g^S, touching the planes $x = 0$ and $x = 1$. We obtain the equations of the tangents by substituting into formulas (5.78) and (5.79) the values zero and unity for x.

For $x_2^L = x_2^S = 0$ (pure component 1),

$$g^L(T, 0) = h_1^{\circ\,(L)} - Ts_1^{\circ\,(L)} = \mu_1^{\circ\,(L)}(T),$$
$$g^S(T, 0) = h_1^{\circ\,(S)} - Ts_1^{\circ\,(S)} = \mu_1^{\circ\,(S)}(T), \qquad (5.80)$$

and for $x_2^L = x_2^S = 1$ (pure component 2),

$$g^L(T, 1) = h_2^{\circ\,(L)} - Ts_2^{\circ\,(L)} = \mu_2^{\circ\,(L)}(T),$$
$$g^S(T, 1) = h_2^{\circ\,(S)} - Ts_2^{\circ\,(S)} = \mu_2^{\circ\,(S)}(T). \qquad (5.81)$$

In studying the character of phase equilibria in the system, we may utilize the intersections of Gibbs free energy surfaces g^L and g^S with planes of given constant temperatures. The form and relative position of the projection of the line of intersection on the (g, x) plane will then determine, as we saw in Chapter 3, the concentrations of the coexisting phases. We shall study the behavior of these projections by means of the first and second derivatives of g with respect to x, and the radius of curvature ρ.

Under the given conditions ($T = $ const, $p = $ const) we have

$$(\partial g/\partial x)_{p,\,T}^{L} = \left(1 - 2x^{L}\right)\omega^{L} + RT\ln\left[x^{L}/(1 - x^{L})\right]$$
$$- \mu_{1}^{\circ\,(L)} + \mu_{2}^{\circ\,(L)}. \tag{5.82}$$

This implies that for $x^{L} \to 0$ $(dg/dx)_{p,T}^{L} \to -\infty$, i.e., the curve $g^{L} = f(x^{L})$ (which for convenience we shall denote by g_{x}^{L}) touches the axis $x = 0$ and points downward; for $x^{L} \to 1$ $(dg/dx)_{p,T}^{L} \to \infty$, i.e., the curve g_{x}^{L} touches the ordinate $x = 1$ and points upward. The roots of the equation

$$\left(1 - 2x^{L}\right)\omega^{L} + RT\ln\left[x^{L}/(1 - x^{L})\right] - \mu_{1}^{\circ\,(L)} + \mu_{2}^{\circ\,(L)} = 0 \tag{5.83}$$

give the values of x^{L} for which the curve g_{x}^{L} has a maximum or a minimum. For solving this equation by the geometric method, it is expedient to transform it into the form

$$\left(1 - 2x^{L}\right)\omega^{L}/RT + c_{T}^{L} = \ln\left[(1 - x^{L})/x^{L}\right], \tag{5.84}$$

where

$$c_{T}^{L} = \left(\mu_{2}^{\circ\,(L)} - \mu_{1}^{\circ\,(L)}\right)/RT. \tag{5.85}$$

The left-hand side of Eq. (5.84) is a straight line in x whose slope depends on the sign and magnitude of the energy of mixing, as well as on temperature. Let us denote this slope by Π^{L} and the ordinate of the line by O_{Π}. The right-hand side of Eq. (5.84) as a function of x is a curve which we denote by K, with its ordinate O_{K}.

We consider various possibilities for the position of the straight line Π^{L} with respect to curve K.

Case 1. $\omega^{L} < 0$.

In this case the slope of the line is positive and the straight line intersects the curve K at one point (cf. Fig. 32). Hence, the curve g_{x}^{L} has only one extremal point.

For $\omega^{L} < 0$, the derivative $(d^{2}g/dx^{2})_{p,T}^{L} > 0$ [cf. (5.64)], i.e., the g_{x}^{L} curve has a minimum (see Fig. 6). The derivative $(d^{2}g/dx^{2})_{p,T}^{L}$ equals the difference in the slopes of the straight line Π^{L} and the tangent to the curve K. If $\omega^{L} < 0$, this difference is positive. The position of the minimum $(x_{\min}^{L})$, determined by the point of intersection of Π and K, changes with temperature: for $c_{T}^{L} > 0, x_{\min} < 1/2$; for $c_{T}^{L} < 0, x_{\min} > 1/2$; for $c_{T}^{L} = 0, x_{\min} = 1/2$.

Case 2.

When the energy of mixing is positive, the slope of the straight line Π is negative. We then meet three variants:

A. If ω^{L} is not large, the straight line Π^{L} intersects the curve K at one point (Fig. 33), where, by formula (5.64), the derivative $(d^{2}g/dx^{2})_{p,T}^{L} > 0$.

Consequently, for small positive values of ω^{L}, when condition (5.64) applies, the curve g_{x}^{L} has one minimum whose position is fixed by the point of intersection of Π^{L} with K. As the temperature is lowered, the position of the minimum shifts from $x_{\min} > 1/2$ to $x_{\min} < 1/2$.

B. At some critical temperature T_{c} the line Π^{L} touches the curve K. Then

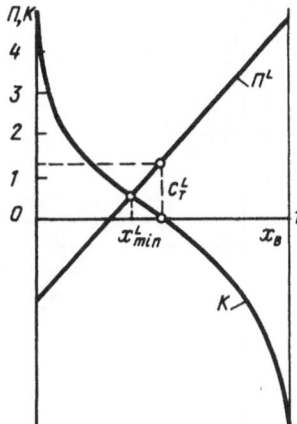

Fig. 32. Geometrical solution of Eq. (5.84) for $\omega^L < 0$.

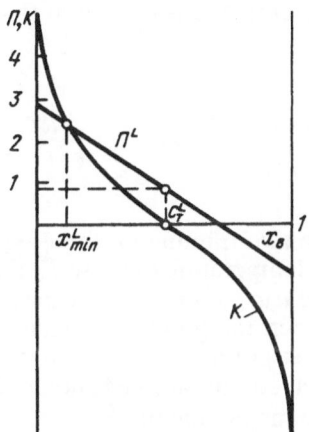

Fig. 33. Geometrical solution of Eq. (5.84) for $0 < \omega^L < 2RT$.

$$T_k = 2x_k(1 - x_k)\,\omega/R. \tag{5.86}$$

On the basis of this expression it is easily verified that the concentration corresponding to the critical point is 0.5. Substituting the value $x^L = 0.5$ into (5.84) we find that $c_{T_c}{}^L = 0$, i.e., at the critical temperature the chemical potentials are identical, in keeping with Eq. (5.85).

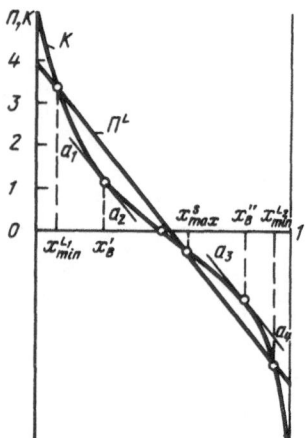

Fig. 34. Geometrical solution of Eq. (5.84) for $\omega^L > 2RT$.

C. When $T < T_c$, the line Π^L intersects the curve K at three points (Fig. 34), with $c_T{}^L > 0$. At the two extreme points condition (5.64) applies, while at the middle one

$$\left(\frac{\partial^2 g}{\partial x_2^2}\right)^L_{T,\,p} = -2\omega^L + \frac{RT}{x_2^L\left(1 - x_2^L\right)} < 0. \tag{5.87}$$

Consequently, there are two minima on the $g_x{}^L$ curve, with one maximum in between (cf. Fig. 10). For temperatures below T_c the system dissociates. If one draws two tangents a_1a_2 and a_3a_4 to the curve K parallel to the line Π^L, then in the concentration interval between the points of contact $x_B{}'$ and $x_B{}''$ condition (5.87) will be fulfilled. The abscissas of the points of contact correspond to points of inflection. Outside this interval condition (5.64) holds. Thus the curve $g_x{}^L$ may have two forms, depending on the magnitude of ω^L:

1. If $\omega^L < 0$, and for all values of ω^L satisfying the condition $0 < \omega^L < RT/[2x^L(1 - x^L)]$, the curve $g_x{}^L$ has one minimum, since the lowest value of the right-hand side of the inequality is equal to $2RT$, so that it may be written in the form $\omega^L < 2RT$.

2. If in some concentration interval including $x = 0.5$ the inequality ($\omega^L > RT)/[2x^L(1 - x^L)]$ applies, i.e., if $\omega^L > 2RT$, then the curve $g_x{}^L$ has two minima separated by a maximum (the case of demixing).

For a more concrete characterization of $g_x{}^L$ one may also determine its radius of curvature

$$\rho^L = \frac{\left\{\sqrt{1 + \left[\left(1 - 2x^L\right)\left(\omega^L/RT\right) + c_T^L + \ln\left[x^L/\left(1 - x^L\right)\right]\right]^2}\right\}^3}{-2\omega^L + RT/[x^L\left(1 + x^L\right)]}. \tag{5.88}$$

If $\omega^L > 0$, then for $T = T_c$, $\rho^L \to \infty$, for $T > T_c$, $\rho^L > 0$, and for $T < T_c$ in some concentration interval, $\rho^L < 0$. If $\omega^L < 0$, we always have $\rho^L > 0$, i.e., the line curves downward at all points. Other conditions being equal, the larger the energy of mixing, the larger the radius of curvature. In order to clarify the dependence of ρ^L on temperature, let us put $x^L = 1/2$. Then

$$\rho^L = [\sqrt{1 + (c_T^L)^2}]^3/(-2\omega^L + 4RT). \qquad (5.89)$$

As the temperature is increased, c_T^L decreases [see formula (5.85)], and so does ρ^L.

The surface $g^s = f(x^s, T)$ may be studied analogously. Knowing the form of the free-energy curves of the coexisting phases, one may treat various types of state diagrams by the method of Rozeboom [175, 176].

We now enumerate the conditions for the formation of state diagrams with a continuous series of solid solutions.

a) $\omega^L = \omega^S = 0$, i.e., the interactions of like and unlike atoms in both phases are identical. In that case the lines Π^L and Π^S are parallel to the x axis and each intersects the curve K at one point. In this case Eq. (5.84) acquires the form

for the liquid phase:

$$c_T^L = \ln[(1 - x^L)/x^L], \qquad (5.90)$$

for the solid phase:

$$c_T^S = \ln[(1 - x^S)/x^S]. \qquad (5.91)$$

Both equations may be solved analytically. The curves g_x^L and g_x^S have in this case one minimum each. The distance between the minima ($x_{min}^L - x_{min}^S$) depends on the difference $c_T^L - c_T^S$. The distance between the points of contact x^L and x^S of the common tangent to the curves g_x^L and g_x^S also depends on this difference: the smaller ($c_T^L - c_T^S$), the nearer are the values of x^L to x^S to each other. If one assumes that c_T^L is larger than c_T^S, then the line Π^L will be above the line Π^S (see Fig. 35a). As the temperature is lowered, the points x_{min}^L and x_{min}^S are displaced to the left, all the while satisfying $x_{min}^L < x_{min}^S$. If we now construct the locus of the projections of the points of contact x^L and x^S of the common tangent to the curves g_x^L and g_x^S onto the (x, T) plane, then we obtain a cigar-type state diagram with a continuous series of solid solutions (cf. Fig. 9).

b) $\omega^L = \omega^S < 0$; $T_{1,m}° \neq T_{2,m}°$. The straight lines Π^L and Π^S will now be parallel to each other, and their slopes will be positive (see Fig. 35b). As the temperature is lowered, as in the preceding case, x_{min}^L is always less than x_{min}^S. Changes in x^L and x^S (the concentrations of the coexisting phases) occur mainly because of the different character of the variations of the curves g_x^L and g_x^S with temperature, and one finally obtains a cigar-type state diagram with continuous series of solid solutions.

c) $0 < \omega^L = \omega^S = 2RT_c < 2RT$; $T_c < T$. As already remarked, for a positive energy of mixing, the lines Π^L and Π^S have negative slopes. For small values of ω each of these lines intersects the curve K at only one point (as shown in Fig. 35c).

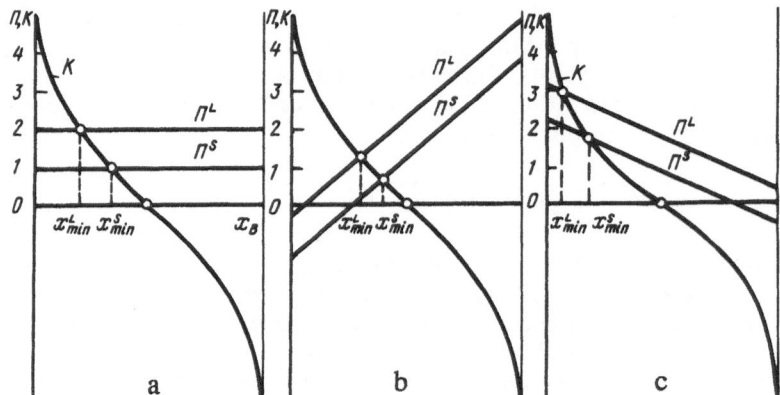

Fig. 35. Geometrical solution of Eq. (5.84) for: a) $\omega^L = \omega^S = 0$; b) $\omega^L = \omega^S < 0$; c) $0 < \omega^L = \omega^S = 2RT_c < 2RT$.

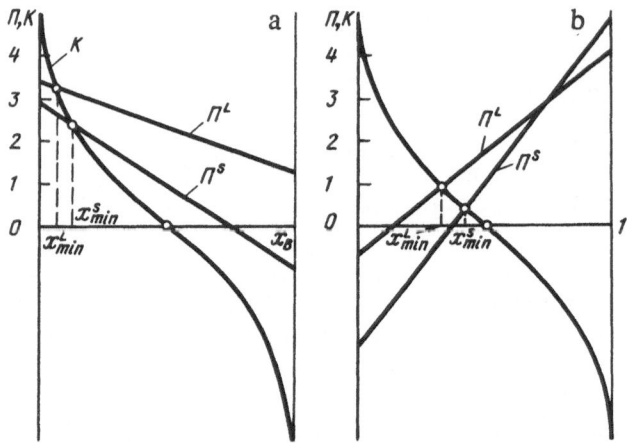

Fig. 36. Geometrical solution of Eq. (5.84) for a state diagram with a continuous series of solid solutions in the presence of a common point on the liquidus and solidus curves: a) minimum; b) maximum.

On lowering the temperature, $x_{\min}{}^L < x_{\min}{}^S$ and $x^L < x^S$. One finally obtains a cigar. On further lowering of the temperature, the line Π^S touches the curve K at some temperature T_K. As the temperature is lowered still further, Π^S intersects the curve K at three points, which correspond to the formation of two minima on the curve $g_x{}^S$ and, consequently, to the dissociation of the solid phase.

On the basis of relations (5.62) and (5.63), the equations for the cigars in the various cases may be written in the form

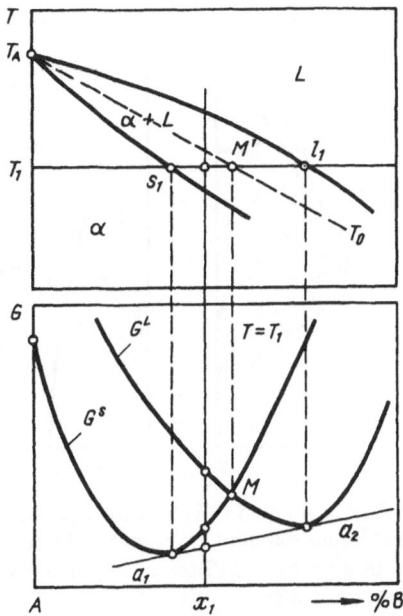

Fig. 37. Scheme for constructing the temperature lines of the equal Gibbs free energies in the liquid and solid phases, by projecting the point M on the isotherm T_1 onto the point M'.

$$T = \frac{\omega\left[(x_2^L)^2 - (x_2^S)^2\right] + \Delta h_{1,\,m}^\circ}{\Delta s_{1,\,m} - R \ln\left[(1 - x_2^L)/(1 - x_2^S)\right]} \qquad (5.92)$$

or

$$T = \frac{\omega\left[(1 - x_2^L)^2 - (1 - x_2^S)^2\right] + \Delta h_{2,\,m}^\circ}{\Delta s_{2,\,m} - R \ln\left(x_2^L\right)/\left(x_2^S\right)}. \qquad (5.93)$$

For negative energies of mixing the cigar obtained is narrower than for $\omega = 0$. If $\omega > 0$, the character of the curves may become more complicated, especially for large values of ω.

d) $\omega^L < \omega^S < 2RT$. In this case $x_{\min}^L$ and $x_{\min}^S$ are near to each other (Fig. 36a). The radius of curvature ρ^L of the curve g_x^L is smaller than for g_x^S, by Eqs. (5.88) and (5.89). As shown earlier, the state diagram obtained will have points of equal concentrations at the minimum (cf. Fig. 11b).

e) $\omega^S < \omega^L < 2RT$. Now $\rho^L > \rho^S$. The positions of the minima of the curves g_x^L and g_x^S are near, since for $T > T_{2,m}^\circ$ the difference $c_T^L - c_T^S$ is small, and the lines Π^L and Π^S intersect K at nearby points (Fig. 36b). Therefore, when the energies of mixing of both phases are related in this way, we obtain points of equal concentrations at the maximum in the phase diagram. Using similar constructions, one may obtain criteria for the formation of other types of state diagrams.

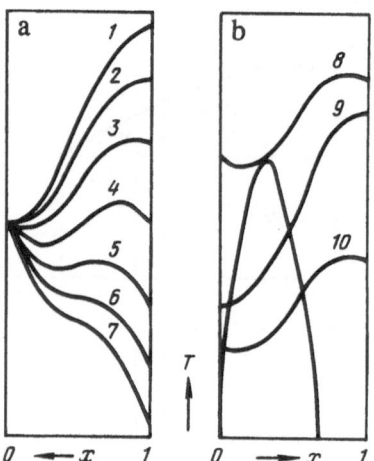

Fig. 38. Curves of equal G values for hypothetical systems in which the liquid solutions are ideal but the solid is a regular solution with $h^E = a_2 = \sin 2\pi x$: a) high-temperature region; b) low-temperature region.

When the conditions $\omega^L < 2RT$ and $\omega^S > 2RT$ are fulfilled (i.e., the melting temperatures of the pure components are small), one obtains a state diagram with a restricted solubility in the solid phase, with a eutectic point.

When $\omega^L < 2RT$ and $\omega^S > 2RT$ $(T_{2,m}° \gg T_{1,m}°)$, one obtains a phase diagram with a restricted solubility in the solid phase, with a peritectic point.

If $\omega^L > 2RT$ and $\omega^S < 2RT$, dissociation should take place in the liquid phase with unrestricted solubility in the solid phase.

If $\omega^L > 2RT$ and $\omega^S > 2RT$ (the solubility of both phases is restricted), one obtains the phase equilibrium diagram illustrated in Fig. 17.

In such a manner, employing the method of Rozeboom and applying geometrical constructions, Kamenetskaya has systematically derived various types of phase diagrams as a function of the magnitude and sign of the energy of mixing of the coexisting phases, taking into account in the separate cases the relation between the melting temperatures of the pure components. It follows from the work of Kamenetskaya that the connection between the type of phase diagram and the character of the intermolecular interaction (the magnitude and sign of ω), as described within the framework of the regular solution approximation, is not single-valued. Such a conclusion is also supported by the results of a thermodynamic argument due to Oonk and Sprenkels [168], based on an analysis of the curves of equal values of the Gibbs free energy of the phases. We follow Oonk and Sprenkels [168, 177] to prove this assertion. Using (5.56), (4.7), (4.17), and (4.84), we write the expression for the line of equal values of the Gibbs free energies of the solid and liquid phase, $T_0(x)$ (cf. Fig. 37), in the general form

$$T_0(x) = \frac{(1-x_2)\,T_{1,\,m}°\,\Delta s_{1,\,m}° + x_2 T_{2,\,m}°\,\Delta s_{2,\,m}° + \Delta h^E(x)}{(1-x_2)\,\Delta s_{1,\,m}° + x_2 \Delta s_{2,\,m}° - \Delta s^E(x)}, \tag{5.94}$$

where

$$\Delta h^E = h^{E\ (L)}(x) - h^{E\ (S)}(x), \quad \Delta s^E = s^{E\ (L)}(x) - s^{E\ (S)}(x). \tag{5.95}$$

If we limit our treatment of $T_0(x)$ to the model of regular solutions, we may rewrite (5.94) in a form convenient for analysis:

$$T_0(x) = \frac{(1-x_2)\, T_{1,\ m}^{\circ}\Delta s_{1,\ m}^{\circ} + x_2 T_{2,\ m}^{\circ}\Delta s_{2,\ m}^{\circ}}{\Delta s_{1,\ m}^{\circ} + x_2\left(\Delta s_{2,\ m}^{\circ} - \Delta s_{1,\ m}^{\circ}\right)}$$

$$+ \frac{\Delta h^E(x)}{\Delta s_{1,\ m}^{\circ} + x_2\left(\Delta s_{2,\ m}^{\circ} - \Delta s_{1,\ m}^{\circ}\right)}. \tag{5.96}$$

If the condition $\Delta h^E(x) = h^{E(L)}(x) - h^{E(S)}(x) = 0$ is satisfied for all values of x (i.e., if both coexisting phases are either ideal solutions or both display the same departure from ideality), then the second term on the right-hand side of Eq. (5.96) vanishes. In this situation the curve $T_0(x)$ is described by the first term in (5.96) and is called the zero line. As a rule, the difference $\Delta s_{2,m}^{\circ} - \Delta s_{1,m}^{\circ}$ is small compared to $\Delta s_{1,m}^{\circ}$, and the zero line hardly deviates from the straight line which joins the melting temperatures of the pure components. This deviation is fixed by the ratio of $\Delta h^E(x)$ to $\Delta s_{1,m}^{\circ}$. Therefore, the character of the $\Delta h^E(x)$ curve is directly reflected in the form of the curve of equal Gibbs free-energy values of the phases.

As an illustration, Fig. 38 shows the successive change of the equal-value G curve of the coexisting phases in model two-component two-phase systems under the influence of varying the difference of the melting temperatures $(T_{2,m}^{\circ} - T_{1,m}^{\circ})$. In calculating the $T_0(x)$ curve, the liquid phase was assumed to be an ideal solution, and the solid phase a regular solution, characterizable by a heat of mixing determined by the formula $h^{E(S)} = a_2 \sin 2\pi x$ with a_2 positive, and by a region of dissociation. Furthermore, it was assumed that $\Delta s_{1,m}^{\circ} = \Delta s_{2,m}^{\circ}$.

According to the results of the calculations carried out by Oonk and Sprenkels [168, 177], the equal G-value curves of the coexisting phases adequately reflect the types of phase diagrams. In other words, the maxima, minima, and points of inflection on the curves of monovariant equilibria coincide exactly with the corresponding properties of the line $T_0(x)$.

Accordingly, the curves of equal G values represented in Fig. 38a, b will correspond to the following types of state diagrams:

a) Curve 4 corresponds to the case when $T_{1,m}^{\circ} = T_{2,m}^{\circ}$. On the corresponding state diagram there should be two extrema: a maximum and a minimum.

b) Curves 1–3 are appropriate to the situation when $T_{2,m}^{\circ} - T_{1,m}^{\circ} > 0$ and illustrate the influence of increasing the melting temperature of the second component of the pure components. On curve 3 the extremal points are still present, and the corresponding type of phase equilibrium will be the same as in the preceding case. On curve 1 the extrema are absent and the derivative $(dT_0(x)/dx)_p > 0$ for all values of x. This case represents a phase equilibrium with a continuous series of solid solutions. Curve 2 shows a case when the extrema are at points $x = 0$ and $x = 1$ [the derivatives $dT_0(x)/dx)_p = 0$], where the liquidus and solidus curves touch. This curve corresponds to a borderline variant, when the type of diagram changes from case (a) to a cigar.

c) Curves 5–7 represent cases when $T_{2,m}^\circ - T_{1,m}^\circ < 0$, and illustrate the influence of lowering the melting temperature of the second component $T_{2,m}^\circ$ on the form of the line $T_0(x)$, as a result of which the positions of the extrema on $T_0(x)$ converge.

The extrema are absent on curve 7 for all values of x $[dT_0(x)/dx)_p < 0]$, and the corresponding type of diagram will be cigarlike. Curve 6 reflects an intermediate case, when the positions of the extrema coincide, and as a result one observes a point of inflection on the $T_0(x)$ line. The presence of this point is appropriate to the case in which the liquidus and solidus curves on the phase equilibrium diagram also have a common point of inflection. The type of phase equilibrium diagram then changes from form (a) to a cigar.

Figure 38b is appropriate to a temperature region in which there is dissociation in the solid phase and, consequently, the equal G-value curves of the coexisting phases are intersected by the spinodal. In this case the intersection of the equal-value G curves – characterizable by the absence of extrema (curve 9) – with the spinodal leads to the formation of a peritectic-type phase equilibrium diagram. For the intersections of the $T_0(x)$ curves having maxima and minima, there are several possible variants:

1. The minimum of the line $T_0(x)$ is inside the region of dissociation, while its maximum is at its boundaries (curve 10). This corresponds to a phase equilibrium diagram with the existence of a eutectic, and a maximum on the liquidus and solidus curves in some definite concentration interval.

2. Both extrema of the $T_0(x)$ line (curves 3–6 in Fig. 38a) lie within the dissociation region; this is appropriate to a peritectic phase equilibrium diagram.

3. The equal G-value curves touch the spinodal, and the maximum and minimum are situated within the region of immiscibility (curve 8); this corresponds to a diagram of phase equilibria with the existence of a peritectic, and of a maximum and minimum on the liquidus and solidus curves in some definite interval of concentration.

We note that curve 2 of the intersection with the spinodal leads to the intermediate case, with a corresponding transition from a peritectic to a diagram of type (3), whereas curve 6 does not indicate an intermediate case.

Using the method expounded above, Oonk and Sprenkels [168] have analyzed the interrelation between the various types of phase equilibria and the character of intermolecular interactions within a regular solution model, excluding from their consideration the case of polymorphism and dissociation in the liquid phase.

For describing the concentration dependence of the difference of the heats of mixing of solid and liquid solutions, they used an expression of the form

$$\Delta h^E(x) = \sum_{k=1}^{n} \Delta a_k \sin k\pi x, \qquad (5.97)$$

where the choice of the magnitude of n was based on results of analyzing real systems.

The results obtained in [168] indicate that one type of diagram of phase equilibria may be associated with various kinds of intermolecular interactions, and, conversely, a particular type of intermolecular interaction may give rise to different types of phase equilibria.

On the basis of the theory of regular solutions, Mager, Lukas, and Petrow [178, 179] have considered the conditions for which the type of the phase diagram may change. They made use of experimental data for 420 systems. The quantity $RT_M = R(T_{1,m}° + T_{2,m}°)/2$ served as an experimental parameter for comparison. Kaufman and Bernstein [180] have developed a method of calculating various phase diagrams by a computer, based on ideal and regular solution models.

Kamenetskaya [167] analyzed the criteria for the occurrence of various types of phase diagrams as a function of the ratio of the fundamental quantities which determine the overall properties of state diagrams at various pressures: $\omega^{(L)}$, $\omega^{(S)}$, $\nu^{E(L)}$, and $\nu^{E(S)}$.

In conclusion, we note that all attempts to predict the possible types of state diagrams on the basis of the model of regular solutions should be considered as purely qualitative, in view of the inherent limitations of this model. The most systematic approach of this kind, put forward in the work of Kamenetskaya, essentially adds very little to the qualitative conclusions obtained on the basis of Rozeboom's method. It does, however, predict qualitatively the evolution of the types of phase equilibria.

Nevertheless, even the qualitative predictions are limited by the principal assumptions, according to which the evolution of a diagram proceeds by combining the dissociation hump in the solid phase with the two-phase equilibrium between the liquid and solid phases. Such an approach is not really justified by any principle. It implies the isomorphism of the structure of the components, which in turn cannot be considered to be a rigorous condition because of its intrinsic limitation.

3. ANALYSIS OF LIQUIDUS AND SOLIDUS CURVES IN SYSTEMS WITH RESTRICTED AND UNRESTRICTED SOLUBILITIES

The thermodynamic analysis of phase equilibria opens up the possibility of clarifying the nature of intermolecular interactions and, consequently, of revealing the physicochemical features of the phases. In this context important information is provided by an estimate of the magnitude and sign of the exchange energy between the components, enabling one to evaluate the predominant features of the several pair interactions in the solution, which in turn leads directly to conclusions concerning the tendency to dissociate or to form compounds. An estimate of the temperature and concentration dependence of ω helps to identify the direction of structural changes due to variations of important state parameters. On the basis of analyzing state diagrams, one may extract information on thermodynamic functions characterizing the formation of one phase or another or, conversely, having available the pertinent thermodynamic data, one may calculate the separate elements of phase diagrams, or estimate the probability of their occurrence.

In general, one may express the equation for the line of monovariant equilibrium in a two-component system in terms of the activity coefficients and the thermodynamic characteristics of the melting components:

$$RT \ln (1 - x_2)^L \gamma_1^L - RT \ln (1 - x_2)^S \gamma_1^S = \Delta \dot{s}_{1,\,m} (T - T_{1,\,m}°) \qquad (5.98)$$

or

$$RT \ln x_2^L \gamma_2^L - RT \ln x_2^S \gamma_2^S = \Delta s_{2,\,m}^{\circ} \left(T - T_{2,\,m}^{\circ}\right).\tag{5.99}$$

In systems with unrestricted solubility of the components in the solid and liquid states, these equations are valid in the whole range of concentrations. In systems with restricted solubility in the solid phase (eutectics and peritectics), the equations are strictly applicable only to solvents, since in that case the standard state of component B dissolved in A is not equivalent to the state of the pure B. Exceptions to this may be systems with restricted solubility formed by isomorphic components.

The explicit form of Eqs. (5.98) and (5.99) depends on the choice of the model employed to analyze intermolecular interactions. Thus, if we use expressions (4.139) and (4.140) for the activity coefficients, we obtain relations characterizing the two-phase equilibrium $S \rightleftharpoons L$ which connect the temperatures and concentrations of the coexisting phases:

$$T = \cfrac{\Delta h_{1,\,m}^{\circ} + \left(x_2^2 \omega_0\right)^L - \left(x_2^2 \omega_0\right)^S - \left(x_2^L\right)^2 \left(1 - 2x_2^L\right) \omega_1^L + \left(x_2^S\right)^2 \left(1 - 2x_2^S\right) \omega_1^S + \left(x_2^L\right)^3 \left(3x_2^L - 2\right) \omega_2^L - \left(x_2^S\right)^3 \left(3x_2^S - 2\right) \omega_2^S}{\Delta s_{1,\,m}^{\circ} + \left(x_2^2 \omega_{01}\right)^L - \left(x_2^2 \omega_{01}\right)^S - \left(x_2^L\right)^2 \left(1 - 2x_2^L\right) \omega_{11}^L + \left(x_2^S\right)^2 \left(1 - 2x_2^S\right) \omega_{11}^S + \left(x_2^L\right)^3 \left(3x_2^L - 2\right) \omega_{21}^L - \left(x_2^S\right)^3 \left(3x_2^S - 2\right) \omega_{21}^S - R \ln \left[\left(1 - x_2^L\right)/\left(1 - x_2^S\right)\right]}.\tag{5.100}$$

$$T = \cfrac{\Delta h_{2,\,m}^{\circ} + \left(x_1^L\right)^2 \omega_0^L - \left(x_1^S\right)^2 \omega_0^S + 2x_2^L \left(x_1^L\right)^2 \omega_1 - 2x_2^S \left(x_1^S\right)^2 \omega_1^S + 3\left(x_1^L x_2^L\right)^2 \omega_2^L - 3\left(x_1^S x_2^S\right)^2 \omega_2^S}{\Delta s_{2,\,m}^{\circ} + \left(x_1^L\right)^2 \omega_{01}^L - \left(x_1^S\right)^2 \omega_{01}^S + 2\left(x_1^L\right)^2 x_2^L \omega_{11}^L - 2\left(x_1^S\right)^2 x_2^S \omega_{21}^S + 3\left(x_1^L x_2^L\right)^2 \omega_{21}^L - 3\left(x_1^S x_2^S\right)^2 \omega_{21}^S - R \ln \left(x_2^L/x_2^S\right)}.\tag{5.101}$$

On the basis of these relations one may easily obtain the equation for the line of monovariant equilibrium $S \rightleftharpoons L$ in various approximate models, such as subregular, quasiregular, strictly regular, or ideal solutions, etc.

For binary systems the most widely used methods of analyzing intermolecular interactions rely on the strictly regular and quasiregular solution models [see Eqs. (5.62) and (5.63)]. For these cases the parameters ω^L and ω^S of a system with a continuous series of solid solutions may be expressed in the form

$$\omega^L = \omega_0^L - \omega_{01}^L T = \frac{RT \left\{\left(1 - x_2^S\right)^2 \ln \left[\left(1 - x_2^L\right)/\left(1 - x_2^S\right)\right] - \left(x_2^S\right)^2 \ln \left(x_2^L/x_2^S\right)\right\}}{\left(1 - x_2^L\right)^2 \left(x_2^S\right)^2 - \left(1 - x_2^S\right)^2 \left(x_2^L\right)^2}$$

$$- \frac{\left(1 - x_2^S\right)^2 \Delta h_{1,\,m}^{\circ} \left(T - T_{1,\,m}^{\circ}\right)/T_{1,\,m}^{\circ} - \left(x_2^S\right)^2 \Delta h_{2,\,m}^{\circ} \left(T - T_{2,\,m}^{\circ}\right)/T_{2,\,m}^{\circ}}{\left(1 - x_2^L\right)^2 \left(x_2^S\right)^2 - \left(1 - x_2^S\right)^2 \left(x_2^L\right)^2},\tag{5.102}$$

$$\omega^S = \omega_0^S - \omega_{01}^S T = \frac{RT \left\{ (1-x_2^L)^2 \ln \left[(1-x_2^S)/(1-x_2^L) \right] - (x_2^L)^2 \ln (x_2^S/x_2^L) \right\}}{(x_2^L)^2 (1-x_2^S)^2 - (x_2^S)^2 (x_2^L)^2}$$

$$+ \frac{(1-x_2^L)^2 \Delta h_{1,\,m}^\circ (T - T_{1,\,m}^\circ) / T_{1,\,m}^\circ - (x_2^L)^2 \Delta h_{2,\,m}^\circ (T - T_{2,\,m}^\circ)/T_{2,\,m}^\circ}{(x_2^L)^2 (1-x_2^S)^2 - (x_2^S)^2 (1-x_2^L)^2}. \qquad (5.103)$$

For systems with a negligible region of solid solutions, one usually turns to equations such as (5.71) and (5.72). Then at points of eutectic composition we obtain

$$\omega^L = \frac{\begin{array}{c} T_e R \ln \left[(1-x_{2\,(e)}^L)/x_{2\,(e)}^L \right] - \Delta s_{1,\,m}^\circ (T_{1,\,m}^\circ - T_e) \\ - \Delta s_{2,\,m}^\circ (T_{2,\,m}^\circ - T_e) \end{array}}{(1 - 2x_{2\,(e)}^L)}. \qquad (5.104)$$

In calculations one frequently substitutes the values of the experimental points in temperature and composition directly into Eqs. (5.102)–(5.104), by means of which one fixes the parameters of the exchange energy. The intrinsic scatter of such experimental data will lead to changes in the energy of mixing, thereby casting doubt on the appropriateness of some of the solution models. The influence of errors in the experimental data entering the quantities ω^L and ω^S is analyzed in the work of Foster and Woods [181]. The optimum calculation of the parameters ω^L and ω^S is based on a selection of values which will lead to the best possible agreement of the theoretically calculated liquidus and solidus curves with the experimental findings. The discrepancy may be characterized by the quantity σ, determined from the relation [182]

$$\sigma^2 = \sigma_{liq}^2 + \sigma_{sol}^2, \qquad (5.105)$$

where

$$\sigma_{liq}^2 = \sum_{j=1}^{M} (T_{liq,calc}^j - T_{liq,exp}^j)^2/M, \qquad (5.106)$$

$$\sigma_{sol}^2 = \sum_{k=1}^{N} (T_{sol,calc}^k - T_{sol,exp}^k)^2/N. \qquad (5.107)$$

Here M and N are the numbers of experimental points. The scanning operation is continued until the deviation σ is minimized.

The largest admissible deviation σ_{max} is established for each particular case, to some extent arbitrarily, depending on the accuracy of the experimental data input. As shown in the analysis carried out in [182], the condition

$$\sigma \leqslant \sigma_{\max} \tag{5.108}$$

may, in fact, be simultaneously satisfied by several models of solutions. The values of the selected parameters determining the temperature and concentration dependence of the exchange energy vary strongly according to the model employed. Consequently, if the experimental errors are large, calculation of thermodynamic properties such as entropy and enthalpy of mixing for alloys, based solely on the analysis of phase diagrams, is likely to be unreliable. Nevertheless, the values obtained for the G^E of alloys and for the activity coefficients of components agree well with experiment.

In the general case the estimate of parameters which determine the temperature and concentration dependence of the exchange energy can be achieved only by a computer calculation.

In the special case when the exchange energy in the liquid and solid phases is the same, Eqs. (5.62) and (5.63) may be transformed into the form

$$
\begin{aligned}
[(x_2^L)^2 - (x_2^S)^2]\, \omega' - Q_1' &= \ln\left[(1 - x_2^S)/(1 - x_2^L)\right], \\
[(1 - x_2^L)^2 - (1 - x_2^S)^2]\, \omega' - Q_2' &= \ln\left(x_2^S/x_2^L\right),
\end{aligned}
\tag{5.109}
$$

where

$$Q_i' = (\Delta h_{i,\,m}^{\circ}/R)\,(1 - T/T_{i,\,m}^{\circ}), \tag{5.109a}$$

and ω' is the reduced exchange energy.

In [183], a simple method was proposed for solving Eq. (5.109) in terms of a universal function, enabling one to calculate readily the state diagram for various values of the exchange energy, and hence to fix the exchange energies of the phases from a best fit of the calculated curve to the experimental points. To this end Eq. (5.109) is transformed into

$$C(x^L) - C(x^S) = Q_2' - Q_1', \quad D(x^L) - D(x^S) = Q_1', \tag{5.110}$$

where the functions

$$
\begin{aligned}
C(z) &= (1 - 2z)\,\omega' + \ln\left[z/(1 - z)\right], \\
D(z) &= z^2\omega' + \ln(1 - z)
\end{aligned}
\tag{5.111}
$$

are tabulated for a very wide range of arguments.

Utilizing (5.98) and (5.99) together with expressions for the activity coefficients as in (4.113) and (4.114), one may analyze the curves of phase equilibrium within the quasichemical approximation. Such an analysis has been carried out for the systems Si(Ge)–metal in [184]. The results are in satisfactory agreement with experimental data.

Jordan [185] proposed a method for analyzing the intermolecular interactions in binary A–B systems within the approximation of a model of a regular associated solution of the type $A + B + B_2$.

For the present case the activities of the components of a binary solution are determined [186] by the equations

$$a_A = \gamma_A x_A = \gamma_{A_1} x_{A_1} = a_{A_1}, \tag{5.112}$$

$$a_B = \gamma_B x_B = \gamma_{B_1} x_{B_1} / \gamma_{B_1}^\circ x_{B_1}^\circ = a_{B_1} / a_{B_1}^\circ. \tag{5.113}$$

The activity coefficients of the monomers A_1 and B_1 may be calculated [186] from

$$RT \ln \gamma_{A_1} \approx \omega^* x_B^2, \quad RT \ln \gamma_{B_1} \approx \omega^* x_A^2, \tag{5.114}$$

where ω^* is an empirical parameter which accounts for the interaction between the structural units of the solution.

In order to find expressions for x_{A_1} and x_{B_1} in terms of experimentally measured quantities, we consider the equilibrium of the dimerization reaction

$$2B_1 \rightleftharpoons B_2,$$

for which the relation

$$K_a(p, T) = (x_{B_1}^2 / x_{B_2}) (\gamma_{B_1}^2 / \gamma_{B_2})$$

is valid. To avoid clumsiness in notation here and in the following, we omit the indices on the equilibrium constants.

The mass conservation condition for the components of the solution in question may be written in the form

$$x_{A_1} = x_A (1 + x_{B_2}), \tag{5.115}$$

$$x_{B_1} = x_B - x_{B_2} (1 + x_A). \tag{5.116}$$

Substituting these equations into the expression for the dissociation constant of the dimer, we obtain

$$K = (\gamma_{B_1}^2 / \gamma_{B_2}) [x_B - x_{B_2} (1 + x_A)]^2 / x_{B_2}. \tag{5.117}$$

With the aid of the Gibbs–Duhem equation in the form

$$x_{A_1} d \ln \gamma_{A_1} + x_{B_1} d \ln \gamma_{B_1} + x_{B_2} d \ln \gamma_{B_2} = 0,$$

and of Eqs. (5.114)–(5.116), we may find $\ln \gamma_{B_2}$. Integrating from $x_B = 1$ to x_B, and assuming the mixture of particles B_1 and B_2 in the pure component B to be ideal, we get

$$RT \ln \gamma_{B_2} = 2\omega^* x_A^2. \tag{5.118}$$

After substituting this expression, and the second of Eq. (5.114), into (5.117), we obtain

$$K = [x_B - x_{B_2} (2 + x_A)]^2 / x_{B_2},$$

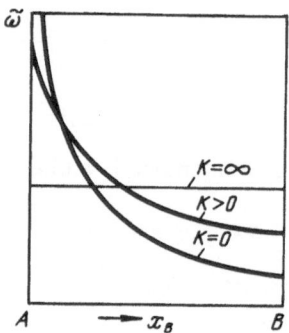

Fig. 39. Concentration dependence of the reduced exchange energy for various degrees of dimerization of component B in the solution (T = const).

from which it is easily verified that

$$x_{B_3} = (1 - x_A)(Q^* - 1)/(1 + x_A)(Q^* + 1), \qquad (5.119)$$

where

$$Q^* = \sqrt{1 + 4(1 - x_A^2)/K}. \qquad (5.120)$$

Substituting expression (5.119) into (5.115) and (5.116), we obtain

$$x_{A_1} = 2x_A(x_A + Q^*)/(1 + x_A)(Q^* + 1), \qquad (5.121)$$

$$x_{B_1} = 2x_B/(1 + Q^*). \qquad (5.122)$$

Finally, substituting (5.121) and (5.122) together with Eqs. (5.114) into (5.112) and (5.113), the activity coefficients of a regular associated solution of the type $A + B + B_2$ will be given by

$$RT \ln \gamma_A = \omega^* x_B^2 + RT \ln [2(x_A + Q^*)/(1 + x_A)(Q^* + 1)], \qquad (5.123)$$

$$RT \ln \gamma_B = \omega^* x_A^2 + RT \ln [(1 + Q^\circ)/(1 + Q^*)], \qquad (5.124)$$

where

$$Q^\circ = \sqrt{1 + 4/K}. \qquad (5.125)$$

For a completely dissociated solution $x_{B_2} \to 0$ and $K \to \infty$. Then Q^* becomes unity, and Eqs. (5.123) and (5.124) become the expressions for the activity coefficients of a regular solution.

Using formulas (5.98), (5.99), (5.123), and (5.124), one may analyze the curves of phase equilibria in a binary A–B system within the regular associated $A + B + B_2$ type model approximation. Thus, when mutual solubility in the second phase is small, relations (5.98) and (5.99) acquire the form

$$T = -\frac{\omega^* \left(x_B^L\right)^2 + \Delta h_{1,\,m}^{\circ}}{\Delta s_{1,\,m}^{\circ} - R \ln \left(1 - x_B^L\right)} \\ - R \ln \{2 \left(Q^* + 1 - x_B^L\right)/[\left(2 - x_B^L\right)\left(Q^* + 1\right)]\} \tag{5.126}$$

$$T = \frac{\omega^* \left(1 - x_B^L\right)^2 + \Delta h_{2,\,m}^{\circ}}{\Delta s_{2,\,m}^{\circ} - R \ln x_B^L - R \ln \left[\left(Q^{\circ} + 1\right)/\left(Q^* + 1\right)\right]}. \tag{5.127}$$

Expression (5.126) was used in [185] to analyze the liquidus line corresponding to the initial crystallization of bismuth in the Bi–Ga system. The result obtained was in good agreement with experimental data.

As a consequence of the considerations of this particular case one may infer that it is not legitimate to employ a model of regular solutions for the thermodynamic analysis of phase equilibrium diagrams in systems whose solutions may contain aggregates of one of the components.

As noted in [185], if isothermal data on γ_i are available, one may easily compute the relative value of the exchange energy by the equation

$$\tilde{\omega} = RT \ln \gamma_i/(1 - x_i)^2, \tag{5.128}$$

where $\tilde{\omega}$ is the relative value of ω.

Meanwhile, on the basis of, say (5.123), we also have

$$\tilde{\omega} = \frac{RT \ln \gamma_1}{x_B^2} = \omega^* + \frac{RT}{x_B^2} \ln \frac{2\left(x_A + Q^*\right)}{\left(1 + x_A\right)\left(Q^* + 1\right)}. \tag{5.129}$$

We may consider this expression in several special cases:
a) If $x_A = 0$, Eq. (5.129) becomes

$$\tilde{\omega}\left(x_A = 0\right) = \omega^* + RT \ln \left[2Q^{\circ}/(1 + Q^{\circ})\right]. \tag{5.130}$$

Hence, in the absence of aggregates in the solution ($K \to \infty$, $Q^{\circ} = 1$) we shall have $\tilde{\omega} \equiv \omega^* \equiv \omega$, i.e., a regular solution. In the case of a strong association of one of the components ($Q^{\circ} \to \infty$, $K \to 0$), with $x_A = 0$, we obtain

$$\tilde{\omega} = \omega^* + RT \ln 2. \tag{5.131}$$

b) If $x_A = 1$, then by l'Hospital's rule we find, via (5.129),

$$\tilde{\omega} = \omega^* + RT/K. \tag{5.132}$$

As in the first case, $\tilde{\omega} = \omega^* = \omega$ for $K \to \infty$.

c) For $K > 0$ we obviously have a situation for the exchange energy which is intermediate between the two preceding extreme cases. The graphical description of the concentration dependence of ω for various degrees of dimerization of the B component of the solution may be illustrated by the scheme of Fig. 39. The preceding equations have been applied extensively to the analysis of the character of the intermolecular interactions within the regular solution model and its various modifications, both in binary and quasibinary systems. Thus, in [182] a systematic thermodynamic analysis was carried out for the curves of phase equilibria in quasibinary systems with a continuous series of solid solutions formed by $A^{III}B^{V}$ type compounds, within the model of quasiregular solutions and its special cases. In our opinion such an approach assumes the virtual absence of association of the compounds in the solutions.

In analyzing the intermolecular interactions in A_mB_n–C_pD_q type quasibinary systems, other methods of describing the phase equilibrium diagrams are possible. Some of these will be considered in the following.

We assume that the compounds/components practically do not dissociate in the solid solution and essentially dissociate in the liquid solution. We shall obtain general thermodynamic equations enabling one to calculate the curves of phase equilibria in this case.

For p = const, we may write for the solid solution

$$\mu^S_{A_mB_n}(T) = \mu^{\circ(S)}_{A_mB_n}(T) + RT\ln\gamma^S_{A_mB_n}\left(1 - x^S_{C_pD_q}\right), \qquad (5.133)$$

$$\mu^S_{C_pD_q}(T) = \mu^{\circ(S)}_{C_pD_q}(T) + RT\ln\gamma^S_{C_pD_q}x^S_{C_pD_q}, \qquad (5.134)$$

while for the liquid solution we have, because of the assumed dissociation,

$$\mu^L_{A_mB_n}(T) = m\mu^L_{A_1}(T) + n\mu^L_{B_1}(T), \qquad (5.135)$$

$$\mu^L_{C_pD_q}(T) = p\mu^L_{C_1}(T) + q\mu^L_{D_1}(T). \qquad (5.136)$$

In equilibrium, we must have

$$\mu^S_{A_mB_n} = \mu^L_{A_mB_n} = m\mu^L_{A_1} + n\mu^L_{B_1}, \qquad (5.137)$$

$$\mu^S_{C_pD_q} = \mu^L_{C_pD_q} = p\mu^L_{C_1} + q\mu^L_{D_1}, \qquad (5.138)$$

where the chemical potential of the monomers is determined by the equation

$$\mu^L_i = \mu^{\circ(L)}_i(T) + RT\ln\gamma^L_i x^L_i, \quad (i = A_1, B_1, C_1, D_1). \qquad (5.139)$$

In view of the fact that according to the present model the compounds/components almost completely dissociate in the liquid phase, the quantities γ_i and x_i appearing in (5.139) should be interpreted as the activity coefficients and analytic mole fractions of the substances A, B, C, and D in the solution. Therefore, to a good approximation, the liquid phase may be considered as a four-component solution. Hence the index 1 on the monomeric forms will be omitted in the following.

At the points associated with the pure components we have, via Eqs. (5.137) and (5.138),

$$\mu^{\circ(S)}_{A_mB_n} = \mu^{\circ(L)}_{A_mB_n} = m\mu^{*(L)}_A + n\mu^{*(L)}_B, \qquad (5.140)$$

$$\mu_{C_pD_q}^{\circ\,(S)} = \mu_{C_pD_q}^{\circ\,(L)} = p\mu_C^{*\,(L)} + q\mu_D^{*\,(L)}, \tag{5.141}$$

where the asterisk pertains to the pure compound.

From Eq. (5.33a) we may write, because of the temperature dependence of $\Delta h_{i,m}^\circ$ and $\Delta s_{i,m}^\circ$,

$$\mu_{A_mB_n}^{\circ\,(L)} = \mu_{A_mB_n}^{\circ\,(S)} + \Delta s_{m,\,A_mB_n}^\circ (T_{m,\,A_mB_n}^\circ - T)$$
$$- \Delta c_p^\circ [T_{m,\,A_mB_n}^\circ - T - T \ln (T_{m,\,A_mB_n}^\circ/T)], \tag{5.142}$$

where Δc_p° is the difference in the heat capacities of the solid and supercooled liquid compound.

Substituting (5.142) into (5.140) and neglecting the heat-capacity difference Δc_p°, we obtain

$$\mu_{A_mB_n}^{\circ\,(S)} = m\mu_A^{*\,(L)} + n\mu_B^{*\,(L)} - \Delta s_{m,\,A_mB_n}^\circ (T_{m,\,A_mB_n}^\circ - T). \tag{5.143}$$

Combining this expression with (5.137), we get, finally,

$$\ln \gamma_{A_mB_n}^S \big(1 - x_{C_pD_q}^S\big) = m \ln \left[\frac{\gamma_A^L x_A^L}{\gamma_A^{*\,(L)} m/(m+n)} \right]$$
$$+ n \ln \left[\frac{\gamma_B^L x_B^L}{\gamma_B^{*\,(L)} n/(m+n)} \right] + \Delta s_{m,\,A_mB_n}^\circ (T_{m,\,A_mB_n}^\circ - T)/RT. \tag{5.144}$$

Similarly, we find from (5.138)

$$\ln \gamma_{C_pD_q}^S x_{C_pD_q}^S = p \ln \left[\frac{\gamma_C^L x_C^L}{\gamma_C^{*\,(L)} p/(p+q)} \right]$$
$$+ q \ln \left[\frac{\gamma_D^L x_D^L}{\gamma_D^{*\,(L)} q/(p+q)} \right] + \Delta s_{m,\,C_pD_q}^\circ (T_{m,\,C_pD_q}^\circ - T)/RT. \tag{5.145}$$

If $m = n = p = q = 1$, Eqs. (5.144) and (5.145) become

$$\ln \gamma_{AB}^S \big(1 - x_{CD}^S\big) = \ln [\gamma_A^L \gamma_B^L / \gamma_A^{*\,(L)} \gamma_B^{*\,(L)}]$$
$$+ \ln (4 x_A^l x_B^l) + \Delta s_{m,\,AB}^\circ (T_{m,\,AB}^\circ - T)/RT, \tag{5.146}$$

$$\ln \gamma_{CD}^S x_{CD}^S = \ln [\gamma_C^L \gamma_D^L / \gamma_C^{*\,(L)} \gamma_D^{*\,(L)}]$$
$$+ \ln (4 x_C^L x_D^L) + \Delta s_{m,\,CD}^\circ (T_{m,\,CD}^\circ - T)/RT. \tag{5.147}$$

Equations similar to (5.144)–(5.147) have been obtained by Jordan [187] for ternary and, as a special case, for quasibinary systems of the type $A_m B_n$–$C_q B_r$.

Equations (5.144)–(5.147) are also widely applied to analyzing phase equilibria in quasibinary systems.

The form of Eqs. (5.144)–(5.147) is determined in each concrete case by the choice of the model solution used to carry out the analysis of the intermolecular interaction.

The analysis of isotherms pertaining to the surface of the liquidus of the ternary systems Ga–As–Zn and Ga–P–Zn, including the quasibinary sections GaAs–Zn and GaP–Zn, was carried out in [188] in the regular solution approximation on the basis of relations similar to (5.146).

Based on the method of Guggenheim [189], Stringfellow and Green put forward a method for estimating the activity coefficients of a three-component mixture in the quasichemical approximation [184]. Their results may be used to analyze equilibria by means of Eqs. (5.146) and (5.147) in quasibinary systems of the type AC–BC. In this case one should consider a three-component liquid solution.

For estimating the activity coefficients of the ith component of a k-compound mixture in the quasichemical approximation, one may use the method described in [190].

Laugier [191] considers a method of analyzing the intermolecular interaction in the quasibinary AC–BC system for cases in which the compounds do not dissociate in the solid solution, while the liquid solution is represented in terms of a regular mixture of the type $A + B + C + AC + BC$.

4. THERMODYNAMIC JUSTIFICATION OF THE RETROGRADE SOLIDUS. CALCULATION OF THE RETROGRADE SOLIDUS IN VARIOUS APPROXIMATIONS

The overwhelming majority of systems containing germanium and silicon, as well as a series of other semiconductors, exhibit a retrograde solidus, or so-called negative solubility. The essence of this phenomenon, as already shown in Chapter 3, consists of the fact that the maximal extent of the region of homogeneity on the basis of, say, the A component is located above the temperature of invariant transformation in the two-component system (cf. Fig. 14).

As noted above, at first glance the retrograde character of the solidus is an unusual phenomenon, since the process of melting is accomplished by the rejection of heat, i.e., on cooling the alloy. The opposite of the usual retrograde melting is also possible, accomplished by the emission rather than the absorption of heat. However, this question has not been studied until now. In 1908 van Laar, using a model similar to the model of regular solutions, calculated theoretically some binary phase diagrams and showed the possible existence of state diagrams with negative solubilities [61, 65]. Since then the possibility of retrograde solubilities lay practically forgotten for a long time. This may be ascribed to the fact that van Laar did not distinguish the specific points of maximal solubility on the curve of the retrograde solidus, as well as to the absence of experimental materials up to about 1926. Thus, when the retrograde character of the solidus was detected experimentally on the zinc side of the zinc–cadmium system [192, 193], the authors considered the phenomenon impossible and attempted to explain it in terms of an allotropic transformation in zinc. Only in 1924 and thereafter did Raub and his co-workers [194–196] observe a significant number of state diagrams with retrograde solubility. In 1948 Meejering [64] analyzed the retrograde phenomenon in a general way by using the van der Waals differential equation, and obtained an approximate expression which should predict the retrograde character of the solidus in binary metallic systems using data on the compositions of coexisting phases at the temperature of the eutectic transformation.

In accordance with [64] we may write the van der Waals differential equation for a two-phase two-component system at constant pressure in the form

$$\left(\frac{\partial x}{\partial T}\right)_p^S = \frac{(\partial s/\partial x)_{p,\,T}^S - (s^S - s^L)/(x^S - x^L)}{(\partial^2 g/\partial x^2)_{p,\,T}^S}. \tag{5.148}$$

Bearing in mind that due to the stability condition the derivative $(\partial^2 g/\partial x^2)_{p,T}{}^S$ is always positive, we are led to conclude from Eq. (5.148) that at the point of maximum solubility on the curve of the retrograde solidus the following condition should apply:

$$(\partial x/\partial T)_p^S = 0 \quad \text{if} \quad (\partial s/\partial x)_{p,\,T}^S - (s^S - s^L)/(x^S - x^L) = 0. \tag{5.149}$$

Let us consider a binary metallic system whose components have one and the same entropy of melting $s_{i,m}{}^\circ$. We write the expression for the entropy of the solid and liquid phases by taking the entropy of mixing to be ideal and using Richards' law, according to which the entropy of melting of a metal is given by

$$\Delta s_{i,\,m}^\circ = bR,$$

where b may assume a value between 1 and 2.7. We then have

$$s^S = \left(1 - x_2^S\right) s_1^{\circ\,(S)} + x_2^S s_2^{\circ\,(S)} - R\left[\left(1 - x_2^S\right)\ln\left(1 - x_2^S\right) + x_2^S \ln x_2^S\right], \tag{5.150}$$

$$s^L = \left(1 - x_2^L\right) s_1^{\circ\,(S)} + x_2^L s_2^{\circ\,(S)} + bR - R\left[\left(1 - x_2^L\right)\ln\left(1 - x_2^L\right) + x_2^L \ln x_2^L\right]. \tag{5.151}$$

Substituting these expressions into (5.149), we obtain, after some manipulations, Meejering's formula:

$$\log x_2^S = \log x_2^L - 0.434b/x_2^L + \left[\left(1 - x_2^L\right)/x_2^L\right]$$
$$\times \log\left(1 - x_2^L\right) - \left[\left(1 - x_2^L\right)/x_2^L\right]\log\left(1 - x_2^S\right). \tag{5.152}$$

By means of Eq. (5.152), Meejering constructed the graph represented in Fig. 40, which allows one to predict the sign of the derivative $(\partial T/\partial x)_p{}^S$ and, consequently, the character of the solidus in binary systems. The shaded portions of the graph delineate the region of coexisting components x^S and x^L, where the derivative $|\partial T/\partial x|_p{}^S \approx \infty$. The upper shaded region corresponds to b between 1 and 1.2, the lower one to b ranging from 2.3 to 2.7. Clearly the derivative $(\partial T/\partial x)_p{}^S$ is positive if the coupled values of x^S and x^L are below the shaded strips and negative if they are above the strips.

According to Meejering, in order to be able to classify the solidus curves into retrograde or normal types by means of the graphs shown in Fig. 40, it is sufficient to have available data on the composition of the liquid and solid phases at just the temperature of the eutectic transformation. Indeed, if the derivative $(\partial T/\partial x)_p^S$ is negative at the eutectic temperature, then at higher temperatures it never becomes ∞, since a minimum x^S is not observed on the solidus curve. In this case the form of the solidus curve will be normal. If, however, the derivative $(\partial T/\partial x)_p^S$ is positive at the eutectic temperature, then as the temperature is increased the solidus will have a vertical tangent, provided it passes through the melting point of the pure component. We then deal with a retrograde solidus curve. In this case two exceptions are possible: the first is connected with an allotropic transformation in the pure component, the second with demixing in the liquid phase. We shall not dwell on these in detail.

As a result of his investigations, Meejering came to the conclusion that a retrograde solidus is possible in binary systems whenever the solubility in the solid state is insignificant and the temperature of the eutectic transformation is lower than the melting temperature of the pure solid solvent.

Unfortunately, the work of Meejering was neglected for a long time, and the problem of the retrograde solidus was not discussed in textbooks and monographs. The situation has changed drastically in the last 10–15 years, owing to the widespread detection of retrograde condensation in binary systems containing semiconductors [66]. Studies appeared in which the possibility of this phenomenon was justified afresh [63, 197–199]. We note that Meejering's approach is rigorous only within certain limits. The final equation (5.152) is approximate, owing to the limited validity of Richards' rule, as noted by one of the present authors [200] when analyzing the periodic dependence of the entropy of melting of simple substances on the atomic number. Nevertheless, the qualitative conclusions of Meejering may be assumed to be well established, since even the maximal deviation from Richards' rule, as observed in practice, cannot alter the specific conclusions derived from Eq. (5.152) and the graphs of Fig. 40.

The cause of retrograde solubility becomes better understood when one determines the position of the solidus line by a known liquidus curve [63, 198, 199].

To simplify the treatment of the thermodynamic causes of negative solubilities, a number of authors [63, 198, 199] assumed that the solid phase resembles a regular solution, while the liquid phase in equilibrium with the solid at the eutectic temperature may be taken to be an ideal solution. The difference of the chemical potentials of the dissolved component 2 in the saturated dilute solid solution and the pure component 2 is determined by the relation

$$\mu_2^{*\,(S)} - \mu_2^{\circ\,(S)} = RT \ln \gamma_2^{*\,(S)} x_2^{*\,(S)}. \tag{5.153}$$

Meanwhile,

$$\mu_2^{*\,(S)} - \mu_2^{\circ\,(S)} = \bar{h}_2^{E\,(S)} - T\bar{s}_2^{M\,(S)}. \tag{5.154}$$

From these we find

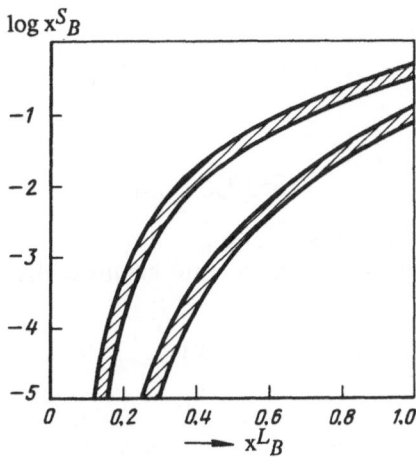

Fig. 40. Toward a thermodynamic justification of the retrograde solidus by Meejering [289].

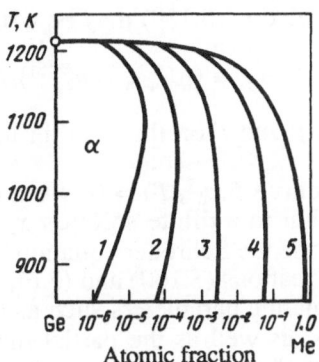

Fig. 41. Solidus curves in Ge–metal systems, calculated for various values of $\overline{h}_{Me}^{E(S)}$ (in J/mole): 1) 92; 2) 114; 3) 46; 4) 23; 5) liquidus curves calculated by Eq. (5.162).

$$RT \ln \gamma_2^{*\,(S)} x_2^{*\,(S)} = \overline{h}_2^{E\,(S)} - T \overline{s}_2^{M\,(S)}. \tag{5.155}$$

Since by the conditions of the problem the solid dilute solution is regular, we have, by definition,

$$\overline{s}_2^{M\,(S)} = -R \ln x_2^{*\,(S)}. \tag{5.156}$$

whence

$$RT \ln \gamma_2^{*\,(S)} = \overline{h}_2^{E\,(S)}. \tag{5.157}$$

Now, since

$$RT \ln \gamma_2^{* \, (S)} = \omega^S x_1^{(S) \, 2} \approx \omega^S,$$

then in the present case

$$\bar{h}_2^{E \, (S)} \approx \text{const};$$

i.e., it does not depend on the composition.

Since the dilute solid solution is in equilibrium with the ideal liquid solution, one may write

$$\mu_2^{\circ \, (S)} + RT \ln x_2^{* \, (S)} + \bar{h}_2^{E \, (S)} = \mu_2^{\circ \, (L)} + RT \ln x_2^{L}, \tag{5.158}$$

or

$$\ln \left(x_2^{* \, (S)}/x_2^{L} \right) = \left(\Delta \mu_2 - \bar{h}_2^{E \, (S)} \right)/RT, \tag{5.159}$$

where $\Delta \mu_2$ is the change in the molar Gibbs free energy in the process of the melting of the second component at temperature T.

Using formula (5.33a), relation (5.159) may be written in the form

$$\ln \left(x_2^{* \, (S)}/x_2^{L} \right) \equiv \ln K_2 = \left(\Delta h_{2, \, m}^{\circ} - \bar{h}_2^{E \, (S)} \right)/RT - \Delta s_{2, \, m}^{\circ}/R. \tag{5.160}$$

Here and below the indices p and T on the equilibrium constant $K_i(T, p)$ will be omitted.

To obtain the solidus curve $f(x_2^S, T) = 0$, we must eliminate x_2^L from Eq. (5.160). If we remember that in a dilute solution $x_2 \lll 1$, whereas the liquid phase is ideal, then we may use the Schröder equation (5.6) for x_1^L.

Solving the combined equations (5.160) and (5.6), we can calculate the solidus curve if we know thermodynamic parameters such as the temperature and heat of melting of both components, as well as the partial molar heat of mixing $\bar{h}_2^{E(S)}$ of the dissolved component in the solid solution. As shown by experiment, the tendency toward retrograde solubility is directly associated with the quantity $\bar{h}_2^{E(S)}$. Thus, for relatively low values of $\bar{h}_2^{E(S)}$, the solidus curve does not reveal any retrograde solubility. The reason that retrograde solubility is not observed for the majority of metallic systems is that the value of $\bar{h}_2^{E(S)}$ is too low. The actual value of $\bar{h}_2^{E(S)}$ needed for the appearance of retrograde phenomena is quite large. Using relations (5.160) and (5.5), and assigning arbitrarily differing values to $\bar{h}_2^{E(S)}$, Thurmond and Struthers [198] constructed four solidus curves in a system containing germanium, with the dissolved second component chosen as an arbitrary substance with a heat of melting of 14.6 kJ/mole, and a melting temperature of 1160 K. These results are presented in Fig. 41.

As seen from the figure, solidus curves corresponding to higher values of $\bar{h}_2^{E(S)}$ exhibit a tendency toward retrograde solubility.

High values of $\bar{h}_2^{E(S)}$ are associated with low solubilities of the component in the solid phase. As a result, the phenomenon of retrograde solubility is observed, as a rule, only in systems with low solubilities. The majority of systems in which the solvents are germanium and silicon are systems of this type. For the example considered above it was necessary to assume that the solid and liquid solutions are, respectively, regular and ideal. This restriction does not, however, contradict the general conclusion that retrograde solubility is directly connected with large values of the quantity $\bar{h}_2^{E(S)}$.

Thurmond and Struthers [198] have also analyzed the phenomenon of retrograde solubility by means of another thermodynamic parameter, namely, the distribution coefficient K_i. Writing the expression for the distribution coefficient K_2° at the melting temperature of the pure solvent in the form

$$\ln K_2^\circ = (\Delta h_{2,\,m}^\circ - \bar{h}_2^{E\,(S)})/RT_{1,\,m}^\circ - \Delta s_{2,\,m}^\circ/R \qquad (5.161)$$

and substituting it into formula (5.160), they obtained the following equation:

$$\ln K_2 = (T_{1,\,m}^\circ/T) \ln K_2^\circ + (\Delta s_{2,\,m}^\circ/R)\,(T_{1,\,m}^\circ/T - 1), \qquad (5.162)$$

where $T_{1,m}^\circ$ is the melting temperature of the pure solvent. The distribution coefficient is determined experimentally. The analysis of systems containing germanium for varying values of K_i showed that the smaller the value of this parameter, the more pronounced retrograde solubility becomes, the maximum probability being shifted to lower temperatures as K_2° grows. It was found [198] that a retrograde solidus is feasible if K_2° is less than 0.1 and $\bar{h}_2^{E(S)}$ is larger than 23,148 kJ/g·mole.

The factors responsible for retrograde behavior were elucidated by Thurmond [198] as follows. As experiment shows, in a system with a retrograde solidus the temperature of the eutectic is usually quite low, and its composition is strongly shifted toward the solute M. In consequence of this, the chemical potential of M in the liquid solution should go through a maximum. This is easily understood if as a measure of the chemical potential one chooses the change in the partial pressure of M along the liquidus curve. At the melting point of the pure solvent, for example germanium, the partial pressure of M vanishes. As more of M is added, the temperature of the liquidus falls slowly, while the concentration increases very sharply; consequently, the partial pressure of M also increases. In the region of large concentrations of the substance M the composition of the liquidus changes with temperature slowly, so that the temperature factor becomes dominant and the partial pressure of M begins to fall. Therefore, the chemical potential or the partial pressure of the solute M along the liquidus curve should pass through a maximum. Since in equilibrium the chemical potentials of the components in the coexisting phases are equal, one should observe a maximum solubility on the solidus curve in the present case.

Since the solid solution is dilute, Henry's law shows that the concentration of the solute M will be directly proportional to its vapor pressure at constant temperature. If Henry's constant does not depend on temperature, then the maximum solubility on the retrograde solidus will occur at a temperature which corresponds to the maximum pressure of substance M above the saturated liquid solution. Because in reality a temperature dependence is observed, the maximum solubility of M should be observed for higher temperatures than the maximum of its partial pressure.

Kirgintsev [201] has given a general method for establishing the retrograde solubility of solid solutions on the basis of the dependence of the logarithm of the equilibrium distribution coefficient on the logarithm of the molar fraction of the solute.

For a two-component system consisting of a solid and a liquid phase, the differential equation of van der Waals at constant pressure has the form (cf. Chapter 2)

$$\left(\frac{\partial T}{dx_2}\right)_p^L = -\frac{x_2^S - x_2^L}{s^S - s^L - (x_2^S - x_2^L)(\partial s/\partial x_2)_{p,T}^L}\left(\frac{\partial^2 g}{\partial x_2^2}\right)_{p,T}^L, \qquad (5.163)$$

$$\left(\frac{\partial T}{\partial x_2}\right)_p^S = -\frac{x_2^S - x_2^L}{s^S - s^L - (x_2^S - x_2^L)(\partial s/dx_2)_{p,T}^S}\left(\frac{\partial^2 g}{\partial x_2^2}\right)_{p,T}^S. \qquad (5.164)$$

In accordance with the form of the state diagram with a retrograde solidus (cf. Fig. 14), and assuming that the stability condition

$$(\partial^2 g/\partial x_2^2)_{p,T}^L > 0 \quad \text{and} \quad (\partial^2 g/\partial x_2^2)_{p,T}^S > 0$$

is satisfied, we find from Eqs. (5.163) and (5.164) that within the interval bracketed by the melting temperature of the solvent and the point of maximum stability m the following holds:

$$\left.\begin{aligned} s^S - s^L - (x_2^S - x_2^L)(\partial s/\partial x_2)_{p,T}^L &> 0, \\ s^S - s^L - (x_2^S - x_2^L)(\partial s/\partial x_2)_{p,T}^S &> 0. \end{aligned}\right\} \qquad (5.165)$$

Thus, at the point m,

$$\left.\begin{aligned} s^S - s^L - (x_2^S - x_2^L)(\partial s/\partial x_2)_{p,T}^L &> 0, \\ s^S - s^L - (x_2^S - x_2^L)(\partial s/\partial x_2)_{p,T}^S &= 0, \end{aligned}\right\} \qquad (5.166)$$

while in the interval between the points m and d,

$$\left.\begin{aligned} s^S - s^L - (x_2^S - x_2^L)(\partial s/\partial x_2)_{p,T}^L &> 0, \\ s^S - s^L - (x_2^S - x_2^L)(\partial s/\partial x_2)_{p,T}^S &< 0. \end{aligned}\right\} \qquad (5.167)$$

Dividing Eq. (5.164) by (5.163) and noting that according to (2.43)

$$(\partial^2 g/\partial x_2^2)_{p,T} = -(\partial \mu_1/\partial x_2)_{p,T}/x_2,$$

we obtain, after some simple transformations, the following expression for the equilibrium distribution coefficient:

$$\left(\frac{\partial \ln K_2}{\partial \ln x_2^L}\right)_{p,T} = \frac{s^S - s^L - (x_2^S - x_2^L)(\partial s/\partial x_2)_{p,T}^S}{s^S - s^L - (x_2^S - x_2^L)(\partial s/\partial x_2)_{p,T}^L}\frac{(\partial \mu_1/\partial x_2)_{p,T}^L}{(\partial \mu_1/\partial x_2)_{p,T}^S} - 1. \qquad (5.168)$$

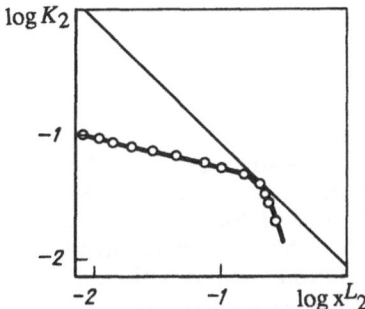

Fig. 42. Illustration of the use of (5.169) and (5.170) to establish the retrograde solidus in the KCl–SrCl₂ system.

Comparing this relation with the conditions (5.166) and (5.167), we see that at the point m

$$\left(\partial \ln K_2 / \partial \ln x_2^L\right)_{p, T} = -1, \tag{5.169}$$

while in the interval between the temperature of invariant equilibrium and the point m (Fig. 14)

$$\left(\partial \ln K_2 / \partial \ln x_2^L\right)_{p, T} < -1. \tag{5.170}$$

Therefore, if the logarithm of the distribution coefficient satisfies (5.169) and (5.170), the solidus curve has a retrograde character and, conversely, if these conditions are not fulfilled, we have a normal solidus. Figure 42 shows the dependence of $\log K_2$ on $\log x_2^L$ for the system KCl–SrCl₂ (SrCl₂ is an impurity) according to the data of [202]. It is seen that for $\log x_2^L = -0.6$ condition (5.169) is realized, so that a retrograde solubility occurs in the KCl–SrCl₂ system.

There are two thermodynamic approaches to estimating the solubility of elements alloyed with semiconductors. The first one relies entirely on treating the heterogeneous equilibrium of solid and liquid phases; the second one also takes into account the ionization of the alloying component atoms in forming the solution. By means of the first method one is able to provide a qualitative description of solidus curves, whereby the behavior of the solid phase is usually described in terms of a regular solution model, and the liquid phase is treated within an ideal [198] or a regular [203] model.

Thus, Thurmond and Struthers [198] calculated the solidus in the binary systems Ge–Sb, Ge–Cu, and Si–Cu, assuming the liquid and solid solutions to be ideal and regular, respectively. In [204] it was shown that the liquid solutions in binary germanium and silicon systems with various elements from the IB, IIIB, IVB, and VB subgroups of the periodic table are described better by means of the theory of regular solutions. This was taken into account by Kozlovskaya and Rubinshtein [205], who, by considering the condition of heterogeneous equilibrium of the solid and liquid phases within the theory of regular solutions, have proposed the following equation for calculating the solidus curve:

$$\ln K_2 = (\Delta h_{2,\,m}^\circ - \omega^S)/RT + (\omega^L/RT)(1 - x_2^L)^2 - \Delta s_{2,\,m}^\circ/R. \quad (5.171)$$

The inadequacy of such an approach, or its modifications, pointed out in [203], is quite obvious for systems with semiconductor components, because of the essential role played by ionization processes in the solid state involving the donor and acceptor alloying elements, as pointed out by Reiss [206–208]. In this context we have considered the possibility of calculating the solidus line in semiconductors by taking into account the degree of ionization of the impurity atoms in the solid solution as their base [209]. The fraction of un-ionized atoms in the solid phase was treated in the regular solution approximation.

From a consideration of heterogeneous equilibrium in the formation of a solid solution based on a semiconductor, one may write

$$J^L \rightleftarrows J^S \rightleftarrows J_i^S + e, \quad (5.172)$$

where J^S and J^L are the un-ionized states of the alloying component in the solid and liquid phases, respectively. J_i^S and e are the ionized state of the alloying component and the carrier charge.

By the condition for heterogeneous equilibrium the chemical potentials of the solute in the coexisting phases are identical, i.e., μ_2^L and μ_2^S. For a real solution we have

$$\begin{aligned} \mu_2^L &= \mu_2^{\circ\,(L)} + RT \ln x_2^L \gamma_2^L, \\ \mu_2^S &= \mu_2^{*\,(S)} + RT \ln x_2^{*\,(S)} \gamma_2^{*\,(S)}. \end{aligned} \quad (5.173)$$

Hence, by the equilibrium condition

$$\mu_2^{\circ\,(L)} + RT \ln x_2^L \gamma_2^L = \mu_2^{*\,(S)} + RT \ln x_2^{*\,(S)} \gamma_2^{*\,(S)}$$

or

$$\begin{aligned} x_2^{*\,(S)} &= x_2^{(L)} \gamma_2^{(L)}/\gamma_2^{*\,(S)} \exp\left[(\mu_2^{\circ\,(L)} - \mu_2^{\circ\,(S)})/RT\right] \\ &\times \exp\left[(\mu_2^{\circ\,(S)} - \mu_2^{*\,(S)})/RT\right] \end{aligned} \quad (5.174)$$

we assume that the difference between $\mu_2^{\circ(S)}$ and $\mu_2^{*(S)}$ is negligible.

Taking into account the well-known assumption of (5.33a), we have

$$x_2^{*\,(S)} = (x_2^{(L)} \gamma_2^{(L)}/\gamma_2^{*\,(S)}) \exp\left[\Delta s_{2,\,m}^\circ (T_{2,\,m}^\circ - T)/RT\right]. \quad (5.175)$$

On the other hand, according to Prigogine and Defay [24], the chemical potential of the dissolved component is equal to the chemical potential of the neutral form, i.e.,

$$\mu_2^{\circ\,(S)} + RT \ln x_2^{*\,(S)} \gamma_2^{*\,(S)} = (\mu_2^{\circ\,(S)})_n + RT \ln (x_2^{*\,(S)})_n (\gamma_2^{*\,(S)})_n, \quad (5.176)$$

where the index n refers to the neutral state. Taking the standard state of the neutral atoms and the dissolved component to be exactly the same, e.g., if they are both taken to be the state of the pure solid component, we may write, on the basis of (5.176),

$$\gamma_2^{* \ (S)} = (x_2^{* \ (S)})_n \, (\gamma_2^{* \ (S)})_n / x_2^{* \ (S)} = F_n \, (\gamma_2^{* \ (S)})_n, \tag{5.177}$$

where F_n is the fraction of un-ionized atoms of the alloying component in the solid solution.

At a higher level of alloying, the quantity F_n may be found from the Fermi–Dirac distribution

$$F_n = \{1 + 1/2 \exp \left[(E_f - E_a)/kT\right]\}^{-1}, \tag{5.178}$$

$$F_n = \{1 + 1/2 \exp \left[(E_d - E_f)/kT\right]\}^{-1}, \tag{5.179}$$

where E_a and E_d are the ionization energies of the acceptors and donors, respectively, reckoned, like the Fermi level R_f, from the upper edge of the valence band.

Allowing for (5.177), we may represent (5.175) in the form

$$x_2^{* \ (S)}/x_2^L = \gamma_2^L/(\gamma_2^{* \ (S)})_n \, F_n \exp \left[(\Delta s_{2, \ m}^{\circ}/RT) \, (T_{2, \ m}^{\circ} - T)\right]. \tag{5.180}$$

In the regular solution approximation we have, in the present case,

$$\gamma_2^L = \exp \left[(\omega^L/RT) \, (1 - x_2^L)^2\right],$$
$$(\gamma_2^{* \ (S)})_n = \exp \{(\omega^S/RT) \, [1 - (x_2^{* \ (S)})_n]^2\}, \tag{5.181}$$

where ω^S is the energy of mixing of the semiconductor and neutral alloying atoms in the solid phase. Substituting these expressions into (5.180), we get

$$K_2 = x_2^{* \ (S)}/x_2^{(L)} = F_n^{-1} \exp \{(\Delta s_{2, \ m}^{\circ}/RT) \, (T_{2, \ m}^{\circ} - T)$$
$$+ (\omega^L/RT) \, (1 - x_2^L)^2 - (\omega^S/RT) \, [1 - (x_2^{* \ (S)})_n]^2\}. \tag{5.182}$$

Since the solubility of the alloying element in the semiconductor is small, we may neglect the quantity $(x_2*^{(S)})_n$. Then

$$K_2 = F_n^{-1} \exp \left[(\Delta s_{2, \ m}^{\circ}/RT) \, (T_{2, \ m}^{\circ} - T)\right.$$
$$+ (\omega^L/RT) \, (1 - x_2^L)^2 - \omega^S/RT]. \tag{5.183}$$

We assume that the behavior of the neutral atoms in the solid phase is accurately described within a regular solution model. We shall then have for the melting temperature of the semiconductor

$$K_2^{\circ} = (F_n^{\circ})^{-1} \exp \left[(\Delta s_{2, \ m}^{\circ}/RT_{1, \ m}^{\circ}) \, (T_{2, \ m}^{\circ} - T_{1, \ m}^{\circ})\right.$$
$$+ \omega^{\circ \ (L)}/RT_{1, \ m}^{\circ} - \omega^S/RT_{1, \ m}^{\circ}], \tag{5.184}$$

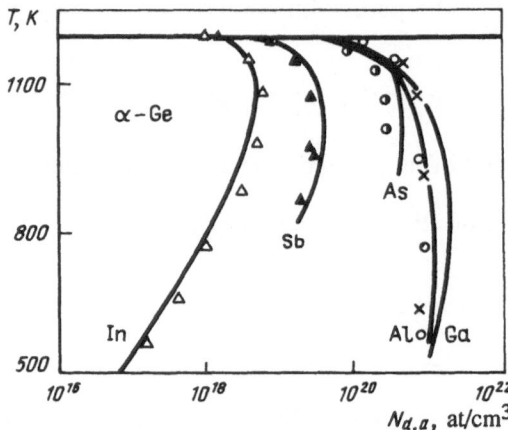

Fig. 43. Solidus curves for some binary systems of germanium with donor and acceptor alloying elements: $N_{d,a}$ is the concentration of donors or acceptors; curves calculated according to Eq. (5.186); points are experimental data [66].

where the superscript "°" on F_n and ω indicates that these quantities are taken at the melting temperature of the semiconductor $T_{1,m}°$.

Solving Eq. (5.184) for ω^S, we obtain

$$\omega^S = \Delta s_{2,\,m}° \left(T_{2,\,m}° - T_{1,\,m}°\right) + \omega^{°\,(L)} - RT_{1,\,m}° \ln\left(F_n°K_2°\right). \qquad (5.185)$$

Substituting ω^S into formula (5.183), we obtain the equation of the solidus:

$$x_2^{*\,(S)} = x_2^L F_n^{-1} \exp\left[(\Delta s_{2,\,m}°/RT)\left(T_{1,\,m}° - T\right)\right.$$
$$\left. + \left(\omega^L/RT\right)\left(1 - x_2^L\right)^2 + \left(T_{1,\,m}°/T\right)\ln\left(F_n°K_2°\right) - \omega^{°\,(L)}/RT\right]. \qquad (5.186)$$

The quantity $F_n°$ may be found by supposing that at the melting temperature the semiconductor is intrinsic, so that the Fermi level is practically at the center of the band gap, i.e.,

$$F_n° = \{1 + 2^{-1}\exp\left[(-E_a + \Delta E/2)/kT\right]\}^{-1}, \qquad (5.187)$$

$$F_n° = \{1 + 2^{-1}\exp\left[(E_d - \Delta E/2)/kT\right]\}^{-1}, \qquad (5.188)$$

where ΔE is the gap width at the semiconductor's melting point.

Thus we may calculate the solidus curve from Eq. (5.186) if we know the following quantities: the position of the Fermi level in the solid phase, the distribution coefficient of the alloying component at the melting temperature of the semiconductor, the change in the width of the band gap with temperature, the entropy of melting and ionization potential of the alloying component, and, finally, the liquidus curve for the system.

Employing the foregoing method, we calculated the solidus curves in binary systems of germanium with aluminum, gallium, indium, arsenic, and antimony. In the calculations we took into account the temperature dependence of the energy of mixing of the components in the melt. Information on the liquidus curves, as well as the data required for the calculation, was taken from [66, 199]. The position of the Fermi level was fixed from Blakemore's curves [210]. The results we obtained are presented in Fig. 43 together with experimental data. The results of the calculations are seen to be in quite good agreement with experiment. The small discrepancies may be ascribed to a somewhat inaccurate choice of the standard state of the alloying component.

The first systematic approach to estimating the solubility, taking into account also the ionization of the alloying atoms, was carried out in the work of Lehovec [211]. If it is assumed that the enthalpy needed to transfer one mole of neutral alloying substance from the solid solution to the liquid melt depends linearly on temperature, and that the behavior of the portion of neutral atoms in the solid phase is ideal, i.e.,

$$(\gamma_2^{*\,(S)})_n = 1, \tag{5.189}$$

then by (5.174) and (5.177) we are led to the expression

$$(x_2^{*\,(S)})_n / x_2^L = (\gamma_2^L / \theta) [F_n^\circ K_2^\circ \theta^\circ / \gamma_2^{\circ\,(L)}]^{T_1^\circ,\,m/T}, \tag{5.190}$$

where $\theta = \exp{[(\Delta s_{2,m}^\circ - a)/R]}$, which is Lehovec's formula.

The foregoing approach to estimating the position of the solidus in systems with semiconductors as the base also enables one to solve the inverse problem, i.e., to establish the position of the Fermi level at a given concentration of the alloying additives at a given temperature. In this instance one uses as initial data information on the character of the heterogeneous equilibrium of the semiconductor–admixture system, as described by the authors previously [212]. Comparison of the results obtained for germanium with the Blakemore graph showed a completely satisfactory agreement.

5. ANALYSIS OF THE LIQUIDUS CURVE OF A CONGRUENTLY MELTING COMPOUND A_mB_n IN VARIOUS APPROXIMATIONS OF IDEAL SOLUTIONS

5.1. Integral Form of the Liquidus Curve Equation in Various Approximations of the Regular Solution Model

On the basis of the van der Waals differential equation we obtained the general thermodynamic expression (5.39), which describes the liquidus curve of a congruently melting compound A_mB_n in differential form. As mentioned, this equation agrees with Prigogine and Defay's formula [24], which was derived using the chemical affinity function. This result underlines the generality of the van der Waals approach, so that one is justified in considering Eq. (5.39) completely rigorous.

The general solution of this equation serves as the basis for using specific models of solutions for the thermodynamic description of the liquidus curve of such systems. Thus we have developed [213] a general method for obtaining the integral form of the liquidus curve for systems with congruently melting compounds, wherein we could incorporate particular models such as the ideal approximation for solving Eq. (5.39). We shall give the general solution of Eq. (5.39) and show that on its basis one may obtain equations for various approximate theories of regular solutions.

We substitute into (5.39) the derivative of the chemical potential with respect to concentration, $(\partial \mu_2 / \partial x_2)_{p,T}{}^{L}$. We find, after separating variables,

$$- (\bar{q}_{A_mB_n}^{L \to S} / T^2) \, dT = - [mR/(1 - x_2^L)] \, dx_2^L$$
$$+ (nR/x_2^L) \, dx_2^L - [mRx_2^L/(1 - x_2^L)] \, d \ln \gamma_2^L + nR d \ln \gamma_2^L. \tag{5.191}$$

In view of Eq. (4.43) we may write for the activity coefficients

$$(1 - x_2^L) \, d \ln \gamma_1 + x_2^L \, d \ln \gamma_2 = 0. \tag{5.192}$$

After substituting $d \ln \gamma_2$ from (5.192) into (5.191), we get

$$\bar{q}_{A_mB_n}^{L \to S} \, d \, (1/T) = mR d \ln (1 - x_2^L) + nR d \ln x_2^L$$
$$+ mR d \ln \gamma_1^L + nR d \ln \gamma_2^L. \tag{5.193}$$

Assuming that $\bar{q}_{A_mB_n}{}^{L \to S}$ does not depend on temperature, and integrating Eq. (5.193) between the limits

$$T = T_{m, A_mB_n}^{\circ}, \quad x_2^L = n \, (m + n)$$

and T, x_2^L, we obtain

$$RT \left[m \ln \frac{\gamma_1^L (1 - x_2^L)}{\gamma_1^{\prime (L)} \, m/(m + n)} + n \ln \frac{\gamma_2^L x_2^L}{\gamma_2^{\prime (L)} \, n/(m + n)} \right]$$
$$= \Delta s_{m, A_mB_n}^{\circ} (T - T_{m, A_mB_n}^{\circ}), \tag{5.194}$$

where the prime denotes the composition of the compound. Equation (5.194) is the general integral form of the liquidus curve corresponding to the initial crystallization of the phase based on the congruently melting compound A_mB_n.

This deduction and the solution of Eq. (5.39) are thermodynamically rigorous if one does not consider the assumption concerning the temperature independence of $\bar{q}_{A_mB_n}{}^{L \to S}$. Equation (5.194) can clearly serve as a point of departure for deriving model forms for the equation of the liquidus curve. To do this, one must express the activity coefficients in the corresponding approximation and substitute them into (5.194).

Using Eqs. (4.139) and (4.140), we obtain from (5.194) the equation of the liquidus curve of a congruently melting compound A_mB_n:

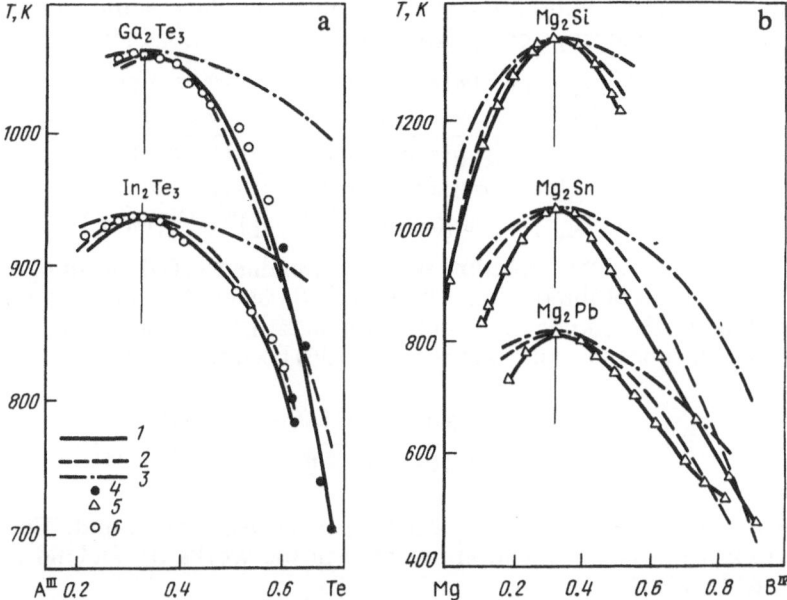

Fig. 44. Liquidus curves corresponding to initial crystallization: a) for the compound $A_2^{III}Te_3$; b) for the compound Mg_2B^{IV}. 1) According to Eq. (5.195); 2) according to Eq. (5.196); 3) according to Eq. (5.47); 4) data from [218, 219]; 5) data from [218, 220–222]; 6) experimental data of present authors.

$$
\begin{aligned}
T = \{\omega_0^L \, [m\,(x_2^L)^2 + n\,(1 - x_2^L)^2 - mn/(m + n)] \\
+ \omega_1^L \, [m\,(x_2^L)^2\,(2x_2^L - 1) + 2nx_2^L\,(1 - x_2^L)^2 - mn^2/(m + n)^2] \\
+ \omega_2^L \, [m\,(x_2^L)^3\,(3x_2^L - 2) + 3n\,(x_2^L)^2\,(1 - x_2^L)^2 - mn^3/(m + n)^3] \\
+ \Delta h_{m,\,A_m B_n}^{\circ}\} \, \{\omega_{01}^L\,[- mn/(m + n) + m\,(x_2^L)^2 + n\,(1 - x_2^L)^2] \\
+ \omega_{11}\,[- mn^2/(m + n)^2 + m\,(x_2^L)^2\,(2x_2^L - 1) + 2nx_2^L\,(1 - x_2^L)^2] \\
+ \omega_{21}^L\,[- mn^3/(m + n)^3 + m\,(x_2^L)^3\,(3x_2^L - 2) + 3n\,(x_2^L)^2 \\
\times\,(1 - x_2^L)^2] - R\ln\,[(m + n)^{m+n}/m^m n^n] \\
- R\ln\,[(1 - x_2^L)^m\,(x_2^L)^n] + \Delta s_{m,\,A_m B_n}^{\circ}\}^{-1}.
\end{aligned}
\tag{5.195}
$$

This includes the most frequently employed particular cases of other transformations (subregular, strictly regular, and quasiregular solutions, etc.). In accordance with the results obtained in [113], it provides a quite accurate prediction of the thermodynamic properties of solutions, in particular of the heat of mixing.

A single equation for the liquidus curve in systems of the type considered here is meaningful for partial dissociation of the compound $A_m B_n$. In the converse case, according to Storonkin [13], the solution divides at the ordinate of the compound into two phase regions, and the left and right branches of the liquidus must be described by independent equations.

In the subregular solution model approximation, which admits a temperature dependence for the exchange energy ($\omega_2 = 0$, $\omega_{21} = 0$), we obtain via Eq. (5.194)

$$T = \frac{\begin{aligned}&\Delta h^{\circ}_{m,\,A_mB_n} + \omega^L_0 \left[m \left(x^L_2\right)^2 + n \left(1 - x^L_2\right)^2 - mn/(m+n)\right] \\ &+ \omega^L_1 \left[m \left(x^L_2\right)^2 \left(2x^L_2 - 1\right) + 2nx^L_2 \left(1 - x^L_2\right)^2 - mn^2/(m+n)^2\right]\end{aligned}}{\begin{aligned}&\Delta s^{\circ}_{m,\,A_mB_n} - \omega^{(L)}_{01} \left[mn/(m+n) - mx^2_2 - n \left(1 - x^{(L)}_2\right)^2\right] \\ &- \omega^L_{11} \left[\left(mn^2/(m+n)^2 - m \left(x^L_2\right)^2 \left(2x^L_2 - 1\right) - 2nx^L_2\right.\right. \\ &\times \left.\left(1 - x^L_2\right)^2\right] - R \ln \left[(m+n)^{m+n} \left(1 - x^L_2\right)^m \left(x^L_2\right)^n/(m^m n^n)\right]\end{aligned}}. \tag{5.196}$$

This is similar to the equation we have obtained [160] by integrating the differential equation of the liquidus curve of A_mB_n using the same approximation. We note that if the molecular interaction parameter ω does not depend on temperature and concentration, expression (5.196) becomes [160]

$$T = \frac{\Delta h^{\circ}_{m,\,A_mB_n} + \omega^L_0 (m+n) \left[x^L_2 - n/(m+n)\right]^2}{\Delta s^{\circ}_{m,\,A_mB_n} - R \ln \left[(m+n)^{m+n} \left(1 - x^L_2\right)^m \left(x^L_2\right)^n/m^m n^n\right]}. \tag{5.197}$$

If $m = n = 1$, i.e., in the simplest case of an AB compound, Eq. (5.197) reduces to a well-known equation obtained in the works of Vieland [214] and Nougaret and Potier [215]:

$$T = \frac{\Delta h^{\circ}_{m,\,AB} + \omega^L_0 \left(2x^L_2 - 1\right)^2/2}{\Delta s^{\circ}_{m,\,AB} - R \ln \left[4x^L_2 \left(1 - x^L_2\right)\right]}. \tag{5.198}$$

Kuznetsov and Leonov [216] obtained a similar result in treating the process of melting of one mole of an A_mB_n crystal having a vanishingly small regime of homogeneity.

We note that the attempt [217] to solve Eq. (5.39) by an approximate integration procedure (the method of varying an arbitrary constant) using a regular solution model has led to results which are in disagreement with the equation of Vieland, rendering its expressions unacceptable.

In conclusion, we would like to note that we have tried practically all of the previously described models of solutions on various groups of congruently melting chemical compounds. It transpired that in a comparatively small range of concentrations around the stoichiometric composition the majority of models give a satisfactory agreement between calculations and the experimental liquidus curve. As the concentration interval is increased, the discrepancy between theory and experiment grows, and it is largest for ideal and strictly regular solutions. The application of the equations is illustrated below for two groups of compounds differing in their compositions (Fig. 44).

We compared the liquidus curves calculated according to Eqs. (5.47), (5.196), and (5.197) with the experimental curves corresponding to the initial crystallization of Ga_2Te_3 and In_2Te_3 in the systems Ga–Te and In–Te (see Fig. 44a). In order to get a more accurate idea of the nature of these curves, we have studied thermographically a series of alloys arranged according to composition to the left and right of the compounds Ga_2Te_3 and In_2Te_3. In each system we studied about 15 alloys. Each alloy composition on the pyrometer signal was plotted on three thermograms, and the mean value of the liquidus temperature was taken [160].

The source materials in preparing the alloys were gallium, indium, and tellurium of semiconductor purity. The synthesis and thermal analysis of the alloys was carried out in a sealed quartz tube evacuated to ~$1.33 \cdot 10^{-3}$ Pa.

The calculation of the liquidus curves corresponding to the initial crystallization of the compounds Ga_2Te_3 and In_2Te_3 was carried out via Eqs. (5.47), (5.196), and (5.197) by means of a computer.

From a comparison of all the calculations with experimental data, it is readily concluded that they agree fairly well with each other in a comparatively narrow concentration regime. Theory and experiment diverge as this concentration range is increased, in the first instance when using the ideal model, and subsequently also for the strongly regular solution. The model of subregular solutions describes the experimental data satisfactorily over the whole range of concentrations.

A wholly similar picture is observed for the case of Mg_2B^{IV} compounds (where B^{IV} is Si, Ge, Sn, Pb). The results of the calculations according to (5.47), (5.196), and (5.197), together with experimental data from [219–222] for comparison, are shown in Fig. 44b.

From an analysis of the results of the calculations it follows that the best description of experimental data on liquidus curves [219–222] is provided by the model of subregular solutions. However, the corresponding calculation for the mixing functions does not always agree well with experiment.

5.2. Integral Thermodynamic Equation of the Liquidus Curve of a Congruently Melting Compound (Brebrick's Equation)

In establishing the intrinsic structure of solutions, an important role is played by thermodynamic relations which describe the liquidus curve in terms of experimentally measured thermodynamic properties of the solutions and their chemical compounds. Such relations were obtained by Brebrick [146, 223].

Using the results of this work we shall derive an equation for the liquidus curve, establishing a direct interrelation between the thermodynamic properties of the liquid and solid phases.

Let us consider the congruent melting of the compound A_mB_n, in equilibrium with a liquid two-component phase of composition x_2^L. We shall assume that the compound has an extremely small region of homogeneity in the vicinity of $x_2 = n/(m + n)$. In equilibrium the chemical potentials of each of the components in the coexisting phases are equal [cf. Eq. (5.54)]. This condition determines the compositions of the equilibrium coexisting solid and liquid phases at any temperature.

Multiplying the first equation of (5.54) by m and the second by n and adding, we get

$$m\mu_1^L + n\mu_2^L = m\mu_1^S + n\mu_2^S. \tag{5.199}$$

Let us write the chemical potentials of the components in the liquid phase in expanded form:

$$\mu_1^L = \mu_1^{\circ\,(L)} + RT \ln \gamma_1^L \left(1 - x_2^L\right), \tag{5.200}$$

$$\mu_2^L = \mu_2^{\circ\,(L)} + RT \ln \gamma_2^L x_2^L. \tag{5.201}$$

Substituting these expressions into (5.199), we obtain

$$mRT \ln \left[\gamma_1^L \left(1 - x_2^L\right)\right] + nRT \ln \left(\gamma_2^L x_2^L\right)$$
$$= m \left(\mu_1^S - \mu_1^{\circ\,(L)}\right) + n\left(\mu_2^S - \mu_2^{\circ\,(L)}\right). \qquad (5.202)$$

Since the region of homogeneity of the compound was assumed to be near $x_2 = n/(m + n)$, the right-hand side of (5.202) may be considered to be the free energy of formation of the solid compound of composition $A_m B_n$ from the liquid components of A and B. Then relation (5.202) will assume the form

$$mRT \ln \left[\gamma_1^L \left(1 - x_2^L\right) + nRT \ln \left[\gamma_2^L x_2^L\right] = \Delta G_f^S \qquad (5.203)$$

where

$$\Delta G_f^S = \Delta H_f^S - T \Delta S_f^S,$$

and ΔG_f, ΔH_f, and ΔS_f are, respectively, the free energy, enthalpy, and entropy of formation of the solid compound from the liquid compounds. We will show in the following that Eq. (5.203) is essentially identical to an equation of the type (5.194), obtained earlier for the liquidus curve of a congruently melting compound $A_m B_n$.

The free energy of mixing in the process of forming a liquid solution of stoichiometric composition is determined by the following:

$$\Delta G_f^{\prime\,(L)} = mRT \ln \left[\gamma_1^{\prime\,(L)} m/(m + n)\right] + nRT \ln \left[\gamma_2^{\prime\,(L)} n/(m + n)\right]. \qquad (5.204)$$

The quantities ΔG_f^S and $\Delta G_f^{\prime(L)}$ are obviously related by

$$\Delta G_f^S = \Delta G_f^{\prime\,(L)} - \Delta G_{m,\,A_m B_n}^{\circ}, \qquad (5.205)$$

where $\Delta G_{m,A_m B_n}^{\circ}$ is the change in the Gibbs free energy of the reaction (5.40) in the process of melting. At the melting temperature $T_{m,A_m B_n}^{\circ}$ of the compound we have already noted that $\Delta G_{m,A_m B_n}^{\circ}$ vanishes [cf. Eq. (5.33a)], so that

$$\Delta G_f^S = \Delta G_f^{\prime\,(L)}. \qquad (5.206)$$

At any other temperature we may write, on the basis of (5.205), and using (5.33a),

$$\Delta G_f^S = \Delta G_f^{\prime\,(L)} + \Delta h_{m,\,A_m B_n}^{\circ} \left[(T/T_{m,\,A_m B_n}^{\circ}) - 1\right], \qquad (5.207)$$

where we have neglected the difference in the heat capacities of the initial substances and the products of the reaction (5.40). Substitution of (5.207) and (5.206) into (5.203) leads to the equation of the liquidus (5.194).

In accordance with the conclusions of Brebrick [223], using Kirchhoff's equation, we may represent the temperature dependence of the enthalpy and the entropy of formation of the compound $A_m B_n$ in the form of a power series. The results up to the second-order term in the expansion are

$$\Delta H_f^S = \Delta H_f^{\bullet\,(S)} + a\left(T - T_{m,\,A_mB_n}^\circ\right) + b(T^2 - T_{m,\,A_mB_n}^{\circ\,2})/2, \tag{5.208}$$

$$\Delta S_f^S = \Delta S_f^{\bullet\,(S)} + a\ln(T/T_{m,\,A_mB_n}^\circ) + b(T - T_{m,\,A_mB_n}^\circ), \tag{5.209}$$

where a and b are constants and the index "$\bullet$" indicates that the quantity so marked refers to the melting temperature of the compound. Remembering that at this melting point

$$\Delta G_{m,\,A_mB_n}^\circ = \Delta h_{m,\,A_mB_n}^\circ - T_{m,\,A_mB_n}^\circ \Delta s_{m,\,A_mB_n}^\circ = 0, \tag{5.210}$$

we may write from (5.206):

$$mRT_{m,\,A_mB_n}^\circ \ln\left[\gamma_1^{\bullet\,(L)}\left(1 - x_2'^{\,(L)}\right)\right] + nRT_{m,\,A_mB_n}^\circ \ln\left[\gamma_2^{\bullet\,(L)}x_2'^{\,(L)}\right]$$

$$= \Delta H_f^{\bullet\,(S)} - T_{m,\,A_mB_n}^\circ \Delta S_f^{\bullet\,(S)} + \Delta h_{m,\,A_mB_n}^\circ - T_{m,\,A_mB_n}^\circ \Delta S_{m,\,A_mB_n}^\circ. \tag{5.211}$$

Substituting (4.180) and (4.181) into (5.211), we obtain from the terms not containing $T_{m,A_mB_n}^\circ$ the expression

$$\Delta H_f^{\bullet\,(S)} = m\bar{h}_1^{\bullet\,EL} + n\bar{h}_2^{\bullet\,EL} - h_{m,\,A_mB_n}^\circ, \tag{5.212}$$

and from the terms with $T_{m,A_mB_n}^\circ$ the relation

$$\Delta S_f^{\bullet\,(S)} = m\bar{s}_1^{\bullet\,EL} + n\bar{s}_2^{\bullet\,EL} - \Delta S_{m,\,A_mB_n}^\circ$$
$$- R\ln\left[(1 - x_2'^{\,(L)})^m (x_2'^{\,(L)})^n\right]. \tag{5.213}$$

These are the additional equations needed to establish the connection between the thermodynamic properties of the solid and liquid phases.

By substituting (4.180), (4.181), (5.208), (5.209), (5.212), and (5.213) into (5.203), we obtain the equation of the liquidus curve of a congruently melting compound:

$$T = \frac{\begin{aligned}&\Delta h_{m,\,A_mB_n}^\circ + m\left(\bar{h}_1^{EL} - \bar{h}_1^{\bullet EL}\right) + n\left(\bar{h}_2^{EL} - \bar{h}_2^{\bullet EL}\right)\\ &+ a\left(T_{m,\,A_mB_n}^\circ - T\right) + aT\ln\left(T/T_{m,\,A_mB_n}^\circ\right)\\ &+ b\left(T_{m,\,A_mB_n}^\circ - T\right)^2/2\end{aligned}}{\begin{aligned}&\Delta s_{m,\,A_mB_n}^\circ + m\left(\bar{s}_1^{EL} - \bar{s}_1^{\bullet EL}\right) + n\left(\bar{s}_2^{EL} - \bar{s}_2^{\bullet EL}\right)\\ &+ R\ln\left\{\left[(1 - x_2'^{\,(L)})/(1 - x_2^L)\right]^m \left[x_2'^{\,(L)}/x_2^L\right]^n\right\}\end{aligned}}. \tag{5.214}$$

This equation may be solved by successive approximations. In contrast to the relations (5.195)–(5.198) obtained earlier, Eq. (5.214) is more general, in that its derivation did not rely on any concrete model for describing the behavior of the liquid phase. Equation (5.214) should always be used together with Eqs. (5.212) and (5.213).

We may simplify expression (5.214) by representing ΔG_f^S as a linear function of temperature in the form

$$\Delta G_f^S = \Delta \overline{H}_f^S - T \,\Delta \overline{s}_f^S + \delta\,(T), \qquad (5.215)$$

where $\delta(T)$ is a term linear in temperature, making a small contribution to $\Delta G_f{}^S$; $\Delta H_f{}^S$ and $\Delta S_f{}^S$ are mean values of the enthalpy and entropy of formation of the compound in a specific temperature interval.

We suppose that this mean value of the enthalpy is related to its value at the melting temperature by the relation

$$\Delta \overline{H}_f^S = \Delta H_f^{\bullet \ (S)} + \Delta, \qquad (5.216)$$

where Δ is a correction term.

Substituting the preceding expression into Eq. (5.215), it is easily verified that

$$\Delta \overline{S}_f^{(S)} = \Delta S_f^{\bullet \ (S)} + \left(\Delta / T_{m,\,A_m B_n}^{\circ}\right) + \delta\left(T_{m,\,A_m B_n}^{\circ}\right) \Big/ T_{m,\,A_m B_n}^{\circ}. \qquad (5.217)$$

Substituting (5.215)–(5.217) into formula (5.203) and using (4.180), (4.181), (5.212), and (5.213), we obtain a simpler equation for the liquidus curve for a congruently melting $A_m B_n$ compound in the inexplicit form:

$$T = \cfrac{\Delta h_{m,\,A_m B_n}^{\circ} - \Delta - \delta\,(T) + m\left(\hbar_1^{EL} - \hbar_1^{\bullet EL}\right) \\ + n\left(\hbar_2^{EL} - \hbar_2^{\bullet EL}\right)}{\left[\Delta h_{m,\,A_m B_n}^{\circ} - \Delta - \delta\left(T_{m,\,A_m B_n}^{\circ}\right)\right] \Big/ T_{m,\,A_m B_n}^{\circ} \\ + m\left(\bar{s}_1^{EL} - \bar{s}_1^{\bullet EL}\right) + n\left(\bar{s}_2^{EL} - \bar{s}_2^{\bullet EL}\right) \\ + R \ln \left\{\left[(1 - x_2'^{\,(L)})/(1 - x_2^{(L)})\right]^m \left(x_2'^{\,(L)}/x_2^{(L)}\right)^n\right\}}. \qquad (5.218)$$

This equation may be solved exactly for the temperature if the terms $\delta(T)$ and $\delta(T_{m,A_m B_n}{}^{\circ})$ are neglected. The resulting equation is similar to Eq. (5.214) and differs from it only by the correction term Δ to the heat of melting of the compound $A_m B_n$. But the premises on which the derivations of these equations are based are different.

Let us now see under what conditions one may justify neglect of the term $\delta(T)$ in Eq. (5.218). Of course, this will depend partly on the accuracy of the calculation and the temperature region employed. Comparing (5.218) and (5.214), we obtain

$$a\left[T - T_{m,\,A_m B_n}^{\circ} + T \ln\left(T/T_{m,\,A_m B_n}^{\circ}\right)\right] + b\left(T - T_{m,\,A_m B_n}^{\circ}\right)^2/2 \\ = \Delta\left(1 - T/T_{m,\,A_m B_n}^{\circ}\right) + \delta\,(T) - T\delta\left(T_{m,\,A_m B_n}^{\circ}\right)/T_{m,\,A_m B_n}^{\circ}. \qquad (5.219)$$

Therefore, comparing the expression

$$\Delta h_{m,\,A_m B_n}^{\circ} + m\left(\hbar_1^{EL} - \hbar_1^{\bullet EL}\right) + n\left(\hbar_2^{EL} - \hbar_2^{\bullet EL}\right) \\ + \Delta\left(1 - T/T_{m,\,A_m B_n}^{\circ}\right) \qquad (5.220)$$

with

$$\Delta h_{m,\, A_mB_n}^{\circ} + m\,(\bar{h}_1^{EL} - \bar{h}_1^{\bullet EL}) + n\,(\bar{h}_2^{EL} - \bar{h}_2^{\bullet EL})$$
$$+ \Delta\,(1 - T/T_{m,\, A_mB_n}^{\circ}) + \delta\,(T)$$
$$- T\delta\,(T_{m,\, A_mB_n}^{\circ})/T_{m,\, A_mB_n}^{\circ}, \qquad (5.221)$$

one may estimate the magnitude of the possible error.

5.3. Equation of the Liquidus Curve of a Congruently Melting AB-Type Compound in the Approximation of a Regular Associated Solution

The temperature and concentration dependence of thermodynamic properties in systems with congruently melting compounds often indicate the existence of nondissociated molecules of the compound in the liquid phase. It is, therefore, of considerable interest to describe the curves of phase equilibria within the model of a regular associated solution consisting of monomers and aggregates.

Let us consider a binary system A–B in which a congruently melting compound AB is formed. We shall suppose that the liquid saturated solution may be represented by a regular mixture of three sorts of particles: the monomers A_1 and B_1, and the aggregates AB.

In order to calculate the liquidus curve for the initial crystallization of the compound AB according to Eq. (5.194), written for the case $m = n = 1$ in the form

$$RT \ln\,[a_1' a_2'/a_1 a_2] = \Delta h_{m,\, AB}^{\bullet} - T\,\Delta s_{m,\, AB}^{\bullet}, \qquad (5.222)$$

one must know the activity coefficients of the components.

According to Jordan [186], we have, for the present case,

$$a_A = a_{A_1} = \gamma_{A_1} x_{A_1}, \quad a_B = a_{B_1} = \gamma_{B_1} x_{B_1}. \qquad (5.223)$$

On the basis of formula (4.216) the activity coefficients of A_1, B_1, and AB of the three-component regular mixture are determined by the equations

$$RT \ln \gamma_{A_1} = x_{B_1}^2 \omega_{A_1-B_1} + x_{AB}^2 \omega_{A_1-AB}$$
$$+ x_{B_1} x_{AB}\,(\omega_{A_1-AB} - \omega_{B_1-AB} + \omega_{A_1-B_1}), \qquad (5.224)$$

$$RT \ln \gamma_{B_1} = x_{AB}^2 \omega_{B_1-AB} + x_{A_1}^2 \omega_{A-B}$$
$$+ x_{A_1} x_{AB}\,(\omega_{B_1-AB} + \omega_{A_1-B_1} - \omega_{A_1-AB}), \qquad (5.225)$$

$$RT \ln \gamma_{AB} = x_{A_1}^2 \omega_{A_1-AB} + x_{B_1}^2 \omega_{B_1-AB}$$
$$+ x_{A_1} x_{B_1}\,(\omega_{A_1-AB} + \omega_{B_1-AB} - \omega_{A_1-B_1}). \qquad (5.226)$$

In using these formulas we require knowledge of their molar fractions x_{A_1}, x_{B_1}, x_{AB}, as well as of the parameters of intermolecular interactions, ω_{ij}.

The law of mass action for the association reaction $A + B = AB$ may be written as

$$K = (x_{A_1} x_{B_1} / x_{AB}) (\gamma_{A_1} \gamma_{B_1} / \gamma_{AB}). \tag{5.227}$$

In the present case the condition of the mass conservation of the components will have the form

$$n_1 = n_{A_1} + n_{AB}, \quad n_2 = n_{B_1} + n_{AB}. \tag{5.228}$$

Using the definition of the molar fractions of the components and the molecular entities, Eq. (5.228) may be easily transformed into the form

$$x_{A_1} = x_A - x_B x_{AB}; \tag{5.229}$$

$$x_{B_1} = x_B - x_A x_{AB}. \tag{5.230}$$

Writing expression (5.227) in the form

$$\ln K = \ln \gamma_{A_1} + \ln \gamma_{B_1} - \ln \gamma_{AB} + \ln (x_{A_1} x_{B_1} / x_{AB}), \tag{5.231}$$

it is simple to verify from Eqs. (5.224)–(5.226) and relations (5.229), (5.230) that

$$\ln K = \ln \{[x_A x_B (1 + x_{AB})^2 - x_{AB}]/x_{AB}\} + (\omega_{A_1-B_1}/RT) [1 - x_A x_B (1 + x_{AB})^2]$$

$$- (\omega_{A_1-AB}/RT) [1 - x_B (1 + x_{AB})^2] - (\omega_{B_1-AB}/RT) [1 - x_A (1 + x_{AB})^2]. \tag{5.232}$$

Unfortunately, Eq. (5.232) is a transcendental equation in x_{AB} and cannot be solved explicitly if $\omega_{ij} \neq 0$. If, however, the parameters K and ω_{ij} are known, one may find the dependence of x_{AB} on the molar fraction of one of the components by solving Eq. (5.232) numerically on the computer. Furthermore, by substituting the value of x_{AB} into Eqs. (5.229) and (5.230), one may estimate the molar fraction of the monomers for given compositions. Substituting the values of x_{A1}, x_{B1}, and x_{AB} for a given composition of the solution x_A and x_B, one may first calculate via (5.224) and (5.225) the activity coefficients of the monomeric forms $\ln \gamma_{A1}$ and $\ln \gamma_{B1}$, and then, via (5.223), the activity coefficients of the components A and B. However, the parameters K and ω_{ij} are not usually known. One therefore resorts to an approximate calculation of the quantities a_A and a_B, proposed by Jordan [186].

According to this method, one expands (5.224) and (5.225) in a Taylor series in x_B and x_A around $x_B = 0$, $x_A = 0$, respectively. Assuming that $\omega_{A_1-AB} = \omega_{B_1-AB}$, and in view of expression (5.232), we obtain the following approximate equation for the activity coefficients of the monomeric forms A_1 and B_1:

$$RT \ln \gamma_{A_1} \approx \{1/[(1 + K_l)^2 \omega_{A_1-AB}]$$
$$+ K_l/[(1 + K_l) \omega_{A_1-B_1}]\} x_B^2 + \cdots = \omega^* x_B^2 + \cdots, \tag{5.233}$$

$$RT \ln \gamma_{B_1} \approx \{1/[(1 + K_l)^2 \omega_{A_1-AB}]$$
$$+ K_l/[(1 + K_l) \omega_{A_1-B_1}]\} x_A^2 + \cdots = \omega^* x_A^2 + \cdots, \tag{5.234}$$

where

$$\omega^* = 1/[(1 + K_l)^2 \, \omega_{A_1-AB}] + K_l/[(1 + K_l) \, \omega_{A_1-B_1}], \qquad (5.235)$$

$$K_l \equiv K \exp{[(\omega_{A_1-B_1} - \omega_{A_1-AB})/RT]}. \qquad (5.236)$$

From (5.236) and (5.235) it is easily verified that in the case of complete association ($K_l = K = 0$)

$$\omega^* = \omega_{A_1 (B_1)-AB}, \qquad (5.237)$$

and for complete dissociation

$$\omega^* = \omega_{A-B}. \qquad (5.238)$$

Using the Gibbs–Duhem equation in the form

$$x_{A_1} d \ln \gamma_{A_1} + x_{B_1} d \ln \gamma_{B_1} + x_{AB} d \ln \gamma_{AB} = 0$$

together with relations (5.229), (5.230), (5.233), (5.234), and the normalization condition $\gamma_{AB} \rightarrow 1$ for $x_B = 0.5$, we obtain for $\ln \gamma_{AB}$:

$$RT \ln \gamma_{AB} = (\omega^*/2) (1 - 4x_A x_B). \qquad (5.239)$$

Substituting (5.233), (5.234), and (5.239) into Eq. (5.227), and using (5.229) and (5.230), we obtain

$$\ln K = (\omega^*/2RT) + \ln [x_A x_B (1 + x_{AB})^2 - x_{AB}] - \ln x_{AB}. \qquad (5.240)$$

Equation (5.240) may easily be solved for x_{AB}:

$$x_{AB} = (1 - P)/(1 + P). \qquad (5.241)$$

Here

$$P \equiv \sqrt{1 - 4x_A x_B/(1 + K')} = \sqrt{1 - 4x_A x_B (1 - \alpha^2)}, \qquad (5.242)$$

where K' is the apparent dissociation constant of the complex AB, related to the actual value of the constant K by the equation

$$K'_x = K \exp{[- \omega^*/2RT]}. \qquad (5.243)$$

In the general case (i.e., $x_A \neq x_B$), P is the fraction of monomeric atoms A and B in the solution, i.e.,

$$P = (n_{A_1} + n_{B_1})/(n_A + n_B). \qquad (5.244)$$

At the same time, $K_x' = \alpha^2/(1 - \alpha^2)$, where α is the degree of ionization of the complex AB at a point corresponding to the composition of the compound $x_B = 0.5$, dependent on temperature.

It is readily verified from (5.242) that at the point $x_B = 1/2$, $P = a°$. If $n_{A_1} = n_A$ and $n_{B_1} = n_B$, then $P = a° = 1$, i.e., the compound AB is completely dissociated in the liquid phase. After substituting Eq. (5.241) into (5.229) and (5.230), we obtain

$$x_{A_1} = (x_A - x_B + P)/(1 + P), \qquad (5.245)$$

$$x_{B_1} = (x_B - x_A + P)/(1 + P). \qquad (5.246)$$

Finally, combining Eqs. (5.245), (5.246), (5.233), and (5.234) with the equations of (5.223), we obtain the following expression for the activity of the components of binary solutions in which interactions lead to the formation of aggregates of type AB:

$$a_A = \gamma_{A_1} x_{A_1} = [(x_A - x_B + P)/(1 + P)] \exp(\omega^* x_B^2/RT), \qquad (5.247)$$

$$a_B = \gamma_{B_1} x_{B_1} = [(x_B - x_A + P)/(1 + P)] \exp(\omega^* x_A^2/RT). \qquad (5.248)$$

Expressions (5.247) and (5.248) satisfy the Gibbs–Duhem equation for a two-component system.

Substituting these equations into Eq. (5.222), we are led, after some algebraic manipulations, to the equation of the liquidus curve for the initial crystallization of a congruently melting compound AB, within the approximation of the theory of a regular associated solution, as obtained by Jordan [186]:

$$T = \frac{\Delta h^°_{m, AB} + 2\omega^* (x_B - 0{,}5)^2}{R \ln S(x_B) + \Delta s^°_{m, AB}}, \qquad (5.249)$$

where

$$S(x_B) = [(1 + P)/(1 + \alpha°)]^2/(4 x_A x_B). \qquad (5.250)$$

We consider two limiting cases:

1. The case of complete dissociation of the compound AB in the liquid phase, when $\alpha° = 1$, $P = 1$. Then (5.250) becomes

$$S(x_B) = 1/(4 x_A x_B),$$

and equality (5.238) applies. Consequently, in this case Eq. (5.198) is valid.

2. The case of complete association of the compound in the liquid phase. This means that the liquid solution consists either of the monomer A_1 and the aggregate AB, or the monomer B_1 and the aggregate AB. In this case we have, on the basis of (5.229), (5.241), and (5.230),

$$\begin{aligned} 1 + P = 2x_B \quad \text{for} \quad x_B > x_A, \\ 1 + P = 2x_A \quad \text{for} \quad x_B < x_A. \end{aligned} \qquad (5.251)$$

Substituting this expression into (5.249) and (5.250), we are led to the following result:

$$T = \frac{\Delta h_{m,\,AB}^{\circ} + 2\omega_{A-AB}\,(x_B - 0.5)^2}{R \ln{(x_A/x_B)} + \Delta S_{m,\,AB}^{\circ}}\ (x_B < x_A); \tag{5.252}$$

$$T = \frac{\Delta h_{m,\,AB}^{\circ} + 2\omega_{B-AB}\,(x_B - 0.5)^2}{R \ln{(x_B/x_A)} + \Delta S_{m,\,AB}^{\circ}}\ (x_B > x_A). \tag{5.253}$$

Thus, in the this case, the phase diagram separates, as it were, into two independent diagrams A–AB and B–AB, with the compound AB serving as the solvent.

In the present approximation the equation of the liquidus curve (5.249) of the congruently melting compound AB contains two parameters ω^* and α°, which under the terms of our treatment are independent of the concentration.

In order to simplify the calculations it is usually assumed that the parameters ω^* and α° are also temperature independent. The values of these parameters may be estimated graphically for not too large values of α°.

Taking into account Eq. (5.253) written for a completely associated solution, we obtain from Eq. (5.249) with $\alpha^{\circ} \neq 0$ and for $x_B > x_A$:

$$2^{-1} \ln{[S\,(x_B)\,(x_A/x_B)]} = [(\omega^* - \omega_{B-AB})/RT]\,(x_B - 0.5)^2. \tag{5.254}$$

Bearing in mind formulas (5.250) and (5.242), and introducing the notation $\Delta x = (x_B - 0.5)$, relation (5.254) may be transformed into the form

$$\ln{(1 + \alpha^{\circ})} + [(\omega^* - \omega_{B-AB})/RT]\,(\Delta x)^2$$
$$= \ln{\{[1 + 2\,\Delta x\,\sqrt{1 + (\alpha^{\circ}/2\,\Delta x)^2 - \alpha^{\circ 2}}]/(1 + 2\,\Delta x)\}}. \tag{5.255}$$

From this it is readily verified that if $\alpha^{\circ} \leq 1$ (for example, $\alpha^{\circ} \leq 0.2$), then for values of $2\Delta x < \alpha^{\circ}$ the right-hand side may be equated to zero to a good approximation, excluding the region near $x = 0.5$. In addition, for α° not large, we may approximate $\ln{(1 + \alpha^{\circ})} \approx \alpha^{\circ}$. Hence,

$$\alpha^{\circ} + [(\omega^* - \omega_{B-AB})/RT]\,(\Delta x)^2 \approx 0. \tag{5.256}$$

Expressing the values of $\omega_{B\text{-}AB}$ from both (5.253) and (5.256), and equating the results, we obtain the following approximate formula:

$$\omega_{B-AB} = [-RT \ln{(x_A/x_B)} + T\,\Delta s_{m,\,AB}^{\circ} - \Delta h_{m,\,AB}^{\circ}]/2\,(\Delta x)^2 = \omega^* + RT\alpha^{\circ}/(\Delta x)^2. \tag{5.257}$$

This implies that $\omega_{B\text{-}AB}$ is hyperbolic in the argument Δx and approaches a constant limiting value ω^* for large Δx. In addition, the dependence of $\omega_{B\text{-}AB}$ on the variable $T/(\Delta x)^2$ is a straight line whose slope provides an estimate of the parameter α°, and whose intercept is ω^*.

The foregoing analysis and approximations were applied by Jordan and Jenkins [186, 192] to the systems $A^{II} - B^{VI}$ and $A^{IV} - B^{VI}$, in which one assumes a strong association in the liquid phase.

In the work of Osamura and Murakami [224], an intermediate model was applied to the example of five binary systems of type $A^{II}B^{V}$. This refers to the model of a regular partially associated solution, with the assumption that the mole fraction of the aggregates in the liquid phase is relatively small, so that to a good approximation one may employ the relations $(1 + x_{AB})^2 \approx 1$, $x_{A_1}x_{B_1} \approx x_A x_B$.

In accordance with the conclusions of [224], and taking into account the assumptions at the basis of (5.223)–(5.225), plus the formula for the constant of the equilibrium reaction

$$a_{AB}/(a_{A_1}a_{B_1}) = C_0 \exp\left[\Delta H^{\circ\,(L)}/RT\right],\qquad(5.258)$$

we have

$$x_{AB} = x_A x_B P^{\circ}_{AB}\left(1 + 2\,\Delta x\omega^-_{AB}/RT\right),\qquad(5.259)$$

where $\Delta H^{\circ(L)}$ is the heat of formation of the complex AB in the liquid phase, and C_0 is a constant relating it to the entropy of formation:

$$P^{\circ}_{AB} = C_0 \exp\left[\left(\Delta H^{\circ\,(L)} - \omega^+_{AB} + \omega_{A_1 - B_1}\right)\right]/RT.\qquad(5.260)$$

The quantities ω_{AB}^- and ω_{AB}^+ are, respectively, $(\omega_{A_1\text{-}AB} - \omega_{B_1\text{-}AB})/2$ and $(\omega_{A_1\text{-}AB} + \omega_{B_1\text{-}AB})/2$, while $\Delta x = 0.5 - x_1$.

If $(\Delta H^{\circ(L)} + \omega_{AB}^+ - \omega_{A_1\text{-}AB})$ is considerably larger than ω_{AB}^-, then after substituting (5.259) into the formula for the excess free energy of the liquid regular solution consisting of the monomers A_1, B_1 and the complex AB, namely,

$$f^E \cong g^E = x_{A_1}x_{B_1}\omega_{A_1\text{-}B_1} + x_{A_1}x_{AB}\omega_{A_1\text{-}AB} + x_{B_1}x_{AB}\omega_{B_1\text{-}AB} + x_{AB}\,\Delta H^{\circ\,(L)},\quad(5.261)$$

we obtain by means of well-known thermodynamic relations the following expression for the activity coefficients of the components:

$$RT\,\ln\gamma_A = x_B^2\omega_{A_1\text{-}B_1} + \left(\Delta H^{\circ\,(L)} + \omega^+_{AB} - \omega_{A_1\text{-}B_1}\right)P^{\circ}_{AB}\left[1 - (1 - 4\,\Delta x)\,\omega^-_{AB}/RT\right],$$

$$(5.262)$$

$$RT\,\ln\gamma_B = x_A^2\omega_{A_1\text{-}B_1} + \left(\Delta H^{\circ\,(L)} + \omega^+_{AB} - \omega_{A_1\text{-}B_1}\right)P^{\circ}_{AB}\left[1 + (1 + 4\,\Delta x)\,\omega^-_{AB}/RT\right],$$

$$(5.263)$$

where the second term is due to the contribution of the aggregates.

In the present approximation we have for ω [224]

$$\omega = \omega_{A_1\text{-}B_1} + \left(\Delta H^{\circ\,(L)} + \omega^+_{AB} - \omega_{A_1\text{-}B_1}\right)P^{\circ}_{AB}\left(1 + 4\omega^-_{AB}\,\Delta x/RT\right).\quad(5.264)$$

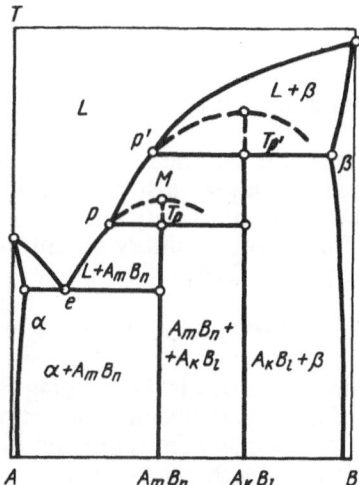

Fig. 45. Basic scheme for state diagram with incongruently melting compounds A_mB_n and A_kB_l.

The quantity $(1 + 4\omega_{AB}{}^{-}\Delta x/RT)$ depends on the concentration. Therefore, when $\omega_{AB}{}^{-} \neq 0$, the parameter ω will depend on composition. The value of ω may be calculated within the model of a regular partially associated solution from data on phase equilibria, in the following manner.

For two points of the liquidus, corresponding to the initial crystallization of the compound AB belonging to the same isotherm, we may write from (5.264):

$$y \equiv RT\,(\omega_{\mathrm{I}} - \omega_{\mathrm{II}})/(\Delta x_{\mathrm{I}} - \Delta x_{\mathrm{II}})$$
$$= 4\left(\Delta H^{\circ\,(L)} + \omega_{AB}^{+} - \omega_{A_1\text{-}B_1}\right)\omega_{AB}^{-}C_0 \exp\left(A/RT\right), \qquad (5.265)$$

where

$$A \equiv \Delta H^{\circ\,(L)} + \omega_{A_1\text{-}B_1} - \omega_{AB}^{+}. \qquad (5.266)$$

The lower indices I and II denote solutions lying to the left and right of the composition of the compound, respectively.

Hence, by plotting the graph $\ln y = f(1/T)$, one may fix the values of the two functions A and $[4(\Delta H^{\circ(L)} + \omega_{AB}{}^{+} - \omega_{A_1\text{-}B_1})\omega_{AB}{}^{-}C_0]$. By plotting $z \equiv \omega - y\Delta x/RT$ as a function of $\exp\,(A/RT)$, one may estimate another two quantities, namely $\omega_{A_1\text{-}B_1}$ and $(\Delta H^{\circ(L)} + \omega_{AB}{}^{+} - \omega_{A_1\text{-}B_1})C_0$. To sum up, from data on the phase equilibria of the binary system we have four equations with five unknown parameters: C_0, $\omega_{A_1\text{-}B_1}$, $\omega_{A_1\text{-}AB}$, $\omega_{B_1\text{-}AB}$, and $\Delta H^{\circ(L)}$.

In the process of the formation of the complex AB in the solution, the most probable values are calculated from the condition of the minimum of the free energy of the solution g^M. Hence, we may use as our fifth equation

$$\partial g^M/\partial x_{A_1} + \partial g^M/\partial x_{B_1} = \partial g^M/\partial x_{AB}, \qquad (5.267)$$

from which we may easily estimate the parameter C_0 by using $g^M = g^{M(id)} + g^E$ and formula (5.261).

As will be discussed later, the parameters of the interactions between monomeric molecules in the liquid phase may also be calculated by other methods, for example by using data on the electronegativity and solubility parameters of the pure components. The heat of formation of binary complexes AB in the liquid phase may be calculated from the equation [224]

$$\Delta H^{\circ(L)} = \text{const} \, f_i D, \qquad (5.268)$$

where f_i is the degree of ionization and D is the alloying factor.

The calculations [224] of the intermolecular interaction parameters from data on phase equilibria, from Eq. (5.268), and from data on the electronegativity of the components are in good agreement with one another.

6. EQUATION OF THE LIQUIDUS CURVE AND THERMODYNAMIC ANALYSIS OF SYSTEMS WITH INCONGRUENTLY MELTING COMPOUNDS

As we have seen above, the thermodynamics of solutions enables one to describe analytically the liquidus curves of systems with congruently melting compounds. However, in many systems chemical compounds are formed which melt incongruently [218, 219].

Figure 45 shows schematically the state diagram of a binary system with a cascade of peritectic transformations, as a consequence of which chemical compounds of various compositions are formed. It follows from this diagram that in general the process of incongruent melting may be represented by the following peritectic reaction:

$$A_m B_n^S \rightleftarrows L_p + A_k B_l^S, \qquad (5.269)$$

where L_p is a melt of composition p, simultaneously saturated with respect to the crystals $A_m B_n$ and $A_k B_l$. If pressure effects are negligible, then, typically for condensed systems, the three-phase equilibrium (5.269) is invariant, as proved in Chapter 3.

There is no rigorous thermodynamic description in the literature of the liquidus curve of a system with a cascade of incongruently melting compounds.

The separate parts of this curve, corresponding to the initial crystallization of one of the compounds with a narrow region of homogeneity, may be described in principle by means of Brebrick's equation (5.214). In its application to incongruent compounds this equation is as accurate as for congruently melting compounds, since no thermodynamic properties characterizing the actual melting process of the compounds enter into it. Nevertheless, Brebrick's equation is not suited for a thermodynamic analysis of the liquidus curve in systems like those indicated in Fig. 45, since, in addition to the physically realized liquidus (e.g., the section ep of Fig. 45), it also describes the hypothetical section pM. Consequently, Brebrick's equation is insensitive to the transition to the peritectic horizontal and does not exhibit the physicochemical essence of incongruent melting, namely the appearance of a completely new form of two-phase equilibrium, encompassing a solid phase of different composition and properties from the compound A_mB_n.

The approach to solving the problem on the basis of Vieland's equation (5.198) would require a knowledge of quantities such as the heat of melting and the melting temperature of the compound. Naturally, in the case of incongruent melting materials, these characteristics are physically unreal. Attempts [225] to overcome this difficulty by means of calculating hypothetical quantities on the basis of experimentally determined enthalpies of the peritectic transformation are, in our opinion, unsuccessful, since the result thus obtained includes a term $[mn/(m + n)]\omega^L$, which is the heat of mixing in forming a physically nonexistent liquid regular solution of composition $n/(m + n)$. Furthermore, it was assumed [225] that the energies of mixing of the components for the nonexistent solution mentioned and for the solution of composition p are equal, which is obviously unjustified. For these reasons the results of [225] cannot be considered correct from a physicochemical viewpoint.

In this context we have attempted to obtain a single equation for the liquidus curve of systems having a cascade of peritectic transformations [226], which would take into account the physicochemical essence of incongruent melting for each compound.

We suppose that the extent of the region of solid solutions based on the compound and the B component is small enough to be ignored in the calculation. We write the peritectic reaction (5.269) by taking account of the mass relations in the following form:

$$A_mB_n^S \rightleftarrows x\,(1 - p)\,A_{\text{sat}}^L + xpB_{\text{sat}}^L + yA_kB_l^S, \tag{5.270}$$

where, according to mass conservation balance,

$$x = [m/(1 - p)] + k\,[n - p\,(m + n)\,]/(1 - p) \\ \times [p\,(k + l) - l], \tag{5.271}$$

$$y = [mp - n\,(1 - p)\,]/[p\,(k + l) - l]. \tag{5.272}$$

The change in the Gibbs free energy in the peritectic reaction (5.270) may then be expressed as

$$\Delta G_p = x\,(1 - p)\,\mu_A^L + xp\mu_B^L + y\mu_{A_kB_l}^{\circ\,(S)} - \mu_{A_mB_n}^{\circ\,(S)} = 0, \tag{5.273}$$

in which the index "$\circ$" refers to the pure material. Substituting the expanded expressions for the chemical potentials μ_i^L into (5.273), and using the model of regular solutions to describe the liquid phase, we obtain after some algebra

$$\Delta G_p = x\,(1 - p)\,\mu_A^{\circ\,(L)} + xp\mu_B^{\circ\,(L)} \\ + x\,(1 - p)\,RT_p \ln\,(1 - p) + xpRT_p \ln p + xp\,(1 - p)\,\omega^L \\ + y\mu_{A_kB_l}^{\circ\,(S)} - \mu_{A_mB_n}^{\circ\,(S)} = 0, \tag{5.274}$$

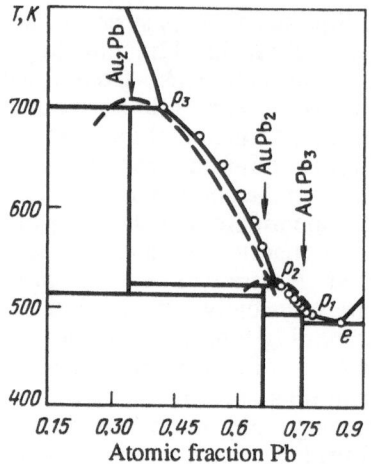

Fig. 46. Comparison of experimental [227] and calculated [Eq. (5.280)] liquidus curves for the gold–lead system.

TABLE 1. Thermodynamic Properties of Incongruently Melting Compounds and the Parameters of Peritectic Reactions in the Au–Pb System

Compound	T_p, K	p_i, Mole fraction Pb	ΔH_p, kJ/mole	$-\Delta H_f^S$, kJ/mole	$-\Delta S_f^S$, J/mole·deg	$-\Delta G_f^S$, kJ/mole	$-\Delta G_f^S$, kJ/mole [230]
Au$_2$Pb	691	0.43	32.6	48.6	37.7	22.6	30.6
AuPb$_2$	527	0.72	23.9 *	38.9	37.7	17.5	20.5
AuPb$_3$	490	0.78	67.4	88.7	136.0	22.2	—

*Experimentally obtained value of ΔH_p for AuPb$_2$ [228, 229].

where ω^L is the energy of mixing of the components in the liquid phase.

For convenience in further calculations, we add and subtract to expression (5.274) the quantity $n\mu_B{}^{\circ(L)}$. Furthermore, we suppose that

$$i\mu_A^{\circ(L)} + j\mu_B^{\circ(L)} - \mu_{A_iB_j}^{\circ(S)} = -\Delta G_f^S{}_{(A_iB_j)}. \qquad (5.275)$$

In view of relations (5.275) and (5.272), we obtain

$$\Delta G_p = -\Delta G_f^S{}_{(A_mB_n)} + y\,\Delta G_f^S{}_{(A_kB_l)} + x\,(1-p)\,RT_p\ln(1-p)$$
$$+ xpRT_p\ln p + xp\,(1-p)\,\omega^L = 0\,, \qquad (5.276)$$

where ΔG_f^S is the free energy of formation of the solid compounds A_mB_n and A_kB_l from the liquid components A and B.

Expanding ΔG_p and ΔG_f by formula (1.113), we obtain from (5.276)

$$\Delta H_p - T_p \, \Delta S_p = -\Delta H_f^S {}_{(A_m B_n)} + T_p \, \Delta S_f^S {}_{(A_m B_n)}$$
$$+ \, y \, \Delta H_f^S {}_{(A_k B_l)} - y T_p \, \Delta S_f^S {}_{(A_k B_l)} + x \, (1 - p) \, R T_p \ln (1 - p)$$
$$+ \, x p R T_p \ln p + x p \, (1 - p) \, \omega^L = 0. \tag{5.277}$$

Combining the terms not containing T_p we find the following expression for the enthalpy of the peritectic reaction:

$$\Delta H_p = -\Delta H_f^S {}_{(A_m B_n)} + y \, \Delta H_f^S {}_{(A_k B_l)} + x p \, (1 - p) \, \omega^L. \tag{5.278}$$

Similarly, from the terms with T_p the entropy of the peritectic reaction is found to be

$$\Delta S_p = -\Delta S_f^s {}_{(A_m B_n)} + y \, \Delta S_f^S {}_{(A_k B_l)} \; - x R \, (1 - p) \ln (1 - p) - x p R \ln p. \tag{5.279}$$

Solving (5.278) and (5.279) for $\Delta H_{f(A_m B_n)}{}^S$ and $\Delta S_{f(A_m B_n)}{}^S$, and substituting the result into the Brebrick equation (5.203), we obtain the following analytic expression for the liquidus curve of the incongruently melting compound (within the regular solution approximation):

$$T = \frac{\begin{array}{c} -\Delta H_p + y \, \Delta H_f^S {}_{(A_k B_l)} + \omega^L \, [x p \, (1 - p) \\ - \, m \, (x_B^L)^2 - n \, (1 - x_B^L)^2] \end{array}}{\begin{array}{c} -\Delta S_p + y \, \Delta S_f^S {}_{(A_k B_l)} - R \ln [(1 - p)^{x \, (1 - p)} p^{x p}] \\ + \, R \ln [(1 - x_B^L)^m \, (x_B^L)^n] \end{array}}. \tag{5.280}$$

Here x_B is the concentration of the liquid solution corresponding to the liquidus at the given temperature. In the present case we take the temperature dependence of the enthalpy and the entropy of the compound $A_m B_n$ to be minimal and neglect it.

The distinguishing feature of the preceding equation is that it contains both the parameters of the peritectic transformation p, ΔH_p, ΔS_p, as well as the thermodynamic characteristics of the formation of the phases whose initial crystallization occurs above the temperature of the peritectic reaction generating the compound $A_m B_n$. Consequently, precisely because of these characteristics, the form of Eq. (5.280) changes in the transition to the peritectic horizontal in the region of higher temperatures.

If above T_p we have crystallization of the incongruently melting compound $A_k B_l$ (Fig. 45), then the liquidus curve pp' will be described by an equation similar to (5.280), except that, instead of p, ΔH_p, and ΔS_p, it will contain the appropriate peritectic transformation parameters p', $\Delta H_p'$, and $\Delta S_p'$, and in place of $\Delta H_{f(A_k B_l)}{}^S$ and $\Delta S_{f(A_k B_l)}{}^S$, the parameters of the generation of the β phase. The parameters x and y in Eqs. (5.271) and (5.272) will change accordingly.

If the compound A_kB_l melts congruently, then clearly the peritectic transformation parameters in Eq. (5.280) will be absent, and in this case

$$T = \frac{\Delta H_f^S{}_{(A_kB_l)} - \omega^L \left[k\left(x_B^L\right)^2 + l\left(1 - x_B^L\right)^2\right]}{\Delta S_f^S{}_{(A_kB_l)} + R \ln\left(1 - x_B^L\right)^k \left(x_B^L\right)^l}, \tag{5.281}$$

i.e., we recover the equation of Brebrick, in which the activity coefficients are expressed in the regular solution approximation.

Finally, if the solid solution based on component B is stable and the solubility is insignificant, then clearly $k = 0$, $l = 1$, so that

$$T = \left[\Delta h_{m,\,B}^{\circ} + \omega^L\left(1 - x_B^L\right)^2\right] / \left(\Delta s_{m,\,B}^{\circ} - R \ln x_B^L\right). \tag{5.282}$$

It is easily seen that this equation is equivalent to Kamenetskaya's equation (5.72).

The foregoing approach to the liquidus curve of incongruently melting compounds enables one to estimate important thermodynamic parameters such as ΔH_p, ΔS_p, $\Delta H_{f(A_kB_l)}{}^S$, and $\Delta S_{f(A_kB_l)}{}^S$ from experimentally constructed state diagrams. The analysis described is quite general and does not depend on the choice of model solutions. Indeed, one may employ any model, including formal power series of any order. The corresponding changes made in Eq. (5.280) will be purely algebraic.

To conclude, let us illustrate the preceding method by the example of the Au–Pb system, in which there is a cascade of three peritectic reactions (Fig. 46) responsible for the generation of the compounds $AuPb_3$, $AuPb_2$, and Au_2Pb [218, 219].

By means of Eq. (5.280) we have calculated sections ep_1, p_1p_2, and p_2p_3 of the liquidus curve, corresponding to the initial crystallization of the above compounds. The results of the calculation are indicated in the diagram by the continuous line. The dashed line denotes results of a calculation from Brebrick's equation, and the points are data from experimental thermal analysis [227]. It is seen from Fig. 46 that there is good agreement between experimental data and the calculations.

From the state diagram we have calculated both the enthalpies of the peritectic reactions and the thermodynamic parameters of the generation of the compounds from the liquid components. The results are presented in Table 1.

In the last column of the table we quote from [229] the values of the free energy of formation of Au_2Pb and $AuPb_2$ from the liquid components at the appropriate temperature T_p. On comparing with the values of ΔG_f^S obtained in our calculation, we see that the agreement is fairly good, indicating that the choice of the model is acceptable and that the general approach is basically correct.

7. VARIOUS METHODS OF ESTIMATING THE EXCHANGE ENERGY FROM EXPERIMENTAL DATA

From our treatment of the various ways of calculating phase-diagram curves in systems with different sorts of interactions among the components, it should be abundantly clear that practically, for any choice of the model of the solution, it is of great importance to form a correct estimate of the exchange energy, the parameter which characterizes the role of the exchange forces between atoms.

The temperature and concentration dependence of ω along the liquidus curve has been constructed within the regular solution approximation, mostly in the study and analysis of experimentally constructed state diagrams [230–237]. It is in the nature of such constructions to provide, in the final analysis, only a qualitative idea of the character of the intermolecular interaction in the system, and of the degree of the latter's departure from regularity in some concentration interval. For more precise estimates and more reliable quantitative conclusions one must have recourse to a combination of different approaches, associated with the development of additional methods for estimating ω.

An analysis of the connection between the exchange energy and various physicochemical properties shows that the quantities ω_{ji} may be estimated from experimental data on the heat of mixing, the vapor pressure, solubility, the composition and temperature at azeotropic and eutectic points, critical temperature of demixing (consolute temperature), excess volume, etc.

We note that the majority of these methods of estimating ω require additional data, in addition to the basic measured quantities mentioned. The exceptions are methods based on measurements of the heat of mixing and vapor pressure.

We now describe briefly the various ways of obtaining ω, the method used being indicated by the section heading.

7.1. Heat of Mixing

Heats of mixing are determined accurately by calorimetric methods [238] which, though laborious, are quite widespread. By formula (4.104)

$$\omega = h^M/x_1 x_2. \tag{5.283}$$

Consequently, a single calorimetric measurement of the heat of mixing suffices to determine the exchange energy.

7.2. Vapor Pressure

For thermodynamic studies strain-gauge (tensometric) measurements are used [239]. If the molecular composition of the vapor phase is known, tensometric data enable one to estimate quite reliably the activity coefficients and, consequently, the exchange energy. From Eqs. (4.97), (4.98), and (4.59) we have

$$p_1 = p_1^\circ x_1 \exp\left(\omega x_2^2 \, RT\right), \tag{5.284}$$

$$p_2 = p_2^\circ x_2 \exp\left(\omega x_1^2 / RT\right). \tag{5.285}$$

Hence,

$$\frac{p_1 x_2}{p_2 x_1} = \frac{p_1^0}{p_2^0} \exp\left[\frac{\omega}{RT}\left(1 - 2x_1\right)\right]. \tag{5.286}$$

One sees from Eq. (5.286) that to determine the exchange energy it is sufficient to measure p_1 and p_2 for any composition excluding $x = 0.5$. Such a determination of the exchange energy has been found to agree well in a series of cases with data calculated from equilibrium phase diagrams [240–242].

7.3. Data on Electronegativity and Atomic Volumes of Components

Stringfellow [243] proposed a method for estimating the parameters of intermolecular interaction ω_{ji} in binary and quasibinary systems, based on the use of data on the electronegativity (X_i), atomic volume (v_i°), and solubility parameters (δ_i) of the components forming the system under study. Such an approach enables one to estimate ω_{ji} on the basis of quantities which are known and tabulated for simple substances. We shall derive a formula for calculating ω_{ji} in terms of these quantities, starting from the expression of Scatchard [91]. For simple liquids whose vapors are nearly ideal, the interaction u_{ii} may be identified with the evaporation energy $\Delta u_{i,v}^\circ$. Thus, one may write

$$\Delta u_{i,\,v}^\circ = c_{ii} v_i^\circ, \tag{5.287}$$

where c_{ii} is the cohesive energy density.

According to Pauling [244], the energy of interaction of opposite polar particles may be found from the equation

$$u_{12} = (u_{11} u_{22})^{1/2} + 3 \cdot 10^4 (\chi_1 - \chi_2)^2. \tag{5.288}$$

Using formula (5.287), we get from (5.288)

$$c_{12} = (c_{11} c_{22})^{1/2} + 3 \cdot 10^4 k (\chi_1 - \chi_2)^2 / (v_1^\circ v_2^\circ)^{1/2}, \tag{5.289}$$

where k is a constant whose value is determined by the difference in the screening constants of various kinds of solutions and by the mutual orientation of the interacting particles, since the contribution to the binding energy due to a difference in electronegativity is determined by Coulomb forces. Substituting (5.289) into Scatchard's equation [91], we find

$$h^M = (x_1 v_1^\circ + x_2 v_2^\circ) \left[(c_{11}^{1/2} - c_{22}^{1/2})^2 - 6 \cdot 10^4 k (\chi_1 - \chi_2)^2 / (v_1^\circ v_2^\circ)^{1/2} \right] \Phi_1 \Phi_2, \tag{5.290}$$

where we take

$$c_{ii} = \delta_i^2 \tag{5.291}$$

(the usual dimensionality of δ is given by $[\delta] = J^{1/2} M^{-5/2}$). Substituting h^M from (5.290) into (5.283) and noting (5.291), we obtain

$$\omega = [v_1^\circ v_2^\circ / (x_1 v_1^\circ + x_2 v_2^\circ)] \, [(\delta_1 - \delta_2)^2 - 6 \cdot 10^4 k \, (\chi_1 - \chi_2)^2 / (v_1^\circ v_2^\circ)^{1/2}]. \quad (5.292)$$

For compositions adjoining either one of the components, i.e., for relatively low values of x_i, relation (5.292) may be put into the form

$$\omega = v_i^\circ (\delta_1 - \delta_2)^2 - 6 \cdot 10^4 k \, (v_i^\circ / v_j^\circ)^{1/2} \, (\chi_1 - \chi_2)^2. \quad (5.293)$$

The value of k in (5.292) and (5.293) cannot be calculated theoretically, but is usually found by comparing (5.292) or (5.293) with values of ω estimated from other data (for example from vapor pressure or phase equilibrium).

The solubility parameter δ_i is usually calculated by means of

$$\delta_i = (\Delta h_{i,\,V}^\circ - RT) / v_i^\circ. \quad (5.294)$$

In estimating $\Delta h_{i,V}^\circ$ using strain-gauge data and the Clausius–Clapeyron equation, one should exercise great caution and use values of the latent heats of the transitions and equations for the temperature dependence of the vapor pressure of simple substances, carefully selected with the aid of a computer to be in accord with data based on the law of Hess and the periodic law of Mendeleev (cf. [245–247]).

By means of Eq. (5.292) we estimated [243] the value of the exchange energy in a series of binary systems comprising germanium and silicon, and compared them with results obtained on the basis of analyzing phase diagrams. The comparison shows that the method of estimating ω from electronegativity and atomic volume data provides quite accurate results.

We now consider the application of this method to quasibinary systems. We tackled such a problem for the particular systems Ge–InP and Ge–GaP [248].

In [249, 250] it was shown that the character of the phase transitions in the quasibinary systems Ge–In(Ga)P is described by eutectic-type state diagrams with insignificant solubilities in the solid state. Analysis of the liquidus curves within the regular solution approximation [249], as well as calculations [251] of the degree of ionization of phosphides along liquidus curves, have shown that the compounds are essentially dissociated in the liquid phase. This provides justification for treating the intermolecular interaction parameter for the systems Ge–In(Ga)P by the method described above (i.e., via electronegativity and atomic volume data, together with the theory of regular solutions), extended to three-component systems.

By analyzing pair interactions in the ternary system A–B–C, one may write the following relations, relating the temperature of the liquidus, the concentration in the ternary system, and the exchange energy in the corresponding binary systems:

$$RT \ln K_{AB} (T) = (\omega_{AB} x_B + \omega_{AC} x_C) \, (x_B + x_C) + (\omega_{AB} x_A + \omega_{BC} x_C)$$
$$\times (x_A + x_C) - \omega_{BC} x_B x_C - \omega_{AC} x_A x_C + RT \ln (x_A x_B), \quad (5.295)$$

$$RT \ln K_C (T) = (\omega_{AB} - \omega_{B\,(A)\,C}) \, x_{B\,(A)}$$
$$+ \omega_{A\,(B)\,C} (x_C - x_{A\,(B)}) + RT \ln (x_{A\,(B)} / x_C). \quad (5.296)$$

In these formulas

$$K_{AB}(T) = \gamma_A \gamma_B x_A x_B, \tag{5.297}$$

$$K_C(T) = \gamma_{A\,(B)} x_{A\,(B)} / \gamma_C x_C. \tag{5.298}$$

Equations (5.295) and (5.296) were written when the condition $x_A + x_B + x_C = 1$ is valid, and correspond to the initial crystallization of AB and C, respectively. The numerical values of $RT \ln K_{AB}(T)$ and $RT \ln K_C(T)$ are fixed from experimental data on the phase diagrams of the corresponding two-component systems by means of the equations

$$RT \ln K_{AB}(T) = \omega_{AB}\left(x_A^2 + x_B^2\right) + RT \ln\left(x_A x_B\right), \tag{5.299}$$

$$RT \ln K_C(T) = \omega_{A\,(B)\,C}\left(x_C - x_{A\,(B)}\right) + RT \ln\left(x_{A\,(B)}/x_C\right), \tag{5.300}$$

which hold with $x_A + x_B = 1$ and $x_{A(B)} + x_C = 1$.

Since in this particular case one considers quasibinary systems of the type C–(AB), one can write

$$x_A = x_B = x. \tag{5.301}$$

Using this condition, we may express Eqs. (5.295) and (5.296) in the form

$$2a_1 x\,(x-1) + a_2 = -2RT \ln x, \tag{5.302}$$

$$a_3 x + a_4 = RT \ln\left[(1-2x)/x\right]. \tag{5.303}$$

The coefficients a_1, a_2, a_3, and a_4 in these equations were calculated from data on the energy of mixing in the corresponding binary systems, as follows:

$$a_1 = 2\left(\omega_{AC} + \omega_{BC}\right) - \omega_{AB}, \tag{5.304}$$

$$a_2 = \omega_{AC} + \omega_{BC} - RT \ln K_{AB}(T), \tag{5.305}$$

$$a_3 = \omega_{AB} - \omega_{BC} - 3\omega_{AC}, \tag{5.306}$$

$$a_4 = \omega_{AC} - RT \ln K_C(T). \tag{5.307}$$

The values of the energy of mixing characterizing the pair interactions in the binary systems Ge–In, (Ga), (P), and In(Ga)–P were calculated by us with the aid of Eq. (5.292). The values of all the quantities entering into (5.292) required for this calculation were taken from [86, 244]. The calculation used binary state diagrams presented in [66, 204, 252, 253]. From data for the energy of mixing calculated by means of Eq. (5.292) we computed the coefficients a_1, a_2, a_3, and a_4. With these coefficients, Eqs. (5.302) and (5.303) were solved graphically for x at a particular temperature on the liquidus. The transition to mole fractions at each temperature of the quasibinary system C–AB (i.e., AB–InP and GaP, C–Ge) was effected by using the relation $x_{AB} = x/(1-x)$.

The results of the calculations are presented in Figs. 47 and 48, together with the experimental data taken from [249, 250]. As seen in Fig. 47, the results for the

Ge–GaP system are quite close to the experimental data obtained by thermal analysis [249].

The picture is somewhat more complicated for the Ge–InP system. The results of the calculation for the liquidus line, which describes the initial crystallization of the solid solution based on indium phosphide, agree well with the data of [249] which, however, differ sharply from the data of [250]. But for the same line appropriate to the solid solution based on germanium, the opposite picture is observed, although the scantiness of data in this temperature and concentration regime does not allow one to draw precise conclusions.

In view of the fact that in both of the experimental papers referred to, the starting materials, the conditions of the experiment, and the apparatus used were practically identical, the only explanation of the observed discrepancy between them lies in the differing degree of supercooling achieved in the corresponding concentration intervals, which may have been caused by differences in the thermal regimes of cooling. The thermodynamic calculations performed by utilizing the reference data on the atomic volumes and the electronegativity of the components is of help in tracing out and understanding the essentials of the equilibrium variant of the state diagram, indicated by the thick lines in Fig. 48.

The observed discrepancy between the calculations and the experimental data may be ascribed to the fact that the gallium and indium phosphides (especially the latter) are not completely dissociated in the solutions. This is indicated by the definite correlation between the amount of discrepancy and the degree of ionization, estimated approximately along the liquidus curves of the systems Ge–In(Ga)P [251].

7.4. ω from Viscosity Data

The possibility of estimating the energy of mixing in the liquid phase ω^L on the basis of data on the viscosity of binary or quasibinary solutions was considered in [254]. In accordance with [255], the diffusion coefficient in a solution is related to the activity coefficient by the following relation:

$$D_i = D_i^\circ (1 + \partial \ln \gamma_i^L / \partial \ln x_i^L), \tag{5.308}$$

where D_i° is the self-diffusion coefficient of the ith component. Assuming, following Moelwyn-Hughes [256], that Stokes' law applies, we shall have, in accordance with (5.308),

$$1/\eta = (1/\eta^{id}) (1 + \partial \ln \gamma_i^L / \partial \ln x_i^L). \tag{5.309}$$

Using the equation of Bachinskii [257], summing the quantity $c_i \eta_i^* x_i$ (where c_i is the constant in Bachinskii's equation, η_i^* is the fluidity, i.e., the inverse of the viscosity), and supposing, in addition, that $c_1 = c_2 = c$, we may write for the case of an ideal solution

$$1/\eta^{id} = x_1^L/\eta_1^\circ + x_2^L/\eta_2^\circ, \tag{5.310}$$

where η_1° and η_2° are the fluidities of the pure components. Substituting this expression into (5.309), we find

$$1/\eta = (x_1^L/\eta_1^\circ + x_2^L/\eta_2^\circ) (1 + \partial \ln \gamma_1^L / \partial \ln x_1^L). \tag{5.311}$$

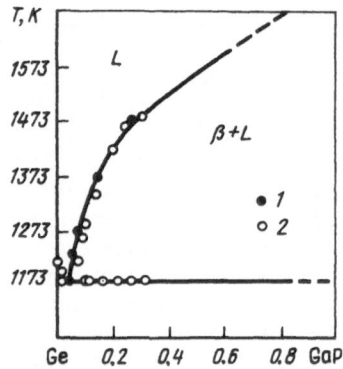

Fig. 47. State diagram of the Ga–GaP system: 1) calculation according to Eqs. (5.292) and (5.295); 2) experimental data of present authors.

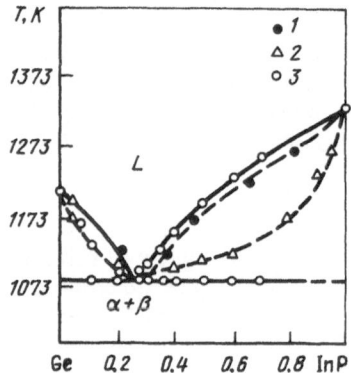

Fig. 48. State diagram of Ge–InP system: 1) calculation according to Eqs. (5.292), (5.295), and (5.296); 2) data from [250]; 3) experimental data of present authors.

In the simplest case, for the model of strictly regular solutions, we may write Eq. (5.311) in the form

$$1/\eta = \left(x_1^L/\eta_1^0 + x_2^L/\eta_2^0\right)\left[1 - 2\left(x_1 x_2 \omega^L\right)/RT\right], \qquad (5.312)$$

where we used Eq. (4.97).

Equation (5.312) enables one to estimate the energy of mixing ω^L on the basis of experimental data for the viscosity. This was demonstrated in the example of a series of sulfide systems in [254]. In conclusion, we note that in the derivation of Eq. (5.311) we made a series of assumptions, among which the least rigorous is the expression for the diffusion coefficient in terms of the viscosity. For a more rigorous approach a more accurate relation between these two quantities should be used (cf. [258, 259]).

The work of Esin et al. [260–262] indicated a way of estimating the energy of mixing from values of the molar volumes and the coefficient of surface tension.

7.5. X-Ray Data

The method of determining the parameter of intermolecular interaction in solid solutions of semiconductor and isovalent substitutes from x-ray data was proposed in the work of Bublik [263], where it was illustrated in the example of a quasibinary system formed by the compounds $A^N B^{B-N}$.

The Fourier component of the concentration fluctuation is connected in a macroscopic theory with the thermodynamic potential of the solid solution, in the following way [264, 265]:

$$1/<\left|\Delta x_{\vec{q}}\right|^2> = (1/RT)\left(\partial^2 g/\partial x^2 + \varphi_{xx}^y\right) \quad \text{for} \quad \vec{q} \to 0, \tag{5.313}$$

where $\vec{q}$ is the wave vector, and φ_{xx}^y is the second derivative with respect to concentration of the thermodynamic potential due to elastic forces. By measuring the intensity of diffuse scattering at the symmetry points of the reciprocal lattice ($+\vec{q}$, $-\vec{q}$), one may separate out the part of the scattering which is caused by statistical displacements and construct the dependence $\langle|\Delta x_{\vec{q}}|^2\rangle = f(\vec{q}^2)$. Extrapolation of the curve thus obtained to $|\vec{q}| = 0$ gives [263]

$$\omega(0) = kT\left[1/<\left|\Delta x_{\vec{q}}\right|^2> - (1/x_1 x_2)\right] \quad \text{for} \quad \vec{q} \to 0, \tag{5.314}$$

where $\omega(0)$ is the Fourier component of the energy of mixing.

Hence, by using (5.313), (5.314), (4.93), and (4.104), it is readily established that

$$\partial^2 g^E/\partial x_2^2 = \omega(0) - \varphi_{xx}^y. \tag{5.315}$$

If the cubic crystal lattice is assumed to be an isotropic elastic continuum, then with some approximation we may use the equation [263]

$$\varphi_{xx}^y = [2K(1-2\sigma)/3(1-\sigma)]\,[(3/a)(\partial a/\partial x)]^2, \tag{5.316}$$

where a is the lattice constant, σ is the Poisson ratio, $K = 1/3(G_{11} + 2G_{12})$, and G_{ij} are the elastic moduli of the solution.

Using for the function g^E the approximation of strictly regular solutions, we obtain an expression connecting the parameter of intermolecular interaction with quantities extracted from x-ray data:

$$\omega = [-\omega(0) + \varphi_{xx}^y]/2. \tag{5.317}$$

However, it should be pointed out that for alloys in which the crystal deformation is significant, the regular solution is not a reliable approximation.

For strictly regular solutions the following relation should apparently hold:

$$\omega = \omega\,(0)/2. \tag{5.318}$$

Therefore, the idea of Bublik, that the quantity ω be conceived as an additive sum of two components (elastic and "chemical"), should be recognized as a very rough approximation. For the more complex behavior of a real solution the connection between ω and x-ray parameters will be presumably more complicated.

8. CALCULATION OF THERMODYNAMIC PROPERTIES OF ALLOYS FROM DATA ON PHASE EQUILIBRIA

An intimate connection exists between the thermodynamic properties of alloys and their state diagrams. It was shown in Chapter 3 that knowledge of the concentration dependence of the free energy of alloys in a wide temperature interval enables one to construct the state diagram. But since the phase diagrams of the majority of binary systems have been well studied, they may be turned to advantageous use for calculating some thermodynamic parameters, for example the heat of melting of the components, the heat and entropy of formation of the alloys, the chemical potentials, or the activity coefficients of the components, etc. Numerous examples are known where one utilizes phase diagrams of alloys to determine their thermodynamic properties [266–270].

We shall now formulate some guidelines. For example, alloys that have a eutectic-type phase diagram, or a diagram with a region of miscibility in the solid or liquid states, will possess, as a rule, positive entropies and heats of formation; negative heats of formation are characteristic of alloys with intermediate phases [271].

These rules cannot be proved rigorously. However, as shown by Voronin [272], using a new class of thermodynamic functions introduced by him (the partial molar functions of heterogeneous mixtures), such correlations can be justified in a series of cases. Referring to this work, we may formulate the following rules:

1. If, in a system without intermediate phases, the region of the initial solid solution based on one of the components is negligibly small, then the solid solution based on the second component has a positive heat and entropy of formation, i.e.,

$$h^{M(\beta)} \geqslant 0, \qquad s^{M(\beta)} \geqslant 0. \tag{5.319}$$

This rule applies to alloys with any type of state diagram, having in the solid phase an immiscibility region adjoining the pure component.

2. In the composition region corresponding to the initial crystallization of the first component, we shall have

$$\bar{h}_1^{M(L)} + \Delta h_{1,\,m}^{\circ} \geqslant 0, \qquad \bar{s}_1^{M(L)} + \Delta s_{1,\,m}^{\circ} \geqslant 0, \tag{5.320}$$

where $\bar{h}_1^{M(L)}$ and $\bar{s}_1^L$ are the partial molar properties of formation of the liquid alloy (melt) pertaining to the liquidus curve with symmetric normalization.

3. In systems with a degenerate eutectic, one can indicate the limiting values of the integral functions of the liquid alloy:

$$h^{M(L)} \geqslant - (1 - x_2)\, \Delta h_{1,\, m}^{\circ}, \qquad s^{M(L)} \geqslant - (1 - x_2)\, \Delta s_{1,\, m}^{\circ}, \qquad (5.321)$$

where the index 1 refers to the more refractory component.

Since the quantity $(h^{M(L)} - Ts^{M(L)})$ is always negative, we may write for the entropy of formation of the alloy, for temperatures which are above the melting temperature of the first component, a "stronger" condition:

$$s^{M(L)} \geqslant - (1 - x_2)\, \Delta h_{1,\, m}^{\circ}/T. \qquad (5.322)$$

If the integral functions of formation of the alloy are positive, then inequalities (5.321) and (5.322) are always satisfied, whereas if they are negative then these inequalities restrict their magnitude.

The inequalities may also be justified for systems with a vanishing region of solid solutions near the first component, and whose concentration at the eutectic point is not far from the second component. This is because the continuation of the branch of the initial crystallization curve of the first component to the eutectic temperature leads to a state diagram which is exactly of the form considered.

4. In the systems where the eutectic point is roughly at mid-position on the concentration axis, and where the solid-state miscibility region is negligible for both components, we have

$$\left. \begin{array}{ll} h^{M(L)} \geqslant - (1 - x_2)\, \Delta h_{1,\, m}^{\circ}, & h^{M(L)} \geqslant - x_2\, \Delta h_{2,\, m}^{\circ}, \\[2mm] s^{M(L)} \geqslant - (1 - x_2)\, \Delta h_{1,\, m}^{\circ}/T, & s^{M(L)} \geqslant - x_2\, \Delta h_{2,\, m}^{\circ}/T. \end{array} \right\} \qquad (5.323)$$

As pointed out in [272], the preceding inequalities should also be obeyed by the thermodynamic functions h and s in systems with an immiscibility region in the liquid phase, as well as for systems with peritectically melting compounds.

The common deficiency of all known solutions to the problem of extracting thermodynamic properties from data based solely on phase equilibria is that some assumption is made about the nature of the temperature and concentration dependence of the thermodynamic functions of the solution. The most widespread approximation is the theory of regular solutions. But, as shown by experiment, many alloys have considerable excess entropies of formation.

In contrast, the method of calculating thermodynamic properties of two-component two-phase systems worked out by the present authors [139, 147, 273–276] does not require any preliminary information on the nature of the solution. The basis of the method, actually developed by Hiskes and Tiller [274–276], is a formal representation of the chemical potentials of the components in the form of a double Taylor series in temperature and concentration, with an arbitrarily chosen number of terms [cf. Eqs. (4.193) and (4.194)]. The calculation is reduced to determining the coefficients for each term of the series.

In order to present clearly the restrictions imposed on the coefficients β_{mn} in Eqs. (4.191) and (4.196), we expand $(T - T_0)^m$ in a binomial series:

$$(T - T_0)^m = \sum_{l=0}^{m} c_m^l\, (-1)^{m-l}\, T_0^{m-l} T^l. \qquad (5.324)$$

Substituting this into (4.196), we obtain

$$\sum_{l=0}^{M} T^l \sum_{m=l}^{M} \sum_{n=0}^{N} (1 - x_0)^n \beta_{mn} c_m^l (-1)^{m-l} T_0^{m-l} = RT. \tag{5.325}$$

Since Eq. (4.196), and hence also (5.325), should apply for all values of T, one may equate the coefficients of like powers of T on both sides of (5.325). We, therefore, obtain a system of $(M + 1)$ equations of the form

$$\sum_{m=l}^{M} \sum_{n=0}^{N} (1 - x_0)^n \beta_{mn} c_m^l (-1)^{m-l} T_0^{m-l} = \begin{cases} R & \text{for} \quad l = 1, \\ 0 & \text{for} \quad l \neq 1, \end{cases} \tag{5.326}$$

where $l = 0, 1, 2, ..., M$.

Analogously, from Eq. (4.191) we have

$$\sum_{m=l}^{M} \sum_{n=0}^{N} (-1)^{n+m-l} x_0^n \beta_{mn} c_m^l T_0^{m-l} = \begin{cases} R & \text{for} \quad l = 1, \\ 0 & \text{for} \quad l \neq 1, \end{cases} \tag{5.327}$$

where $l = 0, 1, 2, ..., M$.

Thus the limitations imposed on the coefficients β_{mn}^i by Eqs. (4.191) and (4.196) lead to a system of $2(M + 1)N$ linear equations containing $(M + 1)(N + 1)$ unknown parameters β_{mn}^i. The missing $(M + 1)(N - 1)$ equations may be obtained from the condition of phase equilibrium.

For each value of the temperature of the equilibrium coexisting phases α and β, we have two equations of phase equilibrium. This determines the minimum number of experimental data required for the model chosen, since determinacy requirements mean that the number of unknowns should not exceed the number of equations.

Substituting expressions (4.193) and (4.195) for the chemical potentials into the condition for phase equilibrium, and introducing the notation

$$f_1^i (m, n) = (-1)^n (T - T_0)^m \sum_{q=0}^{n-1} \frac{(-1)^q}{n - q} c_n^q (1 - x_0)^q [(1 - x_2^i)^{n-q} - 1], \tag{5.328}$$

$$f_2^i (m, n) = (T - T_0)^m \sum_{q=0}^{n-1} \frac{(-1)^q}{n - q} c_n^q x_0^q [(x_2^i)^{n-q} - 1]. \tag{5.329}$$

we obtain for each T the following additional equations:

$$\sum_{n=1}^{N} \sum_{m=0}^{M} (f_1^\alpha (m, n) \beta_{mn}^\alpha - f_1^\beta (m, n) \beta_{mn}^\beta) = RT \ln \left[\frac{(1 - x_2)^\beta}{(1 - x_2)^\alpha} \right] + g_1 (T), \tag{5.330}$$

$$\sum_{n=1}^{N} \sum_{m=0}^{M} (f_2^\alpha (m, n) \beta_{mn}^\alpha - f_2^\beta (m, n) \beta_{m, n}^\beta) = RT \ln \left(\frac{x_2^\beta}{x_2^\alpha} \right) + g_2 (T), \tag{5.331}$$

where

$$g_1(T) = \mu_1^{\circ\,(\beta)}(T, p) - \mu_1^{\circ\,(\alpha)}(T, p),$$

$$g_2(T) = \mu_2^{\circ\,(\beta)}(T, p) - \mu_2^{\circ\,(\alpha)}(T, p).$$

Using the well-known relation for the change of the chemical potential of a pure substance at a phase transition, of the form (5.142), we have

$$g_1(T) = \Delta s_{1,m}^\bullet (T - T_{1,m}^\circ) + \int_T^{T_{1,m}^\circ} \Delta c_{p,1}^\bullet\, dT - T \int_T^{T_{1,m}^\circ} \left(\frac{\Delta c_{p,1}^\circ}{T}\right) dT, \quad (5.332)$$

$$g_2(T) = \Delta s_{2,m}^\circ (T - T_{2,m}^\circ) + \int_T^{T_{2,m}^\circ} \Delta c_{p,2}^\bullet\, dT - T \int_T^{T_{2,m}^\circ} \left(\frac{\Delta c_{p,2}^\circ}{T}\right) dT, \quad (5.333)$$

where $\Delta c_{p,i}^\circ$ is the difference in the heat capacity of the ith component between the α and β phase at temperature T.

Relations (5.332) and (5.333) are only valid for systems having a continuous series of solid solutions, and for some eutectic-type systems. These relations will be replaced by others if the components form crystalline phases differing in structure.

As an example we consider a eutectic-type system in which the solid phase based on component 1 has a different structure from that of 2. Then we should choose the standard state of component 1 in the β phase to be the real state of component 1, and the standard state of component 2 – its state in the infinitely dilute β solution. On the contrary, the standard state of component 2 in the β' phase will be the state of the pure second component, while of component 1 – its state in the infinitely dilute β' state. We are therefore led to assume in this case that

$$\mu_1^{\circ\,(L)}(T, p) - \mu_1^{*\,(\beta')}(T, p) = A + BT + CT \ln T, \quad (5.334)$$

$$\mu_2^{\circ\,(L)}(T, p) - \mu_2^{*\,(\beta)}(T, p) = D + ET + FT \ln T, \quad (5.335)$$

where A, B, C, D, E, F are coefficients to be determined from the condition of chemical equilibrium, together with the β_{mn}^i.

The form of expressions (5.334) and (5.335) was chosen by analogy with the usual expression for $\mu_i^{\circ(L)} - \mu_i^{\circ(S)}$. Therefore, the solution of the system of equations comprising (4.191), (4.196), (5.330), and (5.331), using (5.332), (5.333) or (5.334), (5.335), enables one to calculate the parameters β_{mn}^i, and, consequently, the chemical potentials of the components. The latter serves as the basis for computing all other thermodynamic properties of the alloys of the system under study.

The method of Hiskes and Tiller described above may be generalized somewhat by writing the starting equation (4.186) in the form

$$x_2 [\partial \mu_2^i(x, T)/\partial x_2]_{T, p} = \sum_{m=0}^{M^i} \sum_{n=0}^{N_m^i} \beta_{mn}^i (T - T_0)^m (x_2 - x_0)^n, \quad (5.336)$$

where M^i depends on the nature of the phase i and $N_m{}^i$ depends, in addition, on the value of m. This allows one to use any admissible set of terms in the expansion, and to assume various models of solutions for the different phases.

In this case the number of unknowns is naturally no longer $2(N+1)(M+1)$, and, in addition, through equations of the form (5.326) and (5.327) linear dependences may appear, so that the number of constraints is reduced.

With the exception of a few cases the coefficients $\beta_{mn}{}^i$ do not have a physical meaning. We consider two examples. The coordinates of the initial points of the expansion are taken to be $x_0 = T_0 = 0$, $1 - x_0 = 1$.

A. A strictly regular solution:

$$M = 1, \quad N_0 = 2, \quad N_1 = 0.$$

In order to find relations connecting the various $\beta_{mn}{}^i$, we use Eqs. (5.326) and (5.327). From (5.326) it follows that

for $l = 0$ $\beta_{00} + \beta_{01} + \beta_{02} = 0,$ for $l = 1$ $\beta_{10} = R.$

From Eq. (5.327)

for $l = 0$ $\beta_{00} = 0,$ for $l = 1$ $\beta_{10} = R.$

We therefore have, finally,

$$\beta_{00} = 0, \quad \beta_{10} = R, \quad \beta_{01} = -\beta_{02}. \tag{5.337}$$

Substituting into (4.195) the values $M = 1, N_0 = 2, N_1 = 0, x_0 = T_0 = 0, 1 - x_0 = 1$, and using relation (5.337), we obtain for the chemical potential of the first component

$$\mu_1^i(T, x) = \mu_1^{\circ(i)}(T) + RT \ln(1 - x_2^i)$$
$$+ \sum_{m=0}^{1} \sum_{n=1}^{Nm} \beta_{mn}^i (-1)^n (T)^m \sum_{q=0}^{n-1} (-1)^q c_n^q [(1 - x_2^i)^{n-q} - 1]/(n - q)$$
$$= \mu_1^{\circ(i)}(T) + RT \ln(1 - x_2^i) + \beta_{02}^i x_2^2/2. \tag{5.338}$$

Similarly, it is easily verified from (4.193) and (5.337) that

$$\mu_2^i(T, x) = \mu_2^{\circ(i)}(T) + RT \ln x_2^i$$
$$+ \sum_{m=0}^{1} \sum_{n=1}^{Nm} \beta_{mn}^i T^m \sum_{q=0}^{n-1} (-1)^q c_n^q [(x_2^i)^{(n-q)} - 1]/(n - q)$$
$$= \mu_2^{\circ(i)}(T) + RT \ln x_2^i + \beta_{02}^i (1 - x_2^i)^2/2. \tag{5.339}$$

Equations (5.338) and (5.339) are analogous to the equations for the chemical potentials for strictly regular solutions, (4.97) and (4.98).

B. A quasiregular solution:

$$M = 1, \quad N_0 = 2, \quad N_1 = 2.$$

It is easily verified that in this case Eqs. (5.326) and (5.327) for $l = 0$ are similar to (5.337). For $l = 1$ it follows from (5.326) and (5.327) that

$$\beta_{10} + \beta_{11} + \beta_{12} = R, \quad \beta_{00} + \beta_{01} + \beta_{02} = 0,$$
$$\beta_{00} = 0, \quad \beta_{10} = R, \tag{5.340}$$

whence

$$\beta_{01} = -\beta_{02}, \quad \beta_{11} = -\beta_{12}. \tag{5.341}$$

Since the relations connecting β_{11} with β_{12} and β_{01} with β_{02} are identical, the chemical potentials may be obtained in this case by adding to expressions (5.338) and (5.339) the corresponding terms $T\beta_{12}x_2^2/2$ and $T\beta_{12}(1 - x_2)^2/2$,

$$\mu_1^i = \mu_1^{\circ (i)} (T) + RT \ln (1 - x_2^i) + (x_2^{(i)\,2}/2) (\beta_{02} + \beta_{12}T), \tag{5.342}$$

$$\mu_2^i = \mu_2^{\circ (i)} (T) + RT \ln x_2^i + [(1 - x_2^i)^2/2] (\beta_{02} + \beta_{12}T). \tag{5.343}$$

These equations are analogous to the expressions for the chemical potentials in the quasiregular solution approximation.

Equations (5.330) and (5.331) are ordinary linear equations in the coefficients β_{mn}^i, and to solve them one need only have the same number of equations as the number of unknowns [it should be borne in mind that there are constraints on the β_{mn}^i imposed by Eqs. (5.326) and (5.327)]. Nevertheless, one should utilize the largest possible number of experimental points in the complete temperature regime of the coexisting phases, in order to satisfy the following requirements: we want to be able to study the temperature dependence of the thermodynamic functions in a wide temperature interval, to extrapolate these dependences to the limits of the given interval, and to obtain values of the calculated parameters which are in accord with all known experimental data. Thus the number of equations of the type (5.330) and (5.331) may well exceed the number of unknowns and the solution will be located by one of the methods of mathematical statistics.

Brouwer and Oonk [139] worked out a method based on experimentally constructed $T - x$ state diagrams which enables one to calculate excess thermodynamic functions of mixing like the entropy and the enthalpy, as well as the excess heat capacity of solutions.

The method of Brouwer and Oonk is based, as the preceding one, on a linear regression model ensuing from the condition of the thermodynamic equilibrium of the phases. However, in contrast to the Hiskes–Tiller method, the excess thermodynamic functions are here obtained by expanding the function $g^E(T, x)$ in a Taylor series using the orthogonal polynomials of Redlich and Kister.

In accordance with [139], using relation (4.184) for the excess free energy, we shall obtain, via (4.85) for a eutectic-type system,

$$\mu_1^E (T, \ x) = x_2^2 \sum_{j=1}^{k} \omega_j (T) \{[2j - 1 - 2jx_2] (1 - 2x_2)^{j-2}\}, \tag{5.344}$$

$$\mu_2^E (T, \ x) = (1 - x_2')^2 \sum_{j=1}^{k} \omega_j (T) \, [(1 - 2jx_2') \, (1 - 2x_2')^{j-2}], \tag{5.345}$$

where the prime denotes parameters referring to the solution with a concentration $x_2 \geq x_{2(e)}$.

Here one may distinguish three variants: 1) a eutectic on the side of the first component, in which case Eq. (5.345) may serve as the regression function; 2) a eutectic on the side of the second component, in which case Eq. (5.344) may serve as the regression function; 3) the eutectic is near the center, when the regression functions are obtained by subtracting (5.344) from (5.345).

In the latter case, assuming for example $k = 2$, we obtain the following regression function [139]:

$$\mu_2^E (T, \ x') - \mu_1^E (T, \ x) = \{h_1^E - Ts_1^E + c_1^E \, [T - \theta$$

$$- T \ln (T/\theta)]\} \ [(1 - x')^2 - x^2] + \{h_2^E - Ts_2^E$$

$$+ c_2^E \, [T - \theta - T \ln (T/\theta)]\} \ [(1 - x')^2 \, (1 - 4x') - x^2 (3 - 4x)], \tag{5.346}$$

where h_j^E, s_j^E, and c_j^E are the constants in the expression for the temperature dependence of the parameter ω_j [cf. Eq. (4.185)]. For the demixing curve we have

$$\mu_1^E (x_I) - \mu_1^E (x_{II}) = RT \ln (1 - x_{II})/(1 - x_I),$$
$$\mu_2^E (x_I) - \mu_2^E (x_{II}) = RT \ln (x_{II}/x_I), \tag{5.347}$$

where the subscripts I and II denote the coexisting solutions. The regression function for this curve has the general form

$$f (x, \ T) = [\mu_2^E (x_{II}) - \mu_1^E (x_{II})]_T - [\mu_2^E (x_I) - \mu_1^E (x_I)]_T. \tag{5.348}$$

For a system with a continuous series of solid solutions the regression function has the form of expression (4.185).

The experimental values of the chemical potentials are estimated from the phase equilibrium data via

$$\mu_i^{E \ (S)} (T, \ x_i^S) = RT \ln (x_i^L/x_i^S) - \Delta s_{i, \ m}^\circ (T - T_{i, \ m}^\circ)$$

$$+ \Delta c_{p, \ i}^\circ [T - T_{i, \ m}^\circ - T \ln (T/T_{i, \ m}^\circ)] + \mu_i^{E \ (L)} (T, \ x_i^L). \tag{5.349}$$

This equation shows that in systems with a continuous series of solid solutions the chemical potential of the components in any case may be calculated, provided one knows the excess thermodynamic functions for the other phase.

From a fit of the experimental and calculated values of the quantities g^E or μ_i^E one may estimate the parameters of the model which describes the excess thermodynamic properties of the system under study.

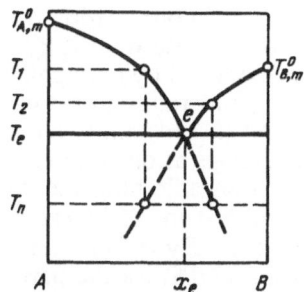

Fig. 49. Scheme for calculating the thermodynamic properties of alloys from phase diagrams by Voronin's method [273].

The calculation of isotherms of the thermodynamic properties of alloys was undertaken by Voronin [273]. Following his work we will show that one may calculate the partial and integral properties of alloys from phase diagrams with simple eutectics. The schematic form of the diagram is shown in Fig. 49, where x_e and T_e are the eutectic point coordinates, and T_1, T_2 are the temperatures of the corresponding branches of the liquidus curve. As the standard state for the solution, we take the state of the pure liquid components. The temperature dependence of the heat and entropy of mixing of the liquid alloys is assumed to be insignificant and is neglected.

Along the whole of the liquidus curve the alloy is in equilibrium with crystals of either the first or the second component, so that the magnitude of μ_i^M for concentrations and temperatures of the liquidus is equal to the change of the chemical potential of the corresponding components in the process of their nonequilibrium solidification. Utilizing this, we shall find expressions for the partial molar Gibbs free energy of mixing of the (saturated) liquid and solid solutions at the temperature T_e.

Recalling the expression

$$\mu_i = \bar{h}_i - T\bar{s}_i, \qquad (5.350)$$

obtained by differentiating (1.113) with respect to n_i at constant T and p, and noting that by the conditions of our treatment $\bar{h}_i^M$ and $\bar{s}_i^M$ do not depend on temperature, it is readily verified from (5.33a) that

$$\mu_1^M(x_2, \ T_e) = \mu_1^M(x_2, \ T_1) - \bar{s}_1^M(x_2) \ (T_e - T_1), \qquad (5.351)$$

$$\mu_2^M(x_2, \ T_e) = \mu_2^M(x_2, \ T_2) - \bar{s}_2^M(x_2) \ (T_e - T_2). \qquad (5.352)$$

Hence, inserting (5.33a), we obtain

$$\mu_1^M(x_2, \ T_e) = \Delta s_{m,\,1}^{\circ}\,(T_1 - T_{m,\,1}^{\circ}) - \bar{s}_1^M(x_2) \ (T_e - T_1), \qquad (5.353)$$

$$\mu_2^M(x_2, \ T_e) = \Delta s_{m,\,2}^{\circ}\,(T_2 - T_{m,\,2}^{\circ}) - \bar{s}_2^M(x_2) \ (T_e - T_2). \qquad (5.354)$$

These equations are valid in different composition regions. Equation (5.353) holds for $x_e \geq x_2 \geq 0$, while Eq. (5.354) is valid for $1 \geq x_2 \geq x_e$. But the limits of usefulness of these equations may be extended by extrapolating the known portions of the liquidus curve from the eutectic temperature to the temperature T_n. Therefore, near the point x_e, T_e, both (5.353) and (5.354) apply simultaneously. The accuracy of extrapolating the liquidus curve drops as one moves away from the composition x_e, so that the portions of the diagram adjoining the pure component are not used in these calculations. Equations (5.353) and (5.354) contain four unknown functions: $\bar{s}_1^M$, $\bar{s}_2^M$, $\mu_1^M(x_2, T_e)$, $\mu_2^M(x_2, T_e)$. By differentiating (5.353) and (5.354) with respect to composition and using (1.16) for the partial molar entropies of mixing, one may eliminate the latter from the system and obtain an expression relating the $\mu_1^M(x_2, T_e)$, $\mu_2^M(x_2, T_e)$, and their compositional derivatives. Furthermore, by using relations (4.12) and (4.13), it is not difficult to replace $\mu_1^M(x_2, T_e)$ and $\mu_2^M(x_2, T_e)$ by $g^M(x_2, T_e)$. As a result of these transformations one obtains the equation [273]

$$
P(x_2) \frac{d^2 g^M(x_2, T_e)}{dx_2^2} + Q(x_2) \left[\frac{dg^M(x_2, T_e)}{dx_2} \right.
$$
$$
\left. - \left(\frac{dg^M(x_2, T_e)}{dx_2} \right)_{x_2 = x_e} \right] + F(x_2) [g^M(x_2, T_e) - g^M(x_e, T_e)] = 0, \qquad (5.355)
$$

where the coefficients

$$
P(x_2) = T_1 - T_2, \qquad (5.356)
$$

$$
Q(x_2) = \frac{(T_2 - T_e)}{(T_1 - T_e)} \frac{dT_1}{dx_2} - \frac{(T_1 - T_e)}{(T_2 - T_e)} \frac{dT_2}{dx_2}, \qquad (5.357)
$$

$$
F(x_2) = - \left[\frac{1}{x_2} \frac{(T_2 - T_e)}{(T_1 - T_e)} \frac{dT_1}{dx_2} + \frac{1}{1 - x_2} \frac{(T_1 - T_e)}{(T_2 - T_e)} \frac{dT_2}{dx_2} \right], \qquad (5.358)
$$

and the quantities

$$
g^M(x_e, T_e) = (1 - x_e)(T_e - T_{m,1}^\circ) \Delta s_{m,1}^\circ + x_e (T_e - T_{m,2}^\circ) \Delta s_{m,2}^\circ, \qquad (5.359)
$$

$$
[dg^M(x_2, T_e)/dx_2]_e = (T_e - T_{m,2}^\circ) \Delta s_{m,2}^\circ - (T_e - T_{m,1}^\circ) \Delta s_{m,1}^\circ \qquad (5.360)
$$

are completely determined by the form of the liquidus curve and the properties of the pure components; Eqs. (5.359), (5.360) are the initial conditions of integration.

In addition to these conditions the integral curve is also given at the boundaries of the interval (0, 1):

$$
g^M(0, T_e) = g^M(1, T_e) = 0, \quad -(dg^M/dx_2)_{x_2=0} = (dg^M/dx_2)_{x_2=1} = \infty. \qquad (5.361)
$$

The foregoing method of estimating thermodynamic properties was illustrated in [273] for the Bi–Cd system, by using a point-sampling method [277]. The calculated mixing functions of the Bi–Cd alloys were shown to be in good agreement with experimental data [273].

Proks [278] proposed a method for calculating the excess thermodynamic functions of mixing g^E and s^E by using isobaric data on phase equilibria in the binary system, together with the heat of mixing. We shall consider two cases.

1. Calculation of the excess entropy of solutions in a eutectic-type system with negligibly small mutual solubility of the components. The condition for the phase equilibrium of the components may be written in this case as

$$\mu_A^L (x, \ T) = \mu_A^{\circ (S)} (T), \qquad \mu_B^L (x', \ T) = \mu_B^{\circ (S)} (T),$$

where the primes denote parameters referring to solutions with concentration $x_B{}^L \geq x_{B(e)}{}^L$.

Hence, for solutions with concentrations $x_B{}^L \leq x_{B(e)}{}^L$ we may write, in view of (5.350),

$$\bar{s}_A^L (T) = [\bar{h}_A^L (x, \ T) - h_A^{\circ (S)} (T)]/T + s_A^{\circ (S)} (T).$$

The partial molar entropy of component A at the temperature T^e, for which experimental heat of mixing data exist, may be estimated by the formula

$$\bar{s}_A^L (T^e) = [\bar{h}_A^L (x, \ T) - h_A^{\circ (S)} (T)]/T + s_A^{\circ (S)} (T)$$
$$+ \int_{T}^{T^e} [\bar{c}_{p,A}^L (x, \ T)/T] \, dT, \tag{5.362}$$

where $\overline{c}_{p,A}$ is the partial heat capacity of component A in the solution. The partial molar entropy of mixing of the A component at temperature T^e is therefore determined from the equation

$$\bar{s}_A^{M (L)} (x, \ T^e) = \bar{s}_A^L (x, \ T^e) - s_A^{\circ (L)} (T^e)$$
$$= [\bar{h}_A^L (x, \ T) - h_A^{\circ (S)} (T)]/T + \int_{T}^{T^e} [\bar{c}_{p, \ A}^L (x, \ T)/T] \, dT$$
$$- \int_{T}^{T^{\circ}_{m, A}} [c_{p, \ A}^{\circ (S)} (T)/T] \, dT - \int_{T^{\circ}_{m, A}}^{T^e} [c_{p, A}^{\circ (L)}/T] \, dT - \Delta h_{m, \ A}^{\circ} (T^{\circ}_{m, A})/T^{\circ}_{m, A}. \tag{5.363}$$

After some manipulations this relation leads to the form [278]

$$\bar{s}_A^{M (L)} (x, \ T^e) = [\bar{h}_A^{E (L)} (x, \ T) + \Delta h_{m, \ A}^{\circ} (T)]/T$$
$$+ \int_{T}^{T^e} [\bar{c}_{p, \ A}^{E (L)} (x, \ T)/T] \, dT + \int_{T}^{T^{\circ}_{m, A}} [\Delta c_{p, \ A}^{\circ} (T)/T] \, dT$$
$$- \Delta h_{m, \ A}^{\circ} (T^{\circ}_{m, A})/T^{\circ}_{m, A}. \tag{5.364}$$

Similarly, one may calculate the partial molar entropy of mixing of the B component for solutions with concentration $x_B{}^L \geq x_{B(e)}{}^L$:

$$\bar{s}_B^{M\,(L)}\left(x',\ T^e\right) = [\bar{h}_B^{E\,(L)}\left(x',\ T\right) + \Delta h_{m,\,B}^{\circ}(T)]/T$$

$$+ \int_{T}^{T^e} \left[\bar{c}_{p,\,B}^{E\,(L)}(T)/T\right] dT + \int_{T}^{T_{m,\,B}^{\circ}} [\Delta c_{p,\,B}^{\circ}(T)/T]\,dT - \Delta h_{m,\,B}^{\circ}\left(T_{m,\,B}^{\circ}\right)/T_{m,\,B}^{\circ}. \quad (5.365)$$

The quantities $\bar{s}_B^{M(L)}(x,\ T^e)$ and $\bar{s}_A^{M(L)}(x,\ T^e)$ are calculated for compositions $x_B^L \leq x_{B(e)}^L$ and $x_B^L \geq x_{B(e)}^L$ from the Gibbs–Duhem relation:

$$\bar{s}_B^{M\,(L)}\left(x,\ T^e\right) = \bar{s}_B^{M\,(L)}\left(x_e\right) - \int_{\bar{s}_A^{M\,(L)}(x_e)}^{\bar{s}_A^{M\,(L)}(x)} (1 - x_B)\,x_B^{-1}\,d\bar{s}_A^{M\,(L)}(x),$$

$$\bar{s}_A^{M\,(L)}\left(x,\ T^e\right) = \bar{s}_A^{M\,(L)}\left(x_e\right) - \int_{\bar{s}_B^{M\,(L)}(x_e)}^{\bar{s}_B^{M\,(L)}(x)} (1 - x_B)^{-1}\,x_B\,d\bar{s}_B^{M\,(L)}(x).$$

The excess entropy of the binary solution A–B at temperature T^e may be determined from the equation

$$s^{E\,(L)}\left(x,\ T^e\right) = \left(1 - x_B^L\right) \bar{s}_A^{M\,(L)}\left(x,\ T^e\right) + x_B^L \bar{s}_B^{M\,(L)}\left(x,\ T^e\right) \\ + R\left[(1 - x_B^L)\ln\left(1 - x_B^L\right) + x_B^L\ln x_B^L\right]. \quad (5.366)$$

2. Calculation of the excess entropy of solutions in eutectic-type systems with mutually soluble components in the solid state. In this case the phase equilibrium conditions for components A and B assume the form

$$\mu_A^L\left(T,\ x^L\right) = \mu_A^S\left(T,\ x^S\right), \qquad \mu_B^L\left(T,\ x'^{(L)}\right) = \mu_B^S\left(T,\ x'^{(S)}\right)$$

or, because of relations (5.350) and (4.2),

$$\bar{h}_A^{M\,(L)}\left(x,\ T\right) - T\bar{s}_A^{M\,(L)}\left(x,\ T\right) = h_A^{\circ\,(S)}(T) - T s_A^{\circ\,(S)}(T) + RT\ln a_A^S;$$
$$\bar{h}_B^{M\,(L)}\left(x,\ T\right) - T\bar{s}_B^{M\,(L)}\left(x,\ T\right) = h_B^{\circ\,(S)}(T) - T s_B^{\circ\,(S)}(T) + RT\ln a_B^S.$$

As in the first case, we may obtain equations for calculating the partial molar entropy of mixing of A and B in the liquid solution, whose form differs from Eq. (5.364) or (5.365) only by the extra terms $-R\ln(1 - x_B^S)$ and $-R\ln x_B'^S$, respectively. The subsequent stages of the analysis in estimating $S^{E(L)}$ are analogous to the exposition in the first variant.

For the calculation of the quantity $g^{E(L)}(T^e)$ in both cases, one may use the well-known thermodynamic relation

$$g^E = h^E - T s^E.$$

For systems with congruently and incongruently melting chemical compounds, one may estimate the values of the thermodynamic functions of the formation of intermediate phases on the basis of phase diagrams.

It follows from Eq. (5.202) for the liquidus curve of the compound $A_m B_n$ that by having available data on the activity coefficients of the components in saturated solutions of corresponding compositions, one may calculate the enthalpy and entropy of formation of the compound, ΔH_f^S and ΔS_f^S.

In the absence of data on the thermodynamic properties of alloys of the given system, the availability of sufficiently accurate data on phase equilibrium enables one to estimate γ_i to a good approximation from the best-fit condition between the theoretical liquidus curve and its experimental construction. In addition, by neglecting the dependence of ΔH_f^S and ΔS_f^S on temperature in the range under study, we may solve simultaneously a set of linear equations such as (5.202) by the method of least squares, for example, and thereby calculate the mean values of the heat and entropy of formation of the compound considered. Thus, for example, using an expression of the form (5.195) for the analysis of the liquidus curve of a congruently melting compound, one may calculate the heat and entropy of formation of the solid compound from the liquid components by means of the following equations:

$$\Delta H_f^S = [mn/(m+n)]\,\omega_0 + [mn^2/(m+n)^2]\,\omega_1 +$$
$$+ [mn^3/(m+n)^3]\,\omega_2 - \Delta h_{m,\;A_m B_n}^\circ, \tag{5.367}$$

$$\Delta S_f^S = [mn/(m+n)]\,\omega_{01} + [mn^2/(m+n)^2]\,\omega_{11} +$$
$$+ [mn^3/(m+n)^3]\,\omega_{21} - \Delta s_{m,\;A_m B_n}^\circ + R \ln m^m n^n/(m+n)^{m+n}. \tag{5.368}$$

These include as particular cases other simpler models of solution.

In conclusion, let us show how one may calculate the thermodynamic characteristics of alloys by constructing the function

$$\ln\,(x_i^L)_{\text{sat}} = f\,(1/T_{\text{liq}}).$$

From the condition of chemical equilibrium we have $\mu_i^S = \mu_i^L$ or, in expanded form,

$$\mu_i^{\circ\,(S)} - \mu_i^{\circ\,(L)} = \Delta s_{m,\;i}^\circ\,(T - T_{m,\;i}^\circ) = RT \ln\,[x_i^L \gamma_i^L / x_i^S \gamma_i^S]. \tag{5.369}$$

For negligible solubilities in the solid phase the activity a_i^S may be taken to be equal to unity. In this case Eq. (5.369) may be separated into entropy and heat components.

Substituting expression (5.350) into Eq. (5.369), we obtain

$$\ln\,(x_i^L)_{\text{sat}} = -\,(\Delta \bar{h}_i^{(S \to L)}/RT) + (1/R)\,(\Delta \bar{s}_i^{(S \to L)}). \tag{5.370}$$

Here

$$\Delta \bar{h}_i^{(S \to L)} = \Delta h_{m,\;i}^\circ + \bar{h}_i^{E\,(L)}, \tag{5.371}$$

$$\Delta \bar{s}_i^{(S \to L)} = \Delta s_{m,\,i}^{\circ} + \bar{s}_i^{E\,(L)}, \tag{5.372}$$

where $\Delta \bar{h}_i^{(S \to L)}$ and $\Delta \bar{s}_i^{(S \to L)}$ are the limiting partial molar heat and entropy of the solution of the ith component.

Equation (5.370) implies that if $\Delta \bar{h}_i^{(S \to L)}$ and $\Delta \bar{s}_i^{(S \to L)}$ are independent of the temperature, then the limiting molar heat of solution of the ith substance may be determined by the slope of the line $\ln(x_i^L)_{\text{sat}} = (1/T_{\text{liq}})$, while from the intercept on the ordinate at $1/T_{\text{liq}} = 0$ one may compute $\Delta \bar{s}_i^{(S \to L)}$. In addition, knowing the heat and entropy of melting of the ith component, it is simple to calculate via (5.371) and (5.372) the partial molar enthalpy of mixing and the excess partial molar entropy of mixing of the ith substance in the liquid phase.

As pointed out by the present authors in [279], the straight-line dependence of $\ln(x_i^L)_{\text{sat}}$ vs. $1/T_{\text{liq}}$ is only observed at low x_i^L concentrations, i.e., for very large dilutions (<2 at. %). This means that the assumed independence of $\Delta \bar{h}_i^{(S \to L)}$ and $\Delta \bar{s}_i^{(S \to L)}$ from temperature is not strictly valid. Therefore, the foregoing calculation is valid only for very dilute solutions ($x_{\text{sat}}^L \to 0$).

As shown by Malinowski [280], the partial molar enthalpy and entropy of the components may also be estimated from the slope of the liquidus curve at the eutectic point in the following manner.

For a eutectic system with negligible mutual solubility of components we have

$$d \left(\mu_i^{\circ\,(S)} / T \right)_p = d \left(\mu_i^L / T \right)_p. \tag{5.373}$$

Expanding the total differentials in Eq. (5.373), we obtain

$$RT^2 \left(\partial \ln a_i / \partial x_i \right)_{p,\,T} = \Delta \bar{h}_i^{(S \to L)} \left(\partial T / \partial x_i \right)_p. \tag{5.374}$$

Writing expression (5.374) for each of the components, we obtain, after some simple transformations,

$$\begin{aligned} &x_1 \Delta \bar{h}_1^{(S \to L)} k_1 - x_2 \Delta \bar{h}_2^{(S \to L)} k_2 \\ &= RT^2 \left[x_1 \left(\partial \ln a_1 / \partial x_1 \right) - x_2 \left(\partial \ln a_2 / \partial x_2 \right) \right]_{p,\,T}, \end{aligned}$$

where $k_i \equiv (\partial T / \partial x_i)p$.

In view of the Gibbs–Duhem relation we have at the eutectic point

$$\left(x_1 \Delta \bar{h}_1^{(S \to L)} k_1 \right)_e = \left(x_2 \Delta \bar{h}_2^{(S \to L)} k_2 \right)_e. \tag{5.375}$$

Using the formula $\Delta \bar{h}_i^{(S \to L)} = T \Delta \bar{s}_i^{(S \to L)}$, a similar expression may be written for the partial molar entropy of mixing of the components at the eutectic point:

$$\left(x_1 \Delta \bar{s}_1^{(S \to L)} k_1 \right)_e = \left(x_2 \Delta \bar{s}_2^{(S \to L)} k_2 \right)_e. \tag{5.376}$$

With the aid of the expression

$$\Delta h_e^{(S \to L)} = \left(x_1 \Delta \bar{h}_1^{(S \to L)} \right)_e + \left(x_2 \Delta \bar{h}_2^{(S \to L)} \right)_e, \tag{5.377}$$

where $\Delta h_e^{(S \to L)}$ is the integral molar heat of mixing of the eutectic alloy, we may obtain yet another relation, which may be applied in practical calculations (cf. [280]).

From Eqs. (5.375) and (5.377) it is easily verified that

$$(\Delta \bar{h}_1^{(S \to L)})_e = [k_2/(x_1 k_1 + x_1 k_2)] \, \Delta h_e^{(S \to L)}; \tag{5.378}$$

$$(\Delta \bar{h}_2^{(S \to L)})_e = [k_1/(x_2 k_1 + x_2 k_2)] \, \Delta h_e^{(S \to L)}. \tag{5.379}$$

At the eutectic temperature

$$\Delta h_{i,m}^\circ (T_e) = \Delta h_{i,m}^\circ (T_{i,m}^\circ) - \int_{T_e}^{T_{i,m}^\circ} \Delta c_{p,i}^{\circ\,(S \to L)} \, dT. \tag{5.380}$$

Bearing in mind (5.380) and (5.371), we obtain via (5.378)–(5.379) expressions enabling one to calculate the partial molar heat of mixing of the components at the eutectic point from the slope of the liquidus curve and the properties of the pure components:

$$(\bar{h}_1^{E\,(L)})_e = [k_2/(x_1 k_1 + x_1 k_2)] \, \Delta h_e^{(S \to L)} - \Delta h_{1,m}^\circ (T_{1,m}^\circ) + \int_{T_e}^{T_{1,m}^\circ} \Delta c_{p,1}^{\circ\,(S \to L)} \, dT,$$

$$(\bar{h}_2^{E\,(L)})_e = [k_1/(x_2 k_1 + x_2 k_2)] \, \Delta h_e^{(S \to L)}$$

$$- \Delta h_{2,m}^\circ (T_{2,m}^\circ) + \int_{T_e}^{T_{2,m}^\circ} \Delta c_{p,2}^{\circ\,(S \to L)} \, dT. \tag{5.381}$$

By the same process as above, we can use Eq. (5.372) and the relation $d\Delta s_{i,m}^\circ = \Delta c_{p,i}^{\circ (S \to L)} d \ln T$ to obtain the following equation for computing the partial molar entropy of mixing of the components at the eutectic point, in terms of the slope of the liquidus curve and the properties of the pure components:

$$(\bar{s}_1^{E\,(L)})_e = [k_2/(x_1 k_1 + x_1 k_2)] \, \Delta s_e^{(S \to L)} - \Delta s_{1,m}^\circ (T_{1,m}^\circ) + \int_{T_e}^{T_{1,m}^\circ} \Delta c_{p,1}^{\circ\,(S \to L)} \, d\ln T,$$

$$(\bar{s}_2^{E\,(L)})_e = [k_1/(x_2 k_1 + x_2 k_2)] \, \Delta s_e^{(S \to L)}$$

$$- \Delta s_{2,m}^\circ (T_{2,m}^\circ) + \int_{T_e}^{T_{2,m}^\circ} \Delta c_{p,2}^{\circ\,(S \to L)} \, d\ln T. \tag{5.382}$$

9. CHOICE OF MODEL SOLUTION BASED ON MATCHING WITH EXPERIMENTAL PHASE EQUILIBRIUM DATA

The mathematical problem of choosing a model solution on the basis of comparing residual dispersions often arises in the modeling of phase transitions, in predicting thermodynamic properties, in studying the character of intermolecular interactions in nonideal systems, etc.

The analysis carried out by Ansara [281] and Bennet et al. [282] showed that the number of systems whose phase equilibria have been studied considerably exceeds the number for which thermodynamic properties of the phases are known. It is therefore of great interest to attempt to determine the parameters on the basis of experimental phase equilibrium data.

In the thermodynamic analysis of phase diagrams wide use is made of the models of ideal, regular and subregular solutions, the quasichemical approximation, and a series of other approaches. Among the criteria allowing one to choose the best agreement between the curves of phase equilibria as calculated by the various models and experimental data, one may pick out two fundamental ones:

1) the mathematical criterion of Fisher (F criterion), which appears to be better from the statistical point of view [283–285];

2) the physicochemical criterion, which includes in the calculation not only the curves of phase equilibria, but also the thermodynamic characteristics of solutions and compounds (for example, the chemical potentials of the components, the enthalpy and excess entropy of solution mixing, the total vapor pressure, the partial vapor pressures for the components, the thermodynamic functions of formation of the compound, etc.). As a result, one can establish for which model the agreement between the calculated and experimental data is most satisfactory.

The problem of determining the parameters reduces to an extremal problem of minimizing some function, which is the criterion enabling one to compare experimentally measured and model-calculated quantities.

The choice of the criterion reflects substantially on estimates of the errors in the physicochemical constants figuring in the original thermodynamic or mathematical model. Since every experiment is accompanied by random errors, one may use various methods of mathematical statistics in selecting the model, for example the most probable value, the least square, Bayes' method, the Chebyshev approximation (minimax method), etc. [283, 285–292].

If the measurements are independent of one another, and if the distribution law is known, the most fundamental form of the criterion is the principle of maximum likelihood. Thus, if the errors are distributed normally, we are led to minimize the sum of squared deviations (the method of least squares) [287–289]. If the distribution is Laplacian, we have to minimize a simple sum of moduli of deviations. In recent years the number of works in which experimental data are treated by the method of least squares (MLS) is on the increase. The method of least squares is realized by searching for those parameters $\vec{\theta}$ for which the "weighted" sum of squared deviations has a minimum:

$$\min_{\vec{\theta},\, \vec{x}_i} SQ = \min_{\vec{\theta},\, \vec{x}_i} \sum_{i=1}^{N} \left(\vec{x}_i^{\,e} - \vec{x}_i \right)^T M_i^{-1} \left(\vec{x}_i^{\,e} - \vec{x}_i \right), \qquad (5.383)$$

where $\vec{x}_i^{\,e}$ is a vector comprising k quantities measured in the ith experiment, and includes the random errors; $\vec{x}_i$ is a vector of the same quantities but free of random errors; M_i is the $k \times k$ covariance matrix of the measured quantities; the index T denotes the transposition operation. One should take into account the functional connection between the actual values of the measured variables for each observation, which include the parameters to be estimated. It is assumed that the matrix M_i is diagonal, and that its elements represent the dispersion of the measured variables.

The existing estimation methods have statistical foundation only for cases in which the model used is a linear function with respect to the unknown parameters [289, 292, 295]. The conditions under which the MLS estimates represent the optimum linear estimates of the parameters [287, 293] (and if the distribution of deviation is close to normal these estimates would also be the most probable values [294]) are the following: 1) there are no experimental errors in the determination of the independent (controlled) variables; 2) the experimental errors of measurement of the independent variables (response functions) have a Gaussian distribution with a zero mean value; 3) the results of observation do not depend on one another for a given experimental point and for different points; 4) the model for describing the response function is adequate, in the sense that it enables one to reproduce the actual values of the variables; 5) the equations of constraint are linear with respect to the selecting parameters $\overline{\theta}$.

For binary, equilibrium, three-phase condensed systems, the measured quantities are the pressure p, the temperature T, and the mole fractions of the components in the liquid, solid, and gaseous phases, x^L, x^S, and x^V, respectively. From the point of view of mathematical statistics one cannot, in the present instance, separate exactly the independent variables and the deviations, since all quantities are defined experimentally, so that they also include the observational errors. Meanwhile, in accordance with the Gibbs phase rule, it is enough to specify only one variable for a complete description of a three-phase binary system. Consequently, from the point of view of thermodynamics, only one of the measured quantities characterizing the three-phase equilibrium in the binary system ought to be considered independent. In condensed two-phase systems, where pressure has little effect on equilibrium, we again have that only one of the measured parameters T, x^L, or x^S can be considered independent. In binary two-phase liquid–vapor systems, two quantities may be considered independent. Therefore, in estimating the parameters of solution models, on the basis of matching them with a complete set of data on phase equilibria in binary two-phase systems, two restrictions are involved.

As independent variables one chooses the molar fractions of the components in the liquid phase, to be determined with the greatest accuracy. In practice, the requirement of determinacy for the controlled variables is frequently not satisfied. If the uncertainty in fixing the independent factors is not large, then use of the usual methods of regression analysis does not lead to serious errors. In the opposite case, the estimated values of the parameters may differ significantly from the actual quantities. In particular, estimates obtained by least squares methods – without accounting for the errors in the definition of the controlled variables – may turn out to be either strongly distorted, or simply untenable [293].

An optimization criterion of the form (5.383) for the estimation of parameters is employed, in principle, in those cases in which the model is nonlinear with respect to the unknown parameters (cf., for example, [283, 288, 284]). In these situations one frequently attempts to reduce the original problem of nonlinear optimization by means of repeated simple transformations to a linear regression analysis. Such a scheme may be accomplished, for example, by using exponential or logarithmic functions.

The formal use of such manipulations of the observational errors of variables usually leads to a situation in which the magnitude of the parameters found by the simplest method is rather different from the values that would be obtained by solving the original problem. For normal measurement errors the parameters obtained by means of such transformations may even have the wrong sign [283]. The most frequently used process of linearization by means of a Taylor series may also lead, because of errors arising out of the truncation of the series, to a substantially different solution from the best estimate. In principle, the scanning of the MLS parameters may also be performed without a preliminary linearization [296–298].

For a practical treatment of experimental results it is important to know which method is the most effective in searching for the SQ extremum [297].

We rewrite the function to be minimized in (5.383) by introducing the concept of dependent and independent variables, as described above. To this end we should first express the dependent variables of the system in terms of the independent ones by means of two constraint equations in explicit form:

$$\vec{Y} = \vec{f}(\vec{x},\ \vec{\theta}); \qquad \vec{Z} = \vec{q}(\vec{x},\ \vec{\theta}), \tag{5.384}$$

where $\vec{Y}$ and $\vec{Z}$ are vectors of the actual values of the dependent variables Y and Z; $\vec{Y} = \{Y_i\}_{i=1}{}^N$ and $\vec{Z} = \{Z_i\}_{i=1}{}^N$; $\vec{x}$ represents a vector of dimensionality $N(k-2)$; $(k-2)$ is the total number of independent variables, which in the present case equals two or one.

Each set of $(k-2)$ elements from $\vec{x}$ may be represented by a vector $\vec{x}_i$, corresponding to the ith experiment.

The assumption that the errors of the measured variables are uncorrelated, i.e., that the matrix $M_i{}^{-1}$ is diagonal, allows one to write the criterion (5.383) in the form

$$\min_{\vec{\theta},\, x_j^i} SQ = \min \sum_{i=1}^N \left\{ \frac{1}{\sigma_{Y_i}^2}(Y_i - Y_i^e)^2 + \frac{1}{\sigma_{Z_i}^2}(Z_i - Z_i^e) \right.$$
$$\left. + \sum_{j=1}^{k-2} \frac{1}{\sigma_{x_j^i}^2}(x_j^i - x_j^{ie})^2 \right\}, \tag{5.385}$$

where $\sigma_{x_j^i}^2$ are the dispersion of errors in the determination of the independent variables, and $\sigma_{Y_i}{}^2$, $\sigma_{Z_i}{}^2$, the corresponding quantities for the responses.

The algorithms in use for selecting the minimum of criterion (5.385) are based, as a rule, on a linearization of the coupled equations by expanding them in a Taylor series of the parameters and the actual values of the variables, and on a transition from a constrained to an unconstrained problem [299].

Certain difficulties arise in selecting the minima of functionals like (5.385). Among these are the determination of the initial approximations for the parameters, the numerical calculation of derivatives with respect to the parameters, and the basic problem of convergence, which plays a decisive role in the quasilinearization procedure [284].

One usually achieves very fast convergence by using good initial estimates of the parameters and the actual values of the measured variables. If these initial choices are poor, convergence is slow or nonexistent.

A series of papers [285, 300] has put forward a method for estimating the parameters of models, based on the principle of maximum likelihood. This method is also general in that it takes into account the errors in the determination of all quantities, but unlike the above method it does not require division of the variables into dependent and independent quantities.

If the measurements are independent and the model is capable of describing the experimental data, while the deviations or discrepancies ϵ_i are distributed normally with some dispersion σ_i^2, then the optimal estimate criterion becomes

$$\min_{\vec{\theta}} SQ = \min_{\vec{\theta}} \sum_{i=1}^{N} \frac{\varepsilon_i^2}{\sigma_i^2} = \vec{\varepsilon}^T M^{-1} \vec{\varepsilon}, \tag{5.386}$$

where

$$\varepsilon_i = f\left(\vec{x}_i^{\,e}, \vec{\theta}\right) = y_i^e - y_i; \tag{5.387}$$

and M is a diagonal matrix with elements σ_i^2.

In accordance with the principle of maximum likelihood, dispersions of the discrepancies of quantities characterizing equilibrium in two-phase binary systems are determined by the relation

$$\sigma_i^2 = \sigma_{y_i}^2 + \left(\frac{\partial y^{\text{calc}}}{\partial x^L}\right)_i^2 \sigma_{x^L}^2 + \left(\frac{\partial y^{\text{calc}}}{\partial x^{S\,(V)}}\right)_i^2 \sigma_{x^{S\,(V)}}^2$$

$$+ \left(\frac{\partial y^{\text{calc}}}{\partial T}\right)_i \sigma_T^2 + \left(\frac{\partial y^{\text{calc}}}{\partial p}\right)_i \sigma_p^2. \tag{5.388}$$

Unfortunately, the methods one meets in the literature for estimating the parameters of models of heterogeneous equilibria do not take into account all – and frequently any – of the errors in the determination of measured variables and, therefore, do not lead to the best statistical estimates of the parameters [110, 139, 146, 147, 274–276, 301–303]. One of the basic aims of choosing a model for the solutions is to obtain an expression that can be used to provide reliable predictions for various thermodynamic properties, phase transitions, etc. The degree of reliability of such expressions will depend on the accuracy of the predictions. This problem is intimately connected with estimating the indeterminacy in the values determined for the parameters.

The application of the method of maximum likelihood is convenient in practice since one does not need to know the absolute values of the dispersions, which are usually unknown, but only their statistical weights W_i:

$$W_i = \sigma^2 / \sigma_i^2 \quad \text{or} \quad W_i = 1/\sigma_i^2, \tag{5.389}$$

where the σ_i are usually taken to be equal to the instrumental errors in measuring the corresponding variables, and σ^2 is the dispersion of observed variables with identical weights.

The estimated parameters of models do not depend on the quantity σ^2 and may be calculated without it. However, to analyze the errors in the values of the parameters calculated by the MLS, one must have a value for σ^2, which may be estimated by means of the relation

$$s^2 = E\left(\sigma^2\right) = s/(N - p). \tag{5.390}$$

General formulas for calculating various characteristics of the accuracy of models are presented in [283] and elsewhere.

The joint probability distribution for the parameter estimates in the nonlinear case is usually an extremely complicated function which is not only difficult to calculate, but also inconvenient to use, even when known. Hence, in practice one does not actually compute the distribution or any of its moments. Instead, one characterizes the errors of the estimates by conditional procedures [288]. The practical use of confidence intervals in estimating the parameters of nonlinear models is treated in [304].

If one uses two different models to describe any set of data, then Eq. (5.390) leads to two different estimates for s^2. If the quantity s^2 proves to be smaller in the second model than in the first, then the second model provides a better description of the experimental data. However, in order to ascertain whether this improvement is statistically significant, one should utilize the F distribution (the Fisher criterion).

$$F\left(\nu,\ \mu\right) = s_1^2/s_2^2, \tag{5.391}$$

which is equal to the ratio of the larger to the smaller quantity if the distribution of errors is taken to be normal, where ν and μ are degrees of freedom for the first and second model, respectively. If $F > F_{1-\alpha}(\nu, \mu)$, the second model should be considered better than the first. If $F < F_{1-\alpha}(\nu, \mu)$, then there is no preference for either model. Here the quantity α characterizes the statistical accuracy in the choice of the model.

Nowak and Flork [305] proposed to solve the problem of choosing from among the various hypotheses one which would best reflect the experimental data on the basis of a probability ratio criterion which requires a minimum amount of information, based on the calculation of simple expressions, and which possesses a series of optimal characteristics.

The problem of the statistical verification of a hypothesis, in those cases where the models are chosen on the basis of matching with certain uniform sets of data, has been considered in [285].

Two of the most essential factors which control the adequacy of the model are the confidence interval at each point of measurement, and the randomness of the residual deviations [283].

If the measured values lie outside the corresponding confidence interval, then this may be due to the following factors: 1) the dispersion of the dependent and independent variables is too small; 2) the model is inadequate.

The most reliable conclusions on the adequacy of a model are arrived at by studying the residual deviations of the variables from the calculated values. If the model is adequate, then these deviations assume purely random values. If, on the other hand, there is a definite regularity in their distribution, then the model was chosen incorrectly. The estimates of the parameters $\theta_1, \theta_2, \ldots, \theta_p$ made by the MLS for different variants of the vector of variable values do not coincide; in addition, the degree of accuracy also varies. The deviations chosen in estimates of the parameters of the model of binary equilibrium for the liquid–solid are the quantities T and γ_i. In [285, 306] a comparative analysis was carried out of the effect of choosing particular deviation functions on the accuracy of the parameter estimates used for analyzing data on the liquid–vapor equilibrium. The most important conclusion of these papers is that for binary two-phase systems the method of maximum likelihood allows one, in principle, to find the most reliable estimate of the parameters by using data for only two parameters, for example p and x^L or T and x^L.

The inclusion of additional information, contrary to the claims in [110, 307], does not improve the accuracy of the parameter estimates, but it does provide the possibility of checking the internal consistency of experimental data [306].

In the studies of Brebrick [146, 301–303] the MLS was used to select a solution model for systems with congruently melting compounds, which led to a quantitative fit between phase equilibrium data and the thermodynamic properties of the corresponding compounds and solutions.

The distinguishing feature of Brebrick's method is the use of supplementary equations connecting the thermodynamic characteristics of the solutions with the properties of the congruently melting compound at the melting point, by means of which one can fix a series of unknown expansion parameters. The method was tested successfully for particular cases of systems in the A^{III}–B^V group [146, 301, 303]. The values of the model parameters were found in the least-squares method by minimizing a function of the form

$$\sigma(T) = \left[\sum_{j=1}^{N} (T_{j,\exp} - T_{j,\text{calc}})^2 / (N - p) \right]^{1/2}, \qquad (5.392)$$

where p is the number of parameters selected and N is the number of experimental points.

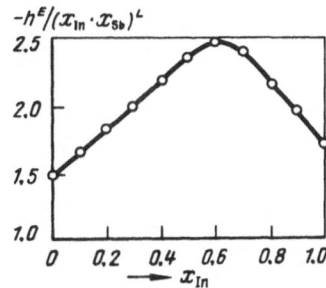

Fig. 50. Composition dependence of the function $h^E/(x_{In}x_{Sb})^L$ in the In–Sb system at 900 K according to data of [311].

TABLE 2. Parameters of Chosen Models and Results of Thermodynamic Calculations Based on Them for Various Systems

Systems	Chosen model**	Model parameters							$\Delta H_f^{(S)}$, kJ/mole	$\Delta S_f^{(S)}$, J/mole-deg
		$-\omega_0$	ω_1	ω_2	ω_{01}	ω_{11}	ω_{21}	$\sqrt{\bar{\sigma}^2}$, deg		
		J/mole			J/mole-deg					
Ga—Sb	6P	10 582	−11 637	4 538	9.1	15.8	4.9	9.3	57.8	151
In—Sb		15 070	−28 779	36 627	−0.046	−12.4	26.8	8.6	57.8	49
In—As		13 060	−15 070	20 930	−0.57	−0.92	0.13	16.3	60.3	33
Ga—Te		83 720	−83 720	−179 998	−15.9	−33.5	−100	4.3	132	60
In—Te		125 580	−115 115	−79 534	−14.2	−92	−125	5.6	137	95
GaTe—Te		68 022	−19 622	−9 209	15.1	−4.3	−36	9.6	167	123
InTe—Te		29 302	−19 883	14 651	1.44	−7.1	3.7	5.3	105	71
Ge—Te *	QR	95 449	—	—	−4.8	—	—	3.1	95	33
Sb—Te	6P	8 372	−63 836	31 395	11.7	−0.42	8.0	9.1	141	65
Bi—Te		42 860	−1 338	22 460	−27.6	−1.0	−0.64	4.1	168	146
Mg—Pb	SR	33 488	21 767	—	−4,2	10.7	—	7.1	56	32
Mg—Sn		56 511	29 302	—	−11.6	4.1	—	15	98	55
Mg—Ge		100 464	−6 820	—	−0.97	6.3	—	20	161	48
Mg—Si		42 948	−41 516	—	−6,3	14.6	—	6.9	132	45
Cd—Sb		9 628	—	—	—	—	—	14.8	36	33

*The model was chosen by analyzing the portion of the liquidus between the concentrations 0.5–0.667 atomic fraction of Te.
**R – regular model; QR – quasiregular model; SR – subregular model, which takes into account the temperature dependence of the exchange energy; 6P – six-parameter model.

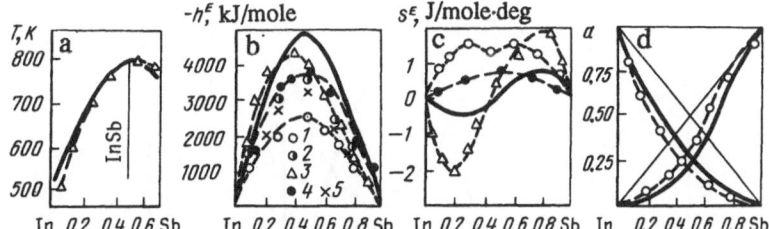

Fig. 51. Results of calculations of the liquidus (a), and of the thermodynamic properties of liquid solutions (b, c, d) in the In–Sb system: a) full line – data of [218, 312, 313]; points – calculation by Eq. (4.195); b) full line – calculation by Eq. (4.135); c) full line – calculation by Eq. (4.137); d) full line – calculation by Eqs. (4.139) and (4.140); in parts b–d the points 1–5 correspond to [311], [315], [308], [314], [229], respectively.

Seven model solutions were analyzed: the ideal, strictly regular and quasiregular solutions, the linear-temperature approximation, a model which took into account the linear dependence of the exchange energy on the temperature and concentration via the Margules series, and the quasichemical approximation.

In each particular case the model parameters were found by condition (5.392) in two ways. In the first variant no restrictions were imposed on any of the model parameters. In the second variant additional constraint equations were formulated, on the basis of (5.212) and (5.213), to compute explicit expressions for the excess partial molar thermodynamic properties. These equations either strictly fixed the values of the separate parameters or imposed definite bounds on them.

Even the best match between the experimental and calculated liquidus curves is not a sufficient criterion for the adequacy of the model and the solution. This conclusion is clearly illustrated by the results of [146, 301]. Five model solutions, namely the quasiregular, the quasichemical, the strictly regular, the linearly temperature-dependent, and the Margules series models, all led to mean square deviations which were within the limits of experimental error. However, the values of the thermodynamic quantities $\Delta H_f S$ and $\Delta S_f S$, calculated for the strictly regular, quasiregular, and quasichemical solutions, were in poor agreement with the experimental values. One may therefore conclude that for these last three the models are inadequate to describe the properties of the $A^{III}B^V$ system. In addition, the other thermodynamic properties computed with the aid of these models also support this conclusion, though the opposite is claimed in many studies [308]. As regards the linear-temperature approximation and the model using the Margules series up to third-order terms, both lead to satisfactory agreement between the calculated curve of monovariant equilibrium and the thermodynamic characteristics of the compound, the agreement being even better in the latter case.

In the work of Brebrick [302] an attempt was made to extend the region of applicability – in terms of concentration – for a single model of a solution, based on eutectic points, which the author considered to be most reliably determined. The influence of errors incurred in the determination of thermodynamic properties of compounds on the values of the parameters was also considered there.

Rao and Tiller [303] illustrate the choice of a model for a two-component solution by adopting the calculation techniques of [309, 310].

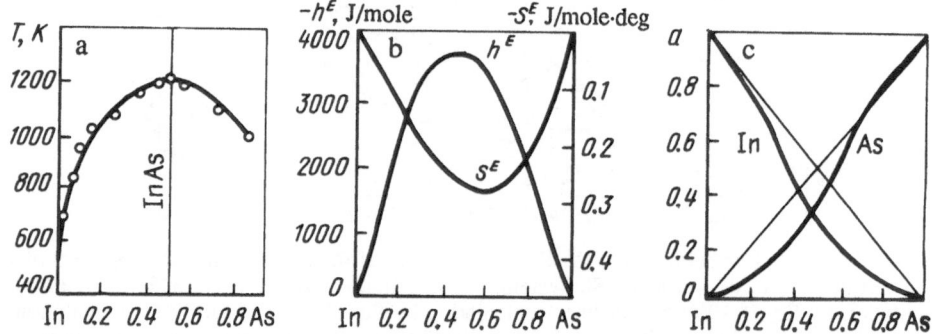

Fig. 52. Results of calculations of the liquidus (a) and thermodynamic properties (b, c) of liquid solutions on the In–As system: a) according to data of [218, 312, 313]; the points on the curve are from the calculation according to Eq. (5.195); b) calculation according to Eqs. (4.135) and (4.137); c) calculation according to Eqs. (4.139) and (4.140).

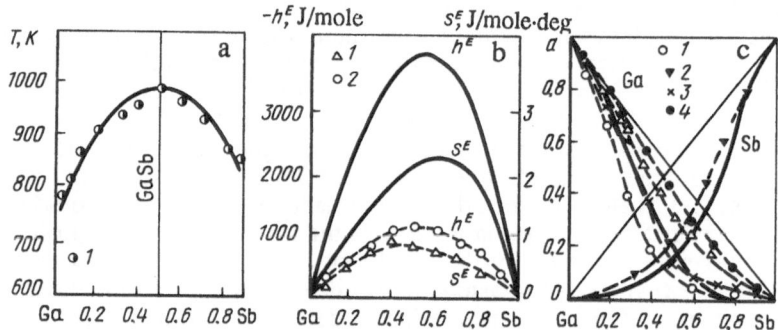

Fig. 53. Results of calculations of the liquidus (a) and thermodynamic properties (b, c) of liquid solutions on the Ga–Sb system: a) full line – according to data of [218, 312, 313]; the points on the curve are from the calculation according to Eq. (5.195); b) full line – calculation according to Eqs. (4.135) and (4.137); 1, 2 – data of [322] and [319], respectively; c) full line – calculation according to Eqs. (4.139) and (4.140); 1–4 – data of [323], [321], [324], and [320], respectively.

The choice of a model in which all possible properties are compared [303, 309, 310] (thus, the liquidus and solidus curves, the thermodynamic properties of solution, etc.) is not practically meaningful since then no parameters remain which would be described on the basis of the given model, at least under the prevailing conditions. Nevertheless, such a choice of a model is very important, since it enables one to estimate the thermodynamic properties of solutions for other values of the state parameters, and to calculate other curves of monovariant equilibrium, for example the solubility limit curve. In addition, by using the periodic table, one may approach with confidence calculations for analog systems formed by anionic or cationic component substitutions, using only the already existing reliable data on phase equilibria.

If the thermodynamic characteristics of solutions and compounds are not available, one can choose the model by successively calculating and comparing the parameters characterizing the exchange energy on the basis of various experimental data (such as viscosity, density, electronegativity, etc.). If the parameters of the exchange energy from the various data agree with one another, then the corresponding model can be considered to be justified within known confidence limits, and one may use it to calculate thermodynamic characteristics of solutions and compounds.

A single model may not be able to describe all the systems. This is especially true for systems with intermediate phases differing in their physicochemical character. One, therefore, has to restrict the limits of concentrations and temperatures within which the chosen model is applicable.

We shall now illustrate the computation of the thermodynamic properties of solutions from phase diagrams for a series of systems containing semiconducting compounds of various groups of simple and complex structures.

9.1. A^{III}–B^V Systems

Let us consider a series of A^{III}–B^V systems in which well-known A^{III}–B^V-type semiconducting phases are formed. The model of subregular solutions is not adequate for these systems, as experiment shows. Thus, for example, the behavior of the function $h^M/(x_1 x_2)$, calculated according to the data of [311] and graphically reconstructed in Fig. 50, clearly indicates that the In–Sb solution is not subregular. We arrived at this conclusion by comparing the excess thermodynamic functions of mixing, calculated via (4.137) and (4.138), with the experimental data of [311]. This comparison also reveals that the concentration dependence of the exchange energy in the indium–antimony system is nonlinear. A six-parameter model was therefore applied in an attempt to specify the quadratic nature of this nonlinearity. The calculation of the parameters of the models employed Eq. (5.196), and was performed by minimizing the sum of squared deviations (5.392) with a direct search method. The parameters computed in this way are presented in Table 2. Figure 51a presents the liquidus curve calculated from Eq. (5.196), compared with the experimental data [312, 313]. The agreement is seen to be quite satisfactory.

The enthalpy and entropy of formation of solid indium antimonide from liquid indium and antimony were calculated on the basis of liquidus–temperature data by means of expressions (5.367) and (5.368), and are presented in Table 2. By comparing these values with the results of [228, 302], one may conclude that the calculated and experimental values of ΔH_f^S and ΔS_f^S agree well with each other.

Thus, at this stage, from the point of view of the work of Brebrick [301, 302], the six-parameter model may be considered to be completely satisfactory. But, as we have already pointed out, the physicochemical criterion based only on a comparison of the heat and entropy of formation is not sufficient to favor a particular model.

Hence, further calculations were carried out for the thermodynamic functions of mixing and the activity coefficients of the components in the solution. The results of these calculations are presented in Fig. 51b, c together with the data available from the literature. It is seen from Fig. 51b that our calculated curve of the concentration dependence of h^E is similar in nature to the experimental curves [229, 311, 314, 315]. The maximum value of h^E corresponds to a composition $x_{Sb} = 0.5$.

The absolute values of the calculated data are within the limits of the experimental errors. Meanwhile, the data computed by Brebrick [302] are distinguished by the concentration dependence of h^E, for which the maximum is shifted to $x_{Sb} \sim 0.35$. Therefore, the choice of the six-parameter model must be judged more apposite. As for the concentration dependence of the excess entropy of mixing, the calculated values do not agree with experiment [311]. In the composition region located on the indium side, the excess entropy of mixing calculated by (4.147) is negative, which is, in principle, impossible when one takes into account the information on the formation of aggregates of In_3Sb and $InSb$ in the liquid phase [315, 316].

One should consider the concentration dependence of the activity coefficients of indium and antimony separately (Fig. 51d). It is seen that the basic character of the data calculated from the phase diagrams is analogous to the experimental values [311]. Some of the quantitative discrepancies are most likely explained by experimental inaccuracies [311]. The data for the activity coefficients of indium and antimony calculated by the six-parameter model are also in accord with a series of other calculations [317].

The conclusion to be drawn from Table 2 and Fig. 51 is that the six-parameter model describes adequately the physicochemical nature of the solutions of the In–Sb system, and that it may be used to calculate the thermodynamic properties of these solutions by employing accurate phase diagram data.

Based on the model chosen for the description of the solution and on data of detailed thermographic studies of the liquidus in composition regions adjacent to the corresponding compounds [312], we have carried out a calculation of the thermodynamic properties of the liquid solutions of In–As and Ga–Sb. The results are presented in Table 2 and in Figs. 52 and 53. Figure 52b shows that the excess entropy of mixing is small compared with the absolute value, which is in qualitative agreement with the conclusions of [318] on the regular behavior of In–As solutions near the melting temperature of indium arsenide. In every case one may state unequivocally that the exchange energy is practically independent of temperature and is a complicated (to be more exact, a quadratic) function of concentration.

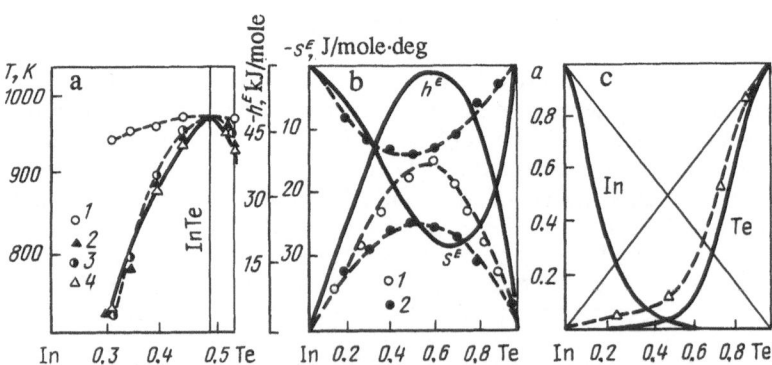

Fig. 54. Results of calculations of the liquidus curve (a) and thermodynamic properties (b, c) of liquid solutions on the In–Te system: a) full line – according to data of [218, 219]; 1–4 – calculation by Eqs. (5.47), (5.198), (5.196), and (5.195), respectively; b) full line – calculation by Eqs. (4.135) and (4.137); 1, 2 – data of [329] and [330], respectively; c) full line – calculation by Eqs. (4.139) and (4.140); the points are according to [330].

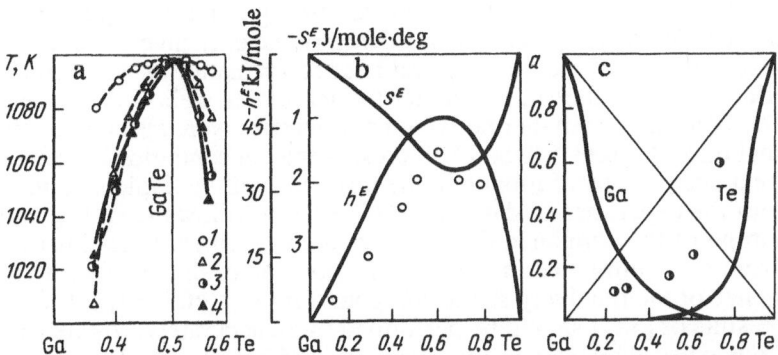

Fig. 55. Results of calculations of the liquidus curve (a) and thermodynamic properties (b, c) of liquid solutions on the Ga–Te system: a) full line – according to data of [218, 219]; 1–4 – the same as in Fig. 54a; b) full line – calculation by Eqs. (4.135) and (4.137); points – data of [328]; c) full line – calculation by Eqs. (4.139) and (4.140); points – data of [330].

The general concentration dependences of the excess heat and entropy of solutions of Ga–Sb, presented in Fig. 53b, correlate well with the available experimental data, but the actual values agree much more poorly. Data on the heat of mixing [319–321] differ from those presented in the figure by the order of 2 kJ/mole. The excess entropy is positive and differs from the data given in [322] by no more than 2.5 J/mole K. The values of the activity coefficients (Fig. 53c), calculated on the basis of the parameter estimates, assume a position intermediate between the experimental data of [322–324]. The values of the heat and entropy of formation of solid indium arsenide and gallium antimonide from the liquid components, calculated in the chosen approximation, are considerably different from the data of the literature [228, 307, 308, 317], which in turn agree poorly among themselves. The size of the dispersion in the comparison of the calculated and the experimental liquidus of the In–As system is quite large (cf. Table 2), while for the Ga–Sb system it is within the experimental error of the thermographic studies.

9.2. A^{III}–B^{VI} Systems

We consider the binary systems of gallium and indium with tellurium as representatives of this group. Available data [218, 219] show the formation of stable, congruently melting mono- and sesquitellurides in these systems.

Owing to the complicated nature of the phase equilibrium in the systems Ga–Te and In–Te, associated with the presence of a demixing region and the incongruently melting compounds Ga_3Te_2, $GaTe_3$, In_9Te_7, In_3Te_5, and In_2Te_5, it is expedient to choose a model of the solution on the basis of monotellurides, encompassing in the initial crystallization the largest temperature and concentration regions. Accordingly, we considered in succession the ideal, strictly regular, and subregular solutions and the six-parameter model. The results of calculations for the liquidus, and the experimental data for comparison, are presented in Figs. 54a and 55a. These indicate that the smallest discrepancy between theory and experiment is observed for the six-parameter model (cf. also Table 2). A reasonably good agreement with experiment is also provided by calculations according to the subregular model which takes into account the temperature dependence of the exchange energy.

The calculation of the entropy and heat of formation of gallium and indium monotelluride within the six-parameter model agrees well with the data of [228, 325–327]. The calculated thermodynamic properties were also compared with experimental data. Figures 54b and 55b present the concentration dependence of the excess heat and entropy of mixing in the formation of the liquid solutions, as compared with the experimental data of [328–330]. It follows from Fig. 54b that the general trend of the computed data follows the experimental results of [329], but differs from those presented in [330] by the position of the maximum. Because of the existence of the relatively stable compound In_2Te_3 (cf. [331]), the computation and the results of [329] should be performed more accurately. As for the quantitative discrepancies, they may be considered to be relatively small in view of the considerable errors in the experimental estimates of the heat of mixing. Figure 55b shows that the heats of mixing – which we have calculated by formula (4.145) – are in sufficiently good agreement with the data of Castanet and Bergman [328], both in terms of absolute values and as far as the nature of the concentration dependence is concerned.

Figures 54c and 55c illustrate the concentration dependence of the activity coefficients calculated from formulas (4.149) and (4.150). Comparison of these curves with the data of Predel et al. [330], obtained on the basis of vapor pressure measurements, shows significant discrepancies. However, if we look at the picture presented by Figs. 54b and 55b, it is clear that the data of [330] are not sufficiently reliable, which may also be due to the failure to take into account the complicated composition of the vapor phase when estimating the activity coefficients.

Within the chosen model of the solution we also carried out the corresponding calculations for the particular systems GaTe–Te and InTe–Te. The transition to a particular system is due to the relative stability of gallium and indium monotelluride in the liquid phase [331, 332]. The parameters of the exchange energy, as well as the entropy and enthalpy of formation of the solid sesquitellurides from liquid tellurium and monotellurides, are presented in Table 2. Comparison of the enthalpy of

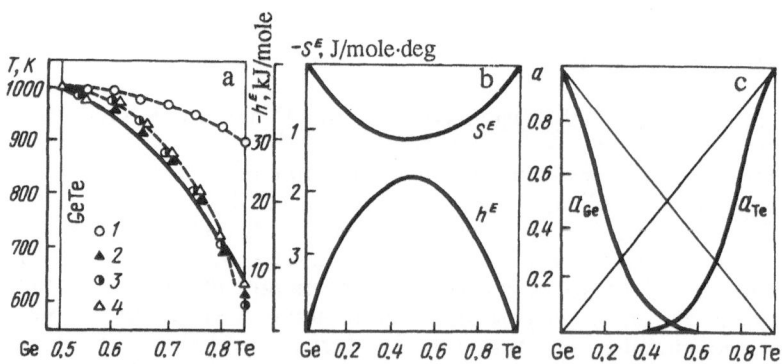

Fig. 56. Results of calculations of the liquidus curve (a) and thermodynamic properties (b, c) of melts in the Ge–Te system: a) full line – data of [218, 219, 336]; 1–4 – the same as in Fig. 54a; b, c) calculation based on Eqs. (4.97) and (4.98).

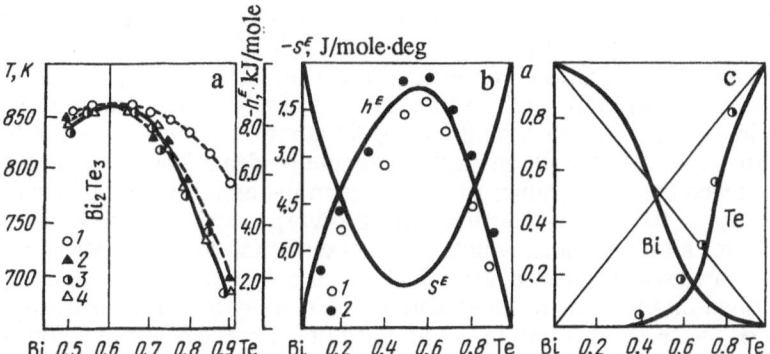

Fig. 57. Results of calculations of the liquidus curve (a) and thermodynamic properties (b, c) of melts in the Bi–Te system: a) full line – experimental; 1–4 – the same as in Fig. 54a; b) full line – calculation by (4.135) and (4.137); 1, 2 – according to the data of [341] and [340], respectively; c) full line – calculation by (4.139) and (4.140); the points are experimental [342].

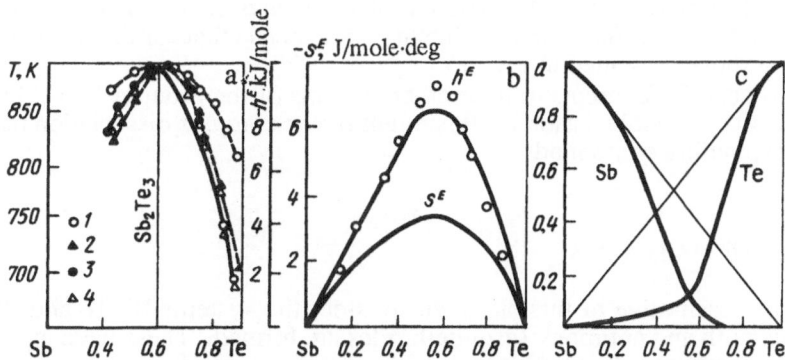

Fig. 58. Results of calculations of the liquidus curve (a) and thermodynamic properties (b, c) of melts in the Sb–Te system: a) full line – experimental, according to data of [163]; 1–4 – the same as in Fig. 54a; b) full line – calculation by (4.135) and (4.137); points are experimental [329]; c) calculation by (4.139) and (4.140).

formation of gallium and indium sesquitelluride with the data of [228, 325, 333, 334] shows good agreement, which attests to the correctness of the six-parameter model.

9.3. A^{IV}–B^{VI} Systems

In our studies we chose the system Ge–Te as the representative member of this group. According to the available information [218, 219], this system has one congruently melting compound and admits the existence of the incongruently melting compound $GeTe_2$ [335]. Germanium monotelluride is distinguished by the skew character of the region of initial crystallization [336] to such an extent that in

[69] this compound was assumed to melt at the transition point. Indeed, the compound melts congruently, but the left branch of the liquidus, corresponding to the initial crystallization of germanium monotelluride, is extremely small. Analysis of the right branch of the liquidus in various approximations shows (Fig. 56a) that in the concentration range 0.5–0.67 atomic fraction of Te the liquidus is satisfactorily described by all the models except the ideal. The amount of dispersion does not exceed the usual values of the errors in the experimental thermal analysis. But when $x_{Te} > 0.67$, one observes strong divergences between calculation and experiment for all the models, which may be ascribed to the existence of germanium ditelluride, or to some other causes. In all cases only one unambiguous conclusion may be drawn on the basis of the calculations, namely that it is necessary to refine the study of phase equilibrium in the vicinity of $x_{Te} = 0.67$. From a comparison of the entropy and enthalpy of formation of germanium monotelluride – as calculated in various approximations – with the experimental data [228, 337, 338], one may conclude that the quasiregular model is the best, which may therefore be used to describe the thermodynamic properties of the solutions in this system. Figures 56b and 56c present, respectively, the results for the excess thermodynamic functions and activity coefficients of germanium and tellurium for the quasiregular model, as a function of composition. The absence of published data does not allow one to compare the calculation with experiment. This means that for the present the only results are those presented in Fig. 56.

No successful attempts to treat other systems of the group $A^{IV}–B^{VI}$ have been reported. One possible reason for this might be the complex dissociation patterns of the corresponding compounds.

9.4. $A^V–B^{VI}$ Systems

As representative of this class we consider the systems Bi–Te and Sb–Te, in which bismuth and antimony sesquitellurides are formed. These have gained widespread use as thermoelectric materials.

The parts of the liquidus in the composition regions adjoining the compounds Bi_2Te_3 and Sb_2Te_3 have been studied by us thermographically. In the analysis we have also made use of the data of other authors, surveyed in [327]. Figures 57a and 58a present the portions of the liquidus curves which correspond to the initial crystallization of bismuth and antimony sesquitelluride, calculated in various approximations. These graphs also present the experimental data, and a comparison shows that the calculation within the six-parameter model approximation gives the least deviation. The mean dispersion (cf. Table 2) falls within the error limits of the thermographic determination of the liquidus curves. The calculated values of the exchange energy parameters are presented in Table 2, which also contains the values of the enthalpy and entropy of formation of bismuth and antimony sesquitelluride, computed at liquidus temperatures by means of Eqs. (5.367) and (5.368). From a comparison of these quantities with values obtained in a thermodynamic calculation in terms of standard thermodynamic functions in the handbook of [339], one may infer that the deviations are within the error limits for estimating the relevant quantities. Therefore, according to [308], the six-parameter model ought to be acceptable for describing the phase equilibrium and thermodynamic properties of alloys in the Bi(Sb)–Te systems. But for a more reliable conclusion concerning the correctness of the chosen model, one should link the thermodynamic properties of solutions calculated in a wide concentration range with experimental data available.

Figure 57b shows our calculations for the curve of the concentration-dependent heat of mixing in the Bi–Te system, agreeing almost perfectly with the experimental data of Maekawa et al. [340] and Blachnik and Enninga [341]. Our results for the activity coefficients of bismuth and tellurium, based on the phase diagram within the six-parameter model, are given in Fig. 57c, which also presents for comparison Brebrick's experimental results, obtained from vapor-pressure measurements [342] via the equation of the temperature dependence of the vapor pressure for pure tellurium [238]. As one can see from the data, the agreement is completely satisfactory. The calculation of the excess entropy of mixing (Fig. 57b) shows that it is negative, and that its absolute value is relatively small.

The calculated values of the thermodynamic functions of mixing in the Sb–Te system are shown in Fig. 58b, together with the experimental data of [329]. It can be seen that the calculation is in excellent agreement with experiment, so that the choice of the model is deemed successful, and the results of the calculation of the activity coefficients of antimony and tellurium presented in Fig. 58c are also quite accurate. The excess entropy of mixing in this instance is extremely small, though positive in sign, which indicates the predominance of the effect of entropy increase on mixing, in comparison to the decrease in the resulting ordering, due to chemical interactions. The picture is reversed in the Bi–Te system considered above. At first sight this may be construed as an inconsistency, especially if one remembers the relatively high thermal stability of antimony telluride compared with bismuth telluride [343, 344]. In fact, however, one must take into account the total effect of the interactions in the liquid phase. This is more pronounced in the Bi–Te system, because of the distinct appearance of bismuth monotelluride as a chemical entity in the liquid phase [343], in spite of the fact that this compound melts incongruently [327], an effect not observed for the Sb–Te system [343].

Summing up the analysis of phase equilibrium and thermodynamic properties of alloys in the Bi(Sb)–Te systems, one may note that the six-parameter model

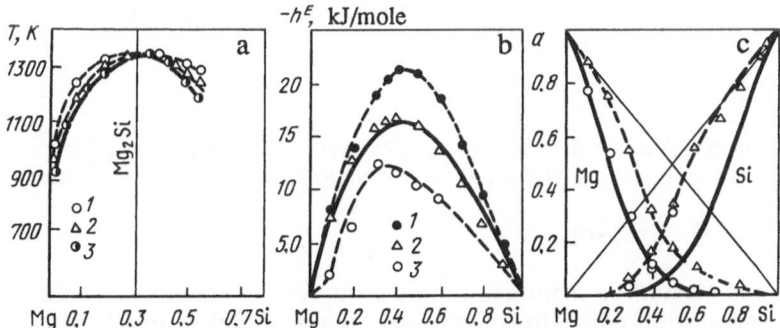

Fig. 59. Results of calculations of the liquidus (a) and thermodynamic properties (b, c) of melts in the Mg–Si system: a) full line – experimental, according to data of [218, 219, 222]; 1–3 – calculation by Eqs. (5.49), (4.196), and (4.197), respectively; b) full line – calculation by Eqs. (4.129) and (4.130); c) full line – calculation by Eqs. (4.131) and (4.132); b, c) 1–3 – experimental points of [222], [352], and [356], respectively.

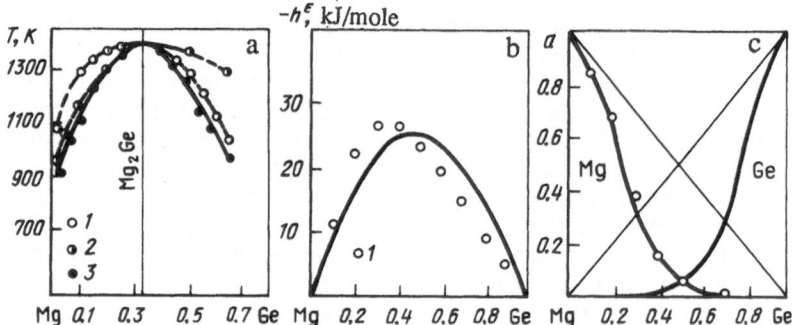

Fig. 60. Results of calculations of the liquidus (a) and thermodynamic properties (b, c) of melts in the Mg–Ge system: a) full line – experimental, according to data of [218, 219]; 1–3 – the same as in Fig. 59a; b, c) full line – the same as in Fig. 59b; c) points are experimental [357].

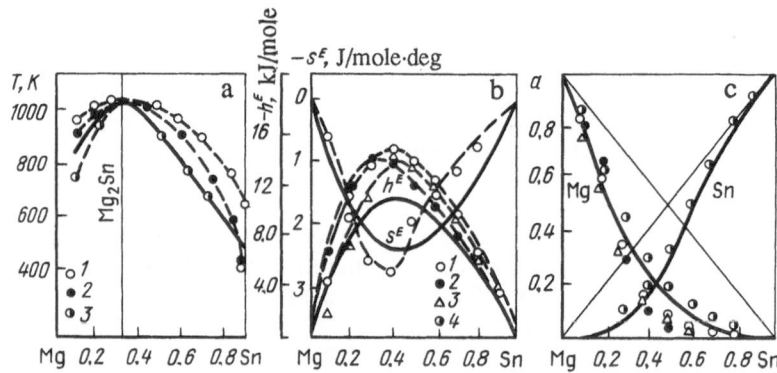

Fig. 61. Results of calculations of the liquidus (a) and thermodynamic properties (b, c) of melts in the Mg–Sn system: a) full line – experimental, according to data of [218, 221]; 1–3 – the same as in Fig. 59a; b, c) full line – the same as in Fig. 59b; c; 1–3 – experimental points of [352], [351], and [354], respectively.

treatment is completely satisfactory and, consequently, that it may be recommended for use on these systems, as well as in the analogous systems of selenium.

9.5. A^{II}–B^{IV} Systems

The representative members of this group in our studies were the binary systems of magnesium with silicon, germanium, tin, and zinc, in all of which congruently melting compounds of the form Mg_2B^{IV} occur.

The approach to choosing the model for these systems is the same as in the preceding cases. Figures 59–62 represent portions of the liquidus curves calculated in various approximations. Also shown are experimental data gathered from [218–222, 313]. Analysis indicates that the model of subregular solutions (which in-

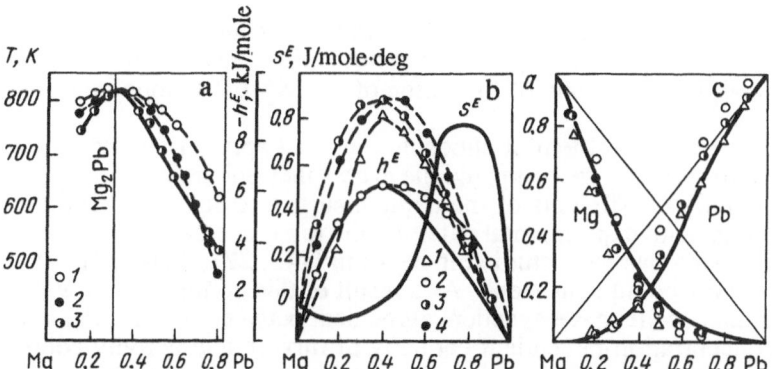

Fig. 62. Results of calculations of the liquidus (a) and thermodynamic properties (b, c) of melts in the Mg–Pb system: a) full line – experimental, according to data of [220]; 1–3 – the same as in Fig. 59a; b, c) full line – the same as in Fig. 59b, c; 1–4 – experimental data according to [354], [356], [350], and [355], respectively.

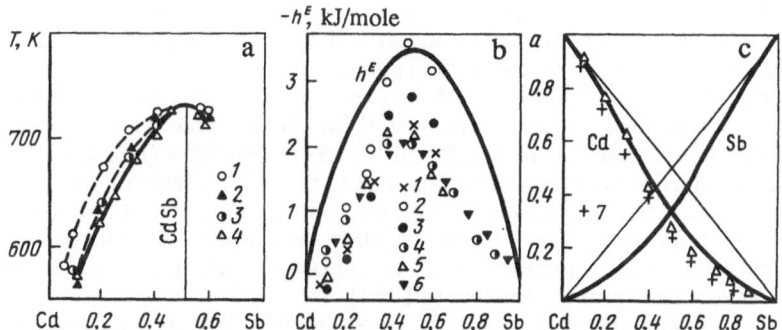

Fig. 63. Results of calculations of the liquidus (a) and thermodynamic properties (b, c) of melts in the Cd–Sb system: a) full line – experimental, according to data of [218, 359]; 1–4 – the same as in Fig. 54a; b) full line – calculation by Eq. (4.104); c) full line – calculation by Eqs. (4.97) and (4.98); b, c) 1–7 – experimental data of [360], [363], [364], [365], [361], [362], and [366], respectively.

cludes the temperature dependence of the exchange energy) and the six-parameter model describe practically identically well the liquidus curves corresponding to the initial crystallization of the compounds Mg_2B^{IV} (cf. also Fig. 44b).

In our opinion, the enhanced values of the dispersion (cf. Table 2) in the Mg–Sn and Mg–Ge systems are due to the use of inaccurate values of the heat of melting of the corresponding compounds, which were borrowed from [345, 346].

The calculation of the thermodynamic parameters of formation of the compounds Mg_2B^{IV} on the basis of data at the liquidus temperature using the two indicated models gives slightly different results. Comparison of the calculated enthalpy and entropy of mixing of the compounds Mg_2B^{IV} with experimental data [326,

347–350] indicates good agreement, so that, as in the case of the $A^{III}B^V$ compounds, the successful description of the liquidus curves and the adequate estimate of the enthalpy and entropy of formation of the Mg_2B^{IV} compounds prove insufficient criteria.

To resolve the problem of choosing the most appropriate model for the solutions in these systems, we compared the excess thermodynamic functions of mixing and the activity coefficients of the components, as calculated within the two approximations, with experimental data. As the reference system we choose Mg–Sn, whose demixing has been studied in detail in [143, 221, 351–354], the data of the various authors being consistent. As a result of this comparison it was shown that the subregular model gave excellent agreement between the calculated and experimental concentration-dependent enthalpy of mixing and activity coefficients (cf. Fig. 61b, c). The six-parameter model, on the other hand, gave substantially different results from the experimental data, especially for the activity coefficients. The final choice fell on the subregular model which also accounts for the temperature dependence of the exchange energy. We note that in the remaining systems of the present group, the comparison with experimental data [143, 220–222, 350, 352, 353, 355–358] favored the subregular solution, although the experimental data for the Mg–Si and Mg–Ge systems is less convincing than for the Mg–Sn. Turning to the first stage of choosing a model, connected with estimating the parameters of the model and the thermodynamic functions of formation of the compounds Mg_2B^{IV}, preference is given to the corresponding quantities calculated according to the model of a subregular solution (cf. Table 2).

The concentration dependence of the excess entropy of mixing calculated in the present approximation is quite complicated for all the systems considered, but because of the very small absolute value of the excess entropy (less than 1 J/mole·K) the clarification of this dependence is scarcely feasible. Nevertheless, on the basis of vapor-pressure data reported in [352], one obtains significant negative values of the entropy of mixing (from –1.7 to –7.5 J/mole·K at the maximum). On account of the complex nature of the evaporation of alloys in these systems, it is difficult to give unequivocal support to the data of [352].

On the other hand, on the basis of rigorous calculations and the data of [143, 347, 351, 353–356], one may neglect, even in principle, the excess entropy of mixing, and calculations may be confined to the subregular model. Therefore, one may conclude, on the basis of the preceding analysis, that the subregular model is to be preferred in performing calculations for the phase equilibria and thermodynamic properties of the $Mg–B^{IV}$ systems.

9.6. A^{II}–B^V Systems

In this group we consider in detail the system Cd–Sb. This system is interesting because one may obtain from it cadmium monoantimonide, which has found wide applications in electronic technology [359]. One learns from a consideration of the cadmium–antimony phase diagram that, apart from the stable cadmium monoantimonide, more complicated metastable alloys may be formed, such metastable equilibria depending essentially on the kinetics of the crystallization process. The analysis of this system relying on the equilibrium diagram should, therefore, be approached with care, taking into account the possible shift in the original phase-transition temperatures, due to the ease of departure from the equilibrium course of this process.

From a systematic calculation – with different models – of the liquidus responsible for the initial crystallization of cadmium monoantimonide, one learns (Fig. 63a) that the results for the subregular and the six-parameter models lie closest to the experimental data. The strictly regular model gives an average deviation of 14.3 K from the experimental liquidus. The calculations of the enthalpy and entropy of formation of cadmium antimonide from the components by various approximations (except the ideal solution) all give consistent results with one another, are in perfect agreement with the data of Hultgren [350], and are close to the data of Glushko's handbook [228]. Consequently, in the first stage of the analysis it is difficult to prefer any single model, since the amount of dispersion on the liquidus cannot be significant, due to the inaccuracy of the experimental data. But the calculations of the excess thermodynamic functions of mixing and the activity coefficients of cadmium and antimony show that the best agreement with the experimental data [360–366] is provided by the model of strictly regular solutions. Thus a paradoxical situation arises. On the one hand, the strictly regular model describes best the thermodynamic properties of the cadmium–antimony melts in the whole concentration region; on the other hand, its description of the experimental liquidus is worse than that of the six-parameter and subregular models. Accordingly, we have carried out an additional study of the vapor pressure above the melts of various compositions by using the boiling-point technique. The results we obtain are in excellent agreement with higher temperature vapor-pressure measurements carried out in [366] by the flux method. The data of our measurements and the results of [367] were processed jointly on a computer, and on the basis of the results obtained we calculated the activity coefficients. The latter were found to agree well with experimental data obtained by the e.m.f. method [361]. Consequently, the results of studying the thermodynamic properties of cadmium–antimony melts should be considered more reliable than the liquidus curve constructed by the method of thermal analysis. In this case one may choose the strictly regular solution and use it for reliable calculations of the phase-equilibrium diagram.

For greater reliability in the use of the strictly regular model, the exchange energy should be estimated on the basis of various experimental data. We utilized the combined data on vapor pressure and heat of mixing [364, 368], as well as data obtained by the e.m.f. method [361]. The values for the exchange energy obtained in all three cases differed little from each other (with an average of ~14.6 kJ/mole). The liquidus calculated on the basis of these data is presented in Fig. 63a. The thermodynamic functions of mixing and activity coefficients computed within the strictly regular model, together with the data of [360–362, 364, 365, 368] and our vapor-pressure study, are presented in Fig. 63b, c.

The calculations we have presented are not to be considered the final word on the choice of the most suitable model solution in the various systems studied. Indeed, what can actually be gleaned from our report is that the attempt undertaken by the authors to calculate the thermodynamic properties of solutions on the basis of phase-equilibrium data in systems with congruently melting compounds is in need of further improvement. This is because the calculations showed that the choosing of a model on the basis of phase-equilibrium data is a multi-extremum problem, whose successful solution depends on the minimization procedure and the form of the functional, as well as other considerations.

Chapter 6

THERMODYNAMIC ANALYSIS OF
THE DISSOCIATION OF
CHEMICAL COMPOUNDS

1. SHAPE AND POSITION OF MAXIMA IN
THE LIQUIDUS AND SOLIDUS CURVES

The immense amount of experimental material accumulated by the school of Kurnakov enabled him to propose that the shape (i.e., radius of curvature) and position of the melting curve maxima are intimately connected with the degree of dissociation of the chemical compounds into their constituent components [44, 369]. Complete nondissociation in a melting compound corresponds to singular cusp-like maxima on the liquidus and solidus curves (where the radius of curvature is zero; cf. Fig. 21a, b). If the compound dissociates only in the liquid phase, the liquidus has a flat maximum, while the solidus maintains a cusp (Fig. 21c, d). If dissociation occurs in both phases, both maxima will be flat (Fig. 21e). As the degree of association increases in one of the phases, the radius of curvature of the corresponding curve increases with it.

Kurnakov's group showed experimentally that for several dissociating compounds the composition at the maximum does not coincide with its chemical formula, the compounds separating into daltonides and berthollides [44] (see below for definitions).

The connection between the shape and position of the liquidus and solidus maxima and the degree of dissociation of a chemical compound was discussed from a thermodynamic point of view in [20]. On this basis Kröger [197] proposed to approximate the liquidus and solidus of compounds dissociating at different levels by explicit mathematical expressions (thus, a hyperbola for partial dissociation, and a parabola for complete dissociation).

Although the nature of monovariant equilibrium curves near the composition of the compounds is important for the theory of physicochemical analysis, until recently it did not get the attention it deserves, apart from the single attempt of Esin [370] to deal with this problem.

We shall discuss the question of whether the shape of the liquidus and solidus follows solely from general thermodynamic theorems, or whether one needs extra data to determine them.

Gibbs showed [1] that flat maxima of temperature–concentration curves follow from general thermodynamic theorems, whereas the existence of cusp maxima cannot be proved thermodynamically but is based on the same experimental results as the theory of limiting dilute solutions. In spite of his work, the problem of the possible thermodynamic justification for the singular maxima has been repeatedly discussed in the literature.

Thus, between 1889 and 1892 there was some discussion between Le Chatelier and Rozeboom [371–373] concerning the form of the maximum for $CaCl_2 \cdot 6H_2O$ on the fusion diagram of the system $CaCl_2$–H_2O. Eventually these authors concluded that the accuracy of experiments was not sufficient to resolve the problem. In addition, Le Chatelier rightly pointed out that general thermodynamic theorems cannot provide an unambiguous answer to this question.

The form of the phase diagram with A_mB_n-type compound formation may be obtained by means of the geometric method by considering the behavior of the Gibbs free-energy curves. As discussed in detail earlier (cf. Chapter 3, Section 3), one must assume *a priori* the presence or complete absence of dissociation for the compounds in the solution, and this will, in turn, determine the particular form of the Gibbs free-energy curve (cf. Figs. 6 and 19). Thus, as pointed out correctly by Patsukova [374], this approach will also fail to provide a rigorous thermodynamic solution to the question of the shape of the maxima for the compounds. The form of the Gibbs free-energy curve in the absence of dissociation of the compound into its components has long been a subject of controversy among researchers (cf. [52, 375–378]).

Van der Waals and Kohnstamm [12] wrote in their polemics with Planck [2]: "Thermodynamics has as little to say about the composition of specific phases as about the question as to which chemical compounds exist and in which phases they can occur." And, later: "There is no way to predict from our results whether any substance might be in equilibrium with other substances, and whether it might be obtained reversibly from other substances."

Hence, the existence of a singular or cusp-like maximum is not a direct consequence of general thermodynamic statements, but neither is it negated by them. The limiting cases will be assumed to be those compounds which do not dissociate into their components at all (i.e., the degree of dissociation α° vanishes).

The analysis of binary state diagrams with congruently melting chemical compounds indicates that in a number of cases the maximum of the fusion curve deviates from the stoichiometric composition. Compounds with significant deviations from stoichiometry were named berthollides by Kurnakov. He noted that berthollides were intermediate phases with variable composition based on dissociating chemical compounds [44]. Subsequently it was shown, by considering data on the continuous transition from daltonide to berthollide compounds, that there is no fundamental difference between them, and the difference which does exist in their fusion diagrams is due to the properties of interphase (heterogeneous) and intraphase (chemical) equilibrium [370, 379].

A series of studies [20, 380, 381] noted the connection between the shape and position of the fusion-curve maximum and the degree of dissociation of the compound, but did not present an explicit analytic dependence of the deviation from stoichiometry on the degree of dissociation.

An important advance toward a quantitative solution to the problem was made by Mlodzeevskii [382, 383] and Esin [370, 384]. Starting from ideal solution theory, Mlodzeevskii obtained a quantitative connection between the shift of the maximum and the degree of dissociation. Esin refined Mlodzeevskii's solution and generalized it to real systems [370, 384].

Starting from the condition of phase equilibrium (2.80) we may write for a two-component, two-phase system

$$\mu_B^S - \mu_A^S = \mu_B^L - \mu_A^L. \qquad (6.1)$$

Introducing the activity coefficients of the corresponding molecular forms γ_{A_1}, γ_{B_1}, and γ_{AB}, one may write the chemical potential of the ith component and the law of mass action in the form

$$\mu_B = \mu_B^{\bullet}(T,\ p) + RT \ln\left[(x_B - u)/(1 - u)\right]\gamma_{B_1} \qquad (6.2)$$

and a similar relation for the A component:

$$(1 - x_B - u)(x_B - u)/u(1 - u) = K\gamma_{AB}/\gamma_{A_1}\gamma_{B_1} = K', \qquad (6.3)$$

where u is the number of moles of AB required for one gross-mole of the phase. On the basis of (6.2) and (6.3) either of the equalities of (6.1) may be written in the form

$$\mu_B - \mu_A = \mu_B^{\circ} - \mu_A^{\circ} + RT \ln\left[(x_B - u)^2/u(1 - u)K'\right] + RT \ln(\gamma_{A_1}/\gamma_{B_1}). \qquad (6.4)$$

At the maximum

$$x_B = 1/2 + \delta; \quad \bar{u} = (1/2)\left[1 - \sqrt{(K' + 4\delta^2)/(K' + 1)}\right],$$

where δ is the deviation from stoichiometric composition of the maximum of the fusion curve.

Substituting the preceding relation in (6.4) and remembering (6.1), we find

$$\sqrt{\frac{K'^{(S)}}{K'^{(L)}}}\frac{\sqrt{K'^{(L)} + 4\delta^2} + 2\delta\sqrt{K'^{(L)} + 1}}{\sqrt{K'^{(S)} + 4\delta^2} + 2\delta\sqrt{K'^{(S)} + 1}} = \exp\left(\frac{q_A - q_B}{2RT_{max}}\right), \qquad (6.5)$$

where

$$q_A = \left(\mu_B^{\circ(S)} + RT \ln\gamma_{B_1}^S\right) - \left(\mu_A^{\circ(S)} + RT \ln\gamma_{A_1}^S\right),$$
$$q_B = \left(\mu_B^{\circ(L)} + RT \ln\gamma_{B_1}^L\right) - \left(\mu_A^{\circ(L)} + RT \ln\gamma_{A_1}^L\right). \qquad (6.6)$$

In accordance with expression (4.78),

$$K_A \gamma_{A_1}^L / \gamma_{A_1}^S = K_A', \qquad K_B \gamma_{B_1}^L / \gamma_{B_1}^S = K_B', \tag{6.7}$$

where K_i' is the apparent distribution coefficient of the ith component.

In view of relation (6.7), we obtain from (6.5) the final result

$$\sqrt{\frac{K'^{(S)}}{K'^{(L)}}} \frac{\sqrt{K'^{(L)} + 4\delta^2} + 2\delta \sqrt{K'^{(L)} + 1}}{\sqrt{K'^{(S)} + 4\delta^2} + 2\delta \sqrt{K'^{(S)} + 1}} = \sqrt{\frac{K_A'}{K_B'}}. \tag{6.8}$$

If one assumes that the changes in the activity coefficients are small compared to $K'^{(S)}$, $K'^{(L)}$, K_A' and K_B', then it follows immediately from (6.8) that δ is different from zero only when the compound dissociates in both phases ($K'^{(S)} > 0$ and $K'^{(L)} > 0$), or when the distribution coefficients of the components between the phases are not equal (in particular, $K_A' \neq K_B'$). But this feature is only characteristic of berthollides and of compounds intermediate between the latter and daltonides.

If, on the other hand, $K_A = K_B$ for example, and the right-hand side of Eq. (6.8) is equal to unity, then the left-hand side may be equal to it for any value of the dissociation constant (and, in particular, $K'^{(S)}$ and $K'^{(L)}$) only when $\delta = 0$, since $K'^{(L)}$ is always larger than $K'^{(S)}$. In other words, equality of the distribution coefficients of the components is completely sufficient to ensure that there should be no shift in the maximum for any degree of dissociation of the compounds in both of the phases.

Finally, if only one of the dissociation constants is zero, say in the solid phase, then for any value of the distribution coefficients we must take $\delta = 0$, in order not to contradict Eq. (6.8). Consequently, there will be no shift in the maximum provided the compound does not dissociate at all, at least in the solid phase.

Esin [370, 384] obtained the connection between the shift of the maximum and the dissociation constant in a somewhat different form:

$$\delta = x_B - 1/2 = \frac{x_{A_1}^L (K_A - K_B)}{2[2(1 - K_A) + x_{A_1}^L (K_A - K_B)]}. \tag{6.9}$$

This relation implies that the composition at the maximum will coincide with the formula of the chemical compound, either when it does not dissociate in the solid phase, or if the distribution coefficients among the phases are identical. The fulfillment of either of these conditions guarantees the absence of a shift.

A shift will be observed on the simultaneous fulfillment of two conditions: dissociation of the compound in both phases, and the inequality of the distribution coefficients. The larger the absolute value of the shift, the stronger the dissociation of the compound in the liquid and, therefore, also in the solid phase, and the greater the difference $(K_A - K_B)$ between the distribution coefficients of the components among the phases. As follows from Eq. (6.9), the maximum is shifted from 0.5 toward the component whose distribution coefficient is larger.

These conclusions may be generalized to the case of forming any compound of A with B with a more complicated composition [370].

Subsequently, Zernike [385] proved, on the basis of a dissociation model coupled with the theory of the lowering of the freezing temperature, that there may be a shift in the fusion curve maximum away from the stoichiometric composition.

Hodgkinson [386, 387] presented a similar explanation of this phenomenon on the basis of a molecular model of disorder.

2. THERMODYNAMIC BASIS OF VARIOUS ESTIMATION PROCEDURES FOR THE DISSOCIATION PARAMETERS OF CONGRUENTLY MELTING COMPOUNDS

There have been a number of attempts in the literature to provide a mathematical justification for the correlation between the liquidus and the degree of dissociation of a congruently melting compound.

Thus, prompted by Kurnakov, Lipin [388, 389] attempted a mathematical description of the fusion curve of a dissociating compound. Using formal mathematical methods, Lipin derived an equation for the fusion of a binary compound, in which the degree of dissociation is a function of temperature, and the dissociation parameter (K) appears only implicitly, so that this equation is not suitable for practical calculations. The problem was solved explicitly for the first time by Stepanov [390], who obtained the equation of the fusion curve of a binary compound, dissociating in the liquid phase, starting from an analysis of the surface of separation of the binary compound in a ternary system.

However, in order to apply the Stepanov equation to the calculation of the dissociation constants of a chemical compound from the fusion diagram, one must have knowledge of two hypothetical quantities: the heat and the temperature of melting in the absence of dissociation. These quantities are not accessible experimentally. Hence, it seems that this equation is usable only for very small dissociations.

In their study of the dependence of the slope of the tangent to the phase equilibrium curves on the degree of dissociation, van der Waals and Kohnstamm [12] made use of Eq. (2.182) (derived by van der Waals) in the following form:

$$\left(\frac{\partial T}{\partial x}\right)_p^L = -\frac{T\,(x^S - x^L)\,(\partial^2 g/\partial x^2)_{p,\,T}^L}{\bar{q}^{(S \to L)}}. \tag{6.10}$$

In order to bypass the difficulties associated with the direct analysis of real systems, the authors split the thermodynamic potential g into two parts. One of these refers to the completely dilute state, and the other portion describes the increment due to "condensation." As a result of this device one obtains

$$\left(\frac{\partial^2 g}{\partial x^2}\right)_{p,\,T} = RT\left[\frac{1}{1-x-u} + \frac{1}{x-u} - \frac{(2x-1)^2\,u\,(x-u)\,(1-x-u)}{(x-u)^2\,(1-x-u)^2\,\Phi}\right], \tag{6.11}$$

where

$$\Phi = u\,(x-u) + u\,(1-x-u) + (x-u)\,(1-x-u).$$

If one now considers the point on the diagram at which the overall compositions of the phases are identical, i.e., where $x^L = x^S$, then according to Eq. (6.10) the derivative $(\partial T/\partial x)_p^L$ will be nonzero only if the second derivative $(\partial^2 g/\partial x^2)_{p,T}^L$ is infinite. From relation (6.11) we find that this occurs either when $1 - x^L - u^L = 0$, or when $x^L - u^L = 0$. What this means is that at the point of the maximum the compound AB does not dissociate into its components at all. The actual number of its moles is equal to the analytic mole fraction of the components $u^L = 1 - x^L = x^L = 1/2$.

In the opinion of van der Waals and Kohnstamm, as long as some dissociation, however small, exists in a crystallizing compound, we have $(\partial T/\partial x)_p^L = 0$ at $x^L = x^S$ and the maximum of the liquidus curve is flat. But this does not mean that the curve $T = f(x)$ will always have the same independence from the degree of dissociation, and that for a completely nondissociating compound this curve will have a sharp peak. On the contrary, from the behavior of the curve $T = f(x)$ near the maximum, one may infer the degree of dissociation.

In order to illustrate the continuity of the transition from flat to cusp maxima as the degree of dissociation falls, van der Waals and Kohnstamm considered, on the basis of Eq. (6.11), a particular example of a compound whose degree of dissociation was zero in the solid phase and very small in the liquid. They obtained the following equation for the tangent near the maximum:

$$(-\bar{q}^{(S \to L)}/RT^2)\,(\partial T/\partial x)^L = 3/2 - (5u_0/4\,\Delta x_{\underline{\cdot}}^L), \tag{6.12}$$

where u_0 is the number of dissociating moles of the AB compound, $\Delta x^L = 1/2 - x^L$, and they satisfy $\Delta x^L \ll 1$ and $u_0 < \Delta x^L$.

It can be seen from (6.12) that for a weakly dissociating compound not too far from the maximum the derivative increases as the degree of dissociation falls. In the limit $u_0 = 0$, Eq. (6.12) gives an expression which is different from that usually obtained from the theory of a limiting dilute solution. This is undoubtedly related to the series of assumptions used in deriving it. For this reason the equation given is approximate and difficult to apply in practice [370].

Schottky, Ulich, and Wagner's [20] investigation of the form of the liquidus and solidus curves, and their mutual relationship, also involved the assumption of a small degree of dissociation of the compounds in the ordering and disordering limits. But neither the dissociation constant nor the degree of dissociation enter explicitly into their equation, and the formal mathematical simplifications they made do not allow one to relate the cases discussed by them to any of the usual classes of solutions (limiting dilution, ideal, perfect). A fairly simple equation involving these constants was given by these authors only in the dilute limit of a liquid solution of the components in a chemical compound, whose dissociation is very small in the melt and zero in the solid phase:

$$\sqrt{(\alpha_0^L)^2 + 4\,(\Delta x^L)^2} - \alpha_0^L = -\bar{q}^{(S \to L)}\,\Delta T/(RT_{\max}^2), \tag{6.13}$$

where α_0 is the degree of dissociation of the compound AB at the melting temperature.

This equation describes well the continuous nature of the shape changes of the liquidus maximum with increasing dissociation. Thus, if $\alpha_0 = 0$, then

$$\Delta T / (2 \, \Delta x^L) = -R T_{m, \, AB}^{\circ 2} / \bar{q}^{(S \to L)}, \tag{6.14}$$

which is identical to the corresponding rule in limiting dilute solutions. The factor 2 in front of Δx^L is the result of using a different scale along the composition axis: the abscissa in Eq. (6.14) has been halved, compared to the usual ones.

Furthermore, if $a_0^L \neq 0$ and $\Delta x^L \to 0$, then for compositions corresponding to the formula of the compound, we have

$$\lim | \Delta T / \Delta x |_{\Delta x \to 0}^L = dT / dx_A = 0,$$

and the maximum is flat. Finally, for a given small Δx^L, the larger a_0^L, the smaller the lowering of ΔT. In other words, the higher the degree of dissociation, the flatter the maximum.

Esin [370] investigated the dependence of the slope of the tangents to the phase equilibrium curves on the degree of dissociation, using the model of ideal associated solutions.

Let the components A and B in a two-component two-phase system form the same chemical compound AB in both phases, according to the scheme

$$\begin{array}{c} A_{sat}^S + B_{sat}^S \rightleftarrows AB^S \\ \updownarrow \qquad \updownarrow \\ A_{sat}^L + B_{sat}^L \rightleftarrows AB^L. \end{array} \tag{6.15}$$

In accordance with the conclusion of Esin [370], we may restrict our consideration to one-half of the state diagram, where x_B^L and x_B^S are smaller than or equal to one-half. From elementary considerations it follows that

$$x_{A_1} = \frac{1 - (2 - \alpha) \, x_B}{1 - (1 - \alpha) \, x_B}, \qquad x_{B_1} = \frac{\alpha x_B}{1 - (1 - \alpha) \, x_B};$$

$$x_{AB} = \frac{(1 - \alpha) \, x_B}{1 - (1 - \alpha) \, x_B}. \tag{6.16}$$

Thus the reciprocal of the dissociation constant of the compound AB is given by

$$\frac{1}{K^L} = \frac{x_{AB}^L}{x_{A_1}^L x_{B_1}^L} = \frac{1 - \alpha^L}{\alpha^L} \cdot \frac{1 - (1 - \alpha^L) \, x_B^L}{1 - (2 - \alpha^L) \, x_B^L}. \tag{6.17}$$

Using (6.16) and (6.17), we obtain for an ideal dissociated solution

$$\begin{aligned} \left(\frac{\partial \mu_B}{\partial x_B} \right)_{p, \, T}^L &= \frac{RT}{\alpha^L x_B^L} \left\{ \frac{\alpha^L}{1 - (1 - \alpha^L) \, x_B^L} + \frac{x_B^L \left(1 - x_B^L \right)}{1 - (1 - \alpha^L) \, x_B^L} \right. \\ &\left. \times \frac{(1 - 2\alpha^L) \, [(K^L)^{-1} + 1] + (\alpha^L)^2 \, [(K^L)^{-1} + 1]^2}{[(K^L)^{-1} + 1] \{ 1 + (\alpha^L)^2 \, [(K^L)^{-1} + 1] - 2\alpha^L \}} \right\}. \end{aligned} \tag{6.18}$$

It follows from this relation that for $\alpha > 0$ the quantity $(\partial \mu_B / \partial x_B)_{p,T}{}^L$ is finite, while for $\alpha = 0$ it is infinite.

Combining Eqs. (6.17) and (6.18) with the derivatives on the liquidus

$$\left(\frac{\partial T}{\partial x_B}\right)_p^S = -\frac{\left(x_B^L - x_B^S\right)\left(\partial \mu_B / \partial x_B\right)_{p,\,T}^S\, T}{\left(1 - x_B^S\right)\bar{q}^{(S \to L)}} \qquad (6.19)$$

and on the solidus

$$\left(\frac{\partial T}{\partial x_B}\right)_p^L = -\frac{\left(x_B^S - x_B^L\right)\left(\partial \mu_B^{\text{''}} / \partial x_B\right)_{p,\,T}^L\, T}{\left(1 - x_B^L\right)\bar{q}^{(L \to S)}} \qquad (6.20)$$

we obtain the required form of the maxima as a function of the degree of dissociation of the chemical compound, provided we can neglect the change in $\overline{q}^{(\alpha \to \beta)}$ with composition in the range under study. In particular, it is easy to obtain thereby [370] all the basic types of maxima for the various chemical compounds classified by Kurnakov [44].

The above method, which is based on an analysis of the slope of the tangent to the fusion curve at the maximum, is inconvenient, in that it does not allow one to trace the continuous transition from a flat to a sharp maximum as the degree of dissociation is lowered. Indeed, the slope of the tangent at the maximum is always zero for any $\alpha > 0$, except for the single case of $\alpha = 0$. The consideration of the slopes of tangents at points near the maximum, under the assumption of slight dissociation [12], is only convenient for limiting dilute solutions [12, 20]. Furthermore, the analysis using an ideal associated solution with arbitrary degrees of dissociation for the compound AB is complicated by the cumbersome initial equation.

In this context Mlodzeevskii's work [43, 382, 383] is of great interest. He gave an expression for the radius of curvature of the maximum as a function of the dissociation constants of an AB-type chemical compound in both phases. The radius of curvature of the maximum is different for each value of α and becomes zero only at $\alpha = 0$.

By means of a geometrical method Mlodzeevskii has derived an equation for the relation between the forms of the Gibbs free-energy curve and the fusion curve. To this end he considered the curve of G in a very small temperature interval $T_m° - T$, so that he could expand G^L and G^S up to first-order terms in the expansion.

Taking the first approximation of the G curve near the maximum to be a parabola, Mlodzeevskii showed [43] that the liquidus and solidus near the maxima are also parabolas. He ultimately obtained an equation relating the radius of curvature of the liquidus and solidus with the dissociation constant of the compound AB and the degree of deviation of the maximum from stoichiometry.

However, as pointed out correctly by Esin [370], the derivation of Mlodzeevskii is not free of formal mathematical assumptions which cannot be specified thermodynamically. The result of this is that the parabolicity of $T = f(x_i)$, justified by Mlodzeevskii in terms of a limiting dilute solution, and the form of the G curves obtained by him, in fact pertain to an ideal associated solution.

Esin obtained [370, 384] the equations of Mlodzeevskii by usual thermodynamic means, without making additional assumptions concerning the form of the curves near the maximum, and generalized them to real systems by introducing the activity coefficients.

The radius of curvature at the maximum of the fusion curve is clearly inversely proportional to the second derivative of the temperature with respect to composition:

$$\rho_{max} = [(\partial^2 T/\partial x_B^2)_{p, \, max}]^{-1}. \tag{6.21}$$

Hence, the whole problem reduces to finding this derivative.

Using the equilibrium condition for the liquid and solid phases in a two-component system in the form (2.182) given by van der Waals and Kohnstamm at $p = $ const, together with expression (5.34a) and its analog for the differential molar heat of formation of the liquid from the solid, we have

$$\left(\frac{\partial T}{\partial x_B}\right)_p^S = -\frac{T\left(x_B^L - x_B^S\right)\left(\partial^2 g/\partial x_B^2\right)_{p, \, T}^S}{\bar{q}^{(S \to L)}}. \tag{6.22}$$

$$\left(\frac{\partial T}{\partial x_B}\right)_p^L = -\frac{T\left(x_B^S - x_B^L\right)\left(\partial^2 g/\partial x_B^2\right)_{p, \, T}^L}{\bar{q}^{(L \to S)}}. \tag{6.23}$$

Dividing the two equations by one another, and using the equality

$$\bar{q}^{(S \to L)} = -\bar{q}^{(L \to S)}, \tag{6.24}$$

valid at the maximum [13], we find

$$\left(\frac{\partial x_B^L}{\partial x_B^S}\right)_p = \frac{\left(\partial^2 g/\partial x_B^2\right)_{p, \, T}^S}{\left(\partial^2 g/\partial x_B^2\right)_{p, \, T}^L}. \tag{6.25}$$

Considering the slopes of the tangents in (6.22) and (6.23) to be functions of x_B^S, x_B^L, T, and $(\partial^2 g/\partial x_B^2)_{p,T}$, and assuming a constant $\bar{q}^{(\alpha \to \beta)}$, we obtain at the maximum for the solidus:

$$\left(\frac{\partial^2 T}{\partial x_B^2}\right)_{p, \, max}^S = -\frac{T_{max}}{\bar{q}^{(L \to S)}}\left(\frac{\partial^2 g}{\partial x_B^2}\right)_{max}^S\left[\left(\frac{\partial x_B^L}{\partial x_B^S}\right)_p - 1\right] \tag{6.26}$$

and similarly for the liquidus

$$\left(\frac{\partial^2 T}{\partial x_B^2}\right)_{p, \, max}^L = -\frac{T_{max}}{\bar{q}^{(L \to S)}}\left(\frac{\partial^2 g}{\partial x_B^2}\right)_{max}^L\left[\left(\frac{\partial x_B^S}{\partial x_B^L}\right)_p - 1\right]. \tag{6.27}$$

Hence, the radii of curvature, in accordance with (6.21) and (6.25), will be given by

$$\rho_{max}^S = -\frac{\left(\partial^2 g/\partial x_B^2\right)_{max}^L \bar{q}^{(S \to L)}}{T_{max}\left(\partial^2 g/\partial x_B^2\right)_{max}^S\left[\left(\partial^2 g/\partial x_B^2\right)_{max}^S - \left(\partial^2 g/\partial^2 x_B\right)_{max}^L\right]}, \tag{6.28}$$

$$\rho_{max}^L = -\frac{(\partial^2 g/\partial x_B^2)_{max}^{(S)}\ \bar{q}^{(S\to L)}}{T_{max}\ (\partial^2 g/\partial x_B^2)_{max}^L\ [(\partial^2 g/\partial x_B^2)_{max}^L - (\partial^2 g/\partial x_B^2)_{max}^S]}. \tag{6.29}$$

In order to obtain the explicit dependence of the radius of curvature on the dissociation constant, Esin used [370, 384] relations of the type (5.3).

Thus, in all the preceding cases the core of the problem is to find an explicit expression for $(\partial \mu_B/\partial x_B)_{p,T}$.

Differentiating Eq. (6.2) with respect to x_B and using (6.3) to find the derivative of u with respect to x_B, we obtain $(\partial \mu_B/\partial x_B)_{p,T}$. Since the composition at the maximum does not normally coincide with the formula for the chemical compound, we shall substitute for x_B the quantity $1/2 + \delta$, where the shift of the maximum δ does not exceed $1/2$. As a result we obtain after some simple algebra

$$\left(\frac{\partial^2 g}{\partial x_B^2}\right)_{max} = \frac{4RT}{(1-4\delta^2)}\ \sqrt{\frac{K'+1}{K'+4\delta}}$$
$$+ \frac{RT(1-2\delta)(\partial \ln \gamma_{AB,\ B_1,\ A_1}/\partial x_B)_{max}}{\sqrt{1+(1/K')}\ \sqrt{1+(4\delta^2/K')}\ [K'+\sqrt{K'+1}\ \sqrt{K'+4\delta^2}+2\delta]}$$
$$+ \frac{2RT}{1-2\delta}\left(\frac{\partial \ln \gamma_{B_1}}{\partial x_B}\right)_{max}, \tag{6.30}$$

where

$$\left(\frac{\partial \ln \gamma_{AB,\ A_1,\ B_1}}{\partial x_B}\right)_{max} = \left\{\frac{\partial \ln [\gamma_{AB}/(\gamma_{A_1}\gamma_{B_1})]}{\partial x_B}\right\}_{max}.$$

For the particular case of an ideal associated solution, Eq. (6.30) becomes

$$\left(\frac{\partial^2 g}{\partial x_B^2}\right)_{max}^S = \frac{4RT}{1-4\delta^2}\ \sqrt{\frac{K^S+1}{K^S+4\delta^2}} = \frac{4RT}{1-4\delta^2}\frac{1}{\varkappa^S}, \tag{6.31}$$

$$\left(\frac{\partial^2 g}{\partial x_B^2}\right)_{max}^L = \frac{4RT}{1-4\delta^2}\ \sqrt{\frac{K^L+1}{K^L+4\delta^2}} = \frac{4RT}{1-4\delta^2}\frac{1}{\varkappa^L}, \tag{6.32}$$

where $\varkappa \equiv \sqrt{(K=4\delta^2)/(K+1)}$.

The radius of curvature of, for example, the liquidus maxima may be found, via (6.29) and (6.32), from the expression

$$\rho_{max}^L = -\frac{(1-4\delta^2)\varkappa^L}{4RT_{max}^2}\frac{\varkappa^L}{\varkappa^L-\varkappa^S}\ \bar{q}^{(S\to L)}. \tag{6.33}$$

If the compound AB is completely nondissociating in the solid phase ($K^S = 0$), then by using (6.24) at the maximum, as well as

$$\bar{q}^{(S \to L)} = (m + n)^{-1} \Delta h^\circ_{m,\, AB} \tag{6.34}$$

we are led, via (6.33), to the equation

$$\rho^L_{max} = -\frac{\Delta h^\circ_{m,\, AB}}{8RT^{\circ 2}_{m,\, AB}} \sqrt{\frac{K^L}{K^L + 1}} = -\frac{\Delta h^\circ_{m,\, AB}}{8RT^{\circ 2}_{m,\, AB}} \alpha^L_0, \tag{6.35}$$

since in this case $\delta = 0$, as shown before.

Equations (6.33) and (6.35) enable one to derive not only all possible combinations of singular and flat maxima, but also show the continuous change of the curvature of the maximum with the degree of dissociation for an ideal dissociated solution. From Eq. (6.35) it is seen immediately that the larger the dissociation constant $K^{(L)}$, the larger the radius of curvature of the maximum, and the flatter it becomes. In the complete absence of dissociation ($K^L = 0$) the radius of curvature is zero and the maximum is a cusp.

The same consequences follow from the more complicated equation (6.33). In order to display the explicit dependence of ρ on K from this equation one must know how δ and K are related. This may be obtained from expression (6.9).

In the ideal associated solution approximation the enthalpy of one gross (overall) mole of any phase may be found from the relation

$$h = (x_B - u)\, h^\circ_B + (1 - x_B - u)\, h^\circ_A + u h^\circ_{AB}, \tag{6.36}$$

where $h_B{}^\circ$, $h_A{}^\circ$, and $h_{AB}{}^\circ$ are the enthalpies of the components A, B, and the compound AB for given p, T and state of aggregation.

From the expression for the dissociation constant of the compound and formula (6.36) it is easy to verify that

$$\bar{q}^{(S \to L)} = (1/2)\, \Delta h_{m,\, AB} + \delta\, (\Delta h^\circ_{m,\, A} - \Delta h^\circ_{m,\, B})$$
$$- (1/2)\, \Delta H^{\circ\, (S)} \varkappa^S + (1/2)\, \Delta H^{\circ\, (L)} \varkappa^L, \tag{6.37}$$

where ΔH° is the molar heat of dissociation of the compound AB at given p and T in the ideal associated solution.

From the set of equations (6.33) and (6.37) one may evaluate the changes in ρ and K. We shall analyze these equations following the treatment of [384].

Since by the condition of our treatment ρ is taken to be negative and the quantity $\bar{q}^{(S \to L)} > 0$, it follows from (6.33) that $K^L > K^S$. In other words, at the maximum the dissociation in the liquid is larger than in the solid. Since the values of both constants change in the same direction (from 0 to ∞), this means that K^L grows faster than K^S. From $K^L > K^S$ it also follows that the distribution coefficients K_A and K_B will be proper fractions, equal to one or zero in the limiting case. It is easy to show that the denominator of formula (6.9) will in this case be always positive and, consequently, that the shift is fixed only by the sign of the difference ($K_A -$

K_B). The absolute value of δ increases monotonically with $x_{A_1}^L$ and K from zero up to the value

$$|\delta| = \frac{1}{2} \frac{1}{1 + 2(1 - K_A)/(K_A - K_B)}.$$

This implies that as K^L grows, the quantities $\sqrt{(K^L + 4\delta^2)/(K^L + 1)}$ and $\sqrt{(K^S + 4\delta^2)/(K^S + 1)}$ increase from zero to unity, the first one faster than the second. Therefore, the second factor in Eq. (6.33) increases continuously with K^L.

Seeing that the quantity $(1 - 4\delta^2)$ decreases more slowly (from 1 to 0) than the increase of x^L (from 0 to 1), the first factor of Eq. (6.33) increases together with K^L. As regards the third factor, its change with x^L according to (6.37) will take the form

$$\frac{d\bar{q}^{(S \to L)}}{dx^L} = \frac{1}{2} \Delta H^{\circ\,(L)} + \left(\Delta h_{m,\,A}^{\circ} - \Delta h_{m,\,B}^{\circ} \right) \frac{d\delta}{dx^L} - \frac{1}{2} \Delta H^{\circ\,(S)} \frac{dx^S}{dx^L}. \quad (6.38)$$

In this approximation the heat of dissociation is always positive, while the difference in the heat of melting of the components may assume any sign. However, since the derivative of δ is less than 1/2, while that of x^S is less than one, and both are quite small, the sign of the whole expression (6.38) is fixed by $\Delta H^{\circ(L)}$, so that it is positive. Consequently, the absolute value of $\bar{q}^{(S \to L)}$ also increases together with K^L. In the unfavorable case in which the compound dissociates practically completely in both phases, and the maximum lies almost on the ordinate of the first component, the distribution coefficients approach one another, $d\delta/dx^L \to 0$, and the derivative of x^S approaches unity. In that case the increment $(d\bar{q}^{(S \to L)}/dx^L)$ very slightly exceeds the quantity $1/2(\Delta h_{m,A}^{\circ} + \Delta h_{m,B}^{\circ} - \Delta h_{m,AB}^{\circ})$. Since under these conditions the heat of melting of the compound AB hardly differs from the sum of the heats of melting of the components, the sign of expression (6.38) remains positive, as before. Hence, all three factors in (6.33), and therefore also $|\rho|$, increase monotonically with K^L. When $K^L = 0$, $\rho_{max}^L = 0$, and the maximum is singular. Accordingly, the rule enunciated by Kurnakov has been given a thermodynamic justification, at least for ideal associated solutions.

Nevertheless, the value of ρ depends not only on K but also on other quantities characterizing the nature of the materials A, B, and AB. In the words of Esin [384]: "While it will not be proved that the influence of these quantities in going from one compound to another will affect ρ to the same degree as a change in K, it is nevertheless impossible to effect a full comparison of the degree of dissociation of various compounds solely on the basis of the curvature of the maxima."

In fact, even in the simplest case of a compound dissociating only in the liquid phase, it is seen from formula (6.35) that the absolute value of the radius of curvature of the maximum may assume different values for one and the same dissociation constant (K^L) as a function of $\Delta h_{m,AB}^{\circ}$.

Application of Eq. (6.30) to the analysis of real solutions at any concentration shows that the various limiting behaviors which apply in ideal associated solutions may also be obtained for real solutions, provided one assumes that the expressions $(\partial \ln \gamma_{AB,A_1B_1}/\partial x_B)_{\max}$ and $(\partial \ln \gamma_{B_1}/\partial x_B)_{\max}$ are finite.

These assumptions, necessary for the existence of a cusp-like maximum, are completely justified and, in fact, follow from the thermodynamic definition of the activity coefficients, namely, from the fact that in going to the limit of a dilute solution of particles A_1 and B_1 in the compound AB these coefficients tend to unity.

But in order to be able to introduce all the basic forms of the state diagrams of two-component two-phase systems forming a chemical compound and, in particular, to show the continuous transition from a flat to a cusp maximum as dissociation increases, it is necessary to assume that the terms in (6.30) containing the activity coefficients should be smaller in absolute value and vary more slowly with changes in the degree of dissociation than the first term. However, the discussion of the reliability of this assumption is beyond the realm of purely thermodynamic studies and should be conducted either on the basis of experiment or of molecular kinetic theory.

Recently the present authors [391, 392] have generalized the method of estimating the degree of dissociation of a congruently melting compound from the liquidus curve to the case when a complicated compound A_mB_n is formed in the system.

Let us consider the state diagram of A–B with a congruently melting compound A_mB_n possessing a narrow region of homogeneity and practically nondissociating in the solid phase. We suppose that only the following equilibria occur in the system:

$$A_mB_n^S \rightleftarrows A_mB_n^L \rightleftarrows m\,(A)_{\text{sat}}^L + n\,(B)_{\text{sat}}^L . \tag{6.39}$$

As was shown in [370], using the theory of ideal associated solutions, when $a_0^S = 0$ the derivative $(\partial^2 g/\partial x_B^2)_{\max}^S$ becomes very large. Using this fact, together with (6.29), (6.34), it is easily verified that in the absence of dissociation in the solid phase the radius of curvature of the liquidus may be found from the relation

$$\rho_{\max}^L = -\frac{\Delta h_{m,\,A_mB_n}^\circ}{(m+n)\,T_{m,\,A_mB_n}^\circ}\,\frac{1}{(\partial^2 g/\partial x_B^2)_{p,\,T}^L} . \tag{6.40}$$

For the present case

$$\mu_B = \mu_B^\circ + RT \ln \{(x_B - nu)/[1 - u\,(m+n-1)]\}. \tag{6.41}$$

Hence, remembering (5.3), it is readily shown that

$$\left(\frac{\partial^2 g}{\partial x_B^2}\right)_{p,\,T} = \frac{1 - u\,(m+n-1) - (du/dx_B)\,[1 - x_B\,(m+n-1)]}{(x_B - nu)\,[1 - u\,(m+n-1)]\,(1-x_B)}\,RT. \tag{6.42}$$

We shall assume that at the point corresponding to the composition of the compound the number of moles $n_{A_mB_n}$ in the melt is a maximum, i.e., the following condition holds:

$$(\partial u/\partial x_B)_{\max} = 0. \tag{6.43}$$

This will give

$$\left(\frac{\partial^2 g}{\partial x_B^2}\right)_{max} = RT^{\circ}_{m,\ A_m B_n} \frac{(m+n)^2}{mn\,[1-(m+n)\,u]}\,.\qquad(6.44)$$

From expressions (6.44) and (6.40) we find for the maximum:

$$u^L_{max} = \frac{1}{m+n} + \frac{(m+n)^2}{mn}\frac{\rho^L_{max}RT^{\circ 2}_{m,\ A_m B_n}}{\Delta h^{\circ}_{m,\ A_m B_n}},\qquad(6.45)$$

where the number of analytic moles of the compound needed in one overall (gross)-mole of the phase of stoichiometric composition is decreased, while the number of dissociating moles of the compound $A_m B_n$ is increased.

Hence, using the definition of α, we find for the maximum:

$$\alpha^L_0 = -\frac{(m+n)^3}{mn}\frac{\rho^L_{max}RT^{\circ 2}_{m,\ A_m B_n}}{\Delta h^{\circ}_{m,\ A_m B_n}}\,.\qquad(6.46)$$

When $m = n = 1$, expression (6.46) leads to an equation of the form (6.35) for a binary compound.

The methods of estimating the degree of dissociation of a compound presented in the work of Mlodzeevskii and Esin were applied for the first time by Patsukova [374] to calculate α_0^L for a series of binary salts. To determine the radius of curvature at the maximum, she used a Lagrangian interpolation scheme.

Wyatt [367, 393–395] developed a method for estimating the degree of dissociation of a compound in the liquid phase from data on crystallization temperatures, or from vapor pressure data of solutions near the composition of the compound.

Following Wyatt [394], we choose AB as the solvent dissociating into atoms of A and B. We add to it substance A, considering the latter as the second component. It is easily verified that

$$x_{AB} = m_1\,(1-\alpha)/n,\qquad(6.47)$$

$$x_{A_1} = (\alpha m_1 + m_2)/n,\qquad(6.48)$$

$$x_{B_1} = \alpha m_1/n,\qquad(6.49)$$

where n is the total number of particles in the mixture, $n = (1-\alpha)m_1 + m_2$; m_i is the analytic concentration of the ith substance, expressed in molalities.

On the basis of Eq.(6.47), the Gibbs–Duhem relation $m_1 d\mu_1 + m_2 d\mu_2 = 0$, and the condition for chemical equilibrium $\mu_{AB} = \mu_{A_1} + \mu_{B_1}$ with $m_1 = $ const, one may derive the following system of equations:

$$\left(\frac{\partial \ln x_{AB}}{\partial m_2}\right)_{m_1} - \left[\frac{\partial \ln (1-\alpha)}{\partial m_2}\right]_{m_1} + \left(\frac{\partial \ln n}{\partial m_2}\right)_{m_1} = 0,$$

$$m_1 \left(\frac{\partial \ln x_{AB}}{\partial m_2} \right)_{m_1} + m_2 \left[\frac{\partial \ln (\alpha m_1 + m_2)}{\partial m_2} \right]_{m_1} - m_2 \left(\frac{\partial \ln n}{\partial m_2} \right)_{m_1} = 0,$$

(6.50)

$$\left(\frac{\partial \ln x_{AB}}{\partial m_2} \right)_{m_1} - \left[\frac{\partial \ln (\alpha m_1 + m_2)}{\partial m_2} \right]_{m_1} - \left(\frac{\partial \ln \alpha}{\partial m_2} \right)_{m_1} + 2 \left(\frac{\partial \ln n}{\partial m_2} \right)_{m_1} = 0.$$

In order to solve these equations we need a fourth equation. For this Wyatt [394] proposed the relation

$$n_1 \, (\partial \ln n_1 / \partial m_2)_{m_1} - m_1 \, (\partial \alpha / \partial m_2)_{m_1} = 1,$$

(6.51)

where $n_1 = \alpha m_1 + m_2$.

From formula (6.50) and (6.51) we get

$$\left(\frac{\partial \ln x_{AB}}{\partial m_2} \right)_{m_1} = - \frac{m_2}{(2\alpha m_1 + m_2)(m_1 + m_2)}.$$

(6.52)

But in the ideal associated solution model we have

$$\frac{1}{RT} \left(\frac{\partial \mu_1}{\partial m_2} \right)_{m_1} = \left(\frac{\partial \ln p_1}{\partial m_2} \right)_{m_1} = \left(\frac{\partial \ln x_{AB}}{\partial m_2} \right)_{m_1},$$

(6.53)

where p_1 is the partial vapor pressure of the compound above the solution. Substituting (6.53) into (6.52), we therefore obtain

$$\alpha = - \frac{m_2}{2m_1} \left[\frac{1}{(m_1 + m_2)(\partial \ln p_1 / \partial m_2)_{m_1}} + 1 \right].$$

(6.54)

Meanwhile, using data on the crystallization temperature of solutions near the composition of the compound, the chemical potential of the solvent may be written in the form [394]

$$\mu_1 = \mu_1^\circ (T, \, p) - (\Delta h_{m, \, 1}^\circ / T_{m, \, 1}^\bullet) \, \Delta T,$$

(6.55)

$$\left(\frac{\partial \mu_1}{\partial m_2} \right)_{m_1} = - \frac{\Delta h_{m, \, 1}^\circ}{T_{m, \, 1}^\circ} \left(\frac{\partial \Delta T}{\partial m_2} \right)_{m_1} - \frac{\Delta T}{T_{m,1}^\circ} \left(\frac{\partial \Delta h_{m, \, 1}^\circ}{\partial m_2} \right)_{m_1},$$

(6.56)

where the derivative $(\partial \Delta h_{m,1}^\circ / \partial m_2)_{m_1}$ describes the change of $\Delta h_{m,1}^\circ$ due to the inhibition of the dissociation of the solvent. If it is assumed that this derivative will be small for small values of m_2, we may write approximately [394]

$$(\partial \mu_1 / \partial m_2)_{m_1} \simeq - (\Delta h_{m, \, 1}^\circ / T_{m, \, 1}^\circ)(\partial \Delta T / \partial m_2)_{m_1}.$$

(6.57)

In view of the expression for the cryoscopic constant [24, 130]

$$\theta_k = R \, (T_{m, \, 1}^\circ)^2 / \Delta h_{m, \, 1}^\circ m_1,$$

one may rewrite relation (6.57) in the form

$$(\partial\mu_1/\partial m_2)_{m_1} = -(RT^\circ_{m,\,1}/\theta_k m_1)(\partial\Delta T/\partial m_2)_{m_1}. \qquad (6.58)$$

The joint solution of Eqs. (6.58) and (6.52) leads, together with (6.53), to the expression

$$\alpha = \frac{m_2}{2m_1}\left[\frac{m_1\theta_k}{(m_1+m_2)(\partial\Delta T/\partial m_2)_{m_1}} - 1\right] \qquad (6.59)$$

or, in an alternative form (cf. [396]),

$$\alpha = \left|\frac{1-2x_2}{2x_2}\right|\left[\frac{RT^{\circ 2}_{m,\,AB}}{\Delta h^\circ_{m,\,AB}(T^\circ_{m,\,AB})\,x_2(1-x_2)\,|\,\partial T/\partial x_2|_p} - 1\right]. \qquad (6.60)$$

As shown by Wyatt [394], one may obtain, on the basis of

$$m_1\left(\partial^2\mu_1/\partial m_2^2\right)^\circ_{m_1} = -\left(\partial\mu_2/\partial m_2\right)^\circ_{m_1},$$

an equation for calculating the degree of dissociation of a compound at the melting point, in terms of data on phase equilibria, or the pressure of the saturated vapor of the compound AB:

$$\alpha_0 = -\frac{1}{2m_1^2\left(\partial^2\ln p_1/\partial m_2^2\right)^\circ_{m_1}}, \qquad (6.61)$$

$$\alpha_0 = \frac{\theta_k}{2m_1\left(\partial^2\Delta T/\partial m_2^2\right)^\circ_{m_1}} = -\frac{\theta_k}{2m_1\left(\partial^2 T/\partial m_2^2\right)^\circ_{m_1}}. \qquad (6.62)$$

The superscript "o" indicates that the quantity so marked refers to the melting point of the compound; Eqs. (6.61)–(6.62) were derived using (6.53) and (6.52) or (6.58), together with the Gibbs–Duhem relation. Simple algebra shows the equivalence of Eqs. (6.62) and (6.35).

In addition Wyatt [394] generalized (6.61) and (6.62) to the case when a compound S dissociates by the following complex scheme:

$$\nu_{AB}S \rightleftharpoons \nu_A A + \nu_B B,$$
$$\downarrow\uparrow \qquad \downarrow\uparrow$$
$$\nu_A a \qquad \nu_B b$$

where a and b are the number of particles which are formed, in turn, from molecules A and B, respectively. The final result in this general case is

$$\alpha_0 = -ab\nu_{AB}\nu_B \left/ \left[\nu_A(a\nu_A+b\nu_B)\,m_1^2\left(\frac{\partial^2\ln p_1}{\partial m_2^2}\right)^\circ_{m_1}\right]\right., \qquad (6.63)$$

$$\alpha_0 = -\theta_k ab\nu_{AB}\nu_B\left/\left[\nu_A(a\nu_A+b\nu_B)\,m_1\left(\partial^2 T/\partial m_2^2\right)^\circ_{m_1}\right]\right.. \qquad (6.64)$$

Dawber and Wyatt [397] have analyzed the consequences of including in the formulas for α^L the activity coefficients of particles forming the associated solution, as estimated in the regular solution approximation. They concluded that the error incurred in neglecting the nonideal behavior of the mixture of A_1, B_1, and AB particles is practically nil, if α^L is between 0 and 0.01, and about 4–5% for $\alpha^L = 0.1$.

In this work on the liquidus curve of the initial crystallization of an AB compound in the regular associated approximation, Jordan [186] obtained an equation of the form of (5.249), which included two unknown parameters: the degree of dissociation of the compound α°, and a function of the energy ω^* of the exchange interaction of the particles in the solution. Since each temperature on the liquidus corresponds to a value α°, one cannot use the equation, in the form proposed by Jordan, for computing the dissociation parameters of compounds from an analysis of the liquidus curves.

3. ESTIMATES OF THE DEGREE OF DISSOCIATION OF CHEMICAL COMPOUNDS IN VARIOUS APPROXIMATIONS OF THE THERMODYNAMICS OF SOLUTIONS

One may obtain a quantitative estimate of the degree of dissociation using the idea of Kurnakov, which connects it with the curvature of the liquidus. This may be accomplished in two ways:

1. By detailed experimental investigation of the liquidus curve corresponding to the initial crystallization of the compound A_mB_n, and the subsequent processing of this data on a computer, with a view to formal approximations in terms of various polynomials. We may call this the formal method, since the description of the liquidus by means of polynomials has nothing to do with the intrinsic character of the intermolecular interactions in the system. At the same time, the formally estimated curvature of the liquidus may be considered to be an experimental estimate, in that the basis of the method is the experimental data, and the accuracy of the estimate will be determined by the accuracy of the experimental method and the number of experimental points on the curve.

2. By model descriptions of the liquidus in terms of various models of the thermodynamics of solutions, and by the derivation of explicit expressions for the dissociation parameters appropriate to the models, on the basis of general thermodynamic relations.

The limitation of the first method is the need for detailed experimental studies of the liquidus curve, especially in the vicinity of the compound A_mB_n, which is not always possible.

The second approach will be limited by the intrinsic inadequacy of whatever model is being used for the solution.

The difficulty of estimating the degree of dissociation is not removed by finding the radius of curvature. For the model approach one still has to introduce and analyze the expressions for the dissociation parameters. We note that this approach was developed by the present authors [312, 391, 392, 398–402], and was substantiated by comparison with experimental estimates of the radius of curvature in a series of concrete systems, including compounds of the type AB and A_mB_n. We shall consider the estimate of the dissociation parameters for congruently melting chemical compounds in a generalized manner, and will then provide some illustrative examples for a series of systems, including semiconducting systems of various types.

It follows from a comparison of Eqs. (6.40) and (6.46) that the degree of dissociation of the compound A_mB_n in the liquid phase is related to the second derivative of the Gibbs free energy with respect to concentration:

$$\alpha_0^L = \frac{(m+n)^2}{mn} \frac{RT^\circ_{m, A_mB_n}}{(\partial^2 g/\partial x_B^2)_{p, T}}. \tag{6.65}$$

Using formula (5.3) for the second derivative and Eq. (4.140), we obtain in the six-parameter model

$$\left(\frac{\partial^2 g}{\partial x_B^2}\right)_{p, T} = \frac{RT}{x_B(1-x_B)} - 2[\omega_0 - \omega_{01}T + (\omega_1 - \omega_{11}T)(3x_B - 1)$$

$$+ 3x_B(\omega_2 - \omega_{21}T)(2x_B - 1)]. \tag{6.66}$$

At the maximum, when $x_B = n/(m+n)$, we get

$$\left(\frac{\partial^2 g}{\partial x_B^2}\right)_{p, T}^\circ = \frac{(m+n)^2}{mn}\left\{RT^\circ_{m, A_mB_n} - \frac{2mn}{(m+n)^2}\right.$$

$$\times \left[\omega_0 - \omega_{01}T^\circ_{m, A_mB_n} + \frac{2n-m}{m+n}(\omega_1 - \omega_{11}T^\circ_{m, A_mB_n})\right.$$

$$\left.\left. + \frac{3n(n-m)}{(m+n)^2}(\omega_2 - \omega_{21}T^\circ_{m, A_mB_n})\right]\right\}. \tag{6.67}$$

Substituting (6.67) into (6.65), we obtain for the degree of dissociation of the compound A_mB_n within the six-parameter model

$$\alpha_0^L = RT^\circ_{m, A_mB_n}\left\{RT^\circ_{m, A_mB_n} - 2\frac{mn}{(m+n)^2}\left[\omega_0 - \omega_{01}T^\circ_{m, A_mB_n}\right.\right.$$

$$\left.\left. + \frac{2n-m}{m+n}(\omega_1 - \omega_{11}T^\circ_{m, A_mB_n}) + \frac{3n(n-m)}{(m+n)^2}(\omega_2 - \omega_{21}T^\circ_{m, A_mB_n})\right]\right\}^{-1}. \tag{6.68}$$

On the basis of this equation one may arrive at expressions for the degree of dissociation in the subregular, quasiregular, and strictly regular solutions.

Thus, in the subregular solution, which takes into account the temperature dependence of the exchange energy, assuming the latter depends linearly on concentration, one finds

$$\omega_2 = 0; \quad \omega_{21} = 0, \tag{6.69}$$

so that for the degree of dissociation one has

$$\alpha_0^L = RT^\circ_{m, A_mB_n}\left\{RT^\circ_{m, A_mB_n} - \frac{2mn}{(m+n)^2}\left[\omega_0 - \omega_{01}T^\circ_{m, A_mB_n}\right.\right.$$

$$\left.\left. + \frac{2n-m}{m+n}(\omega_1 - \omega_{11}T^\circ_{m, A_mB_n})\right]\right\}^{-1}. \tag{6.70}$$

In the quasiregular model the exchange energy is assumed to be independent of concentration, so that, in addition to condition (6.69), one also has

$$\omega_1 = 0; \quad \omega_{11} = 0. \tag{6.71}$$

Then from Eq. (6.68) we obtain

$$\alpha_0^L = RT_{m, \, A_mB_n}^\circ \left[RT_{m, \, A_mB_n}^\circ - \frac{2mn}{(m+n)^2} \left(\omega_0 - \omega_{01} T_{m, \, A_mB_n}^\circ \right) \right]^{-1}. \tag{6.72}$$

Finally, in the model of a strictly regular solution, the exchange energy has no temperature dependence either ($\omega_{01} = 0$), so that

$$\alpha_0^L = RT_{m, \, A_mB_n}^\circ \left[RT_{m, \, A_mB_n}^\circ - \frac{2mn\omega_0}{(m+n)^2} \right]^{-1}. \tag{6.73}$$

For an AB-type compound Eqs. (6.68), (6.70), and (6.72) become, respectively,

$$\alpha_0^L = \frac{4RT_{m, \, AB}^\circ}{T_{m, \, AB}^\circ (4R + 2\omega_{01} + \omega_{11}) - (2\omega_0 + \omega_1)}; \tag{6.74}$$

$$\alpha_0^L = \frac{2RT_{m, \, AB}^\circ}{T_{m, \, AB}^\circ (2R + \omega_{01}) - \omega_0}. \tag{6.75}$$

$$\alpha_0^L = 2RT_{m, \, AB}^\circ / (2RT_{m, \, AB}^\circ - \omega_0). \tag{6.76}$$

Therefore, in the simplest case of an AB compound, the six-parameter and subregular models give identical expressions for the degree of dissociation, which is dictated by the known concentration symmetry of this type of compound.

Rotenberg and Kuznetsov [217] have obtained an equation whose form is close to that of (6.76), by differentiating an equation like (5.77). It differs from (6.76) by the absence of the factor 2 in the numerator. This discrepancy is due to the use of the Mlodzeevskii equation which contains the differential molar heat of formation of a mole of the phase. It was not taken into account in [217], whereas we included it in our exposition.

While on the subject of estimating the degree of dissociation for A_mB_n and AB compounds by means of Eqs. (6.73) and (6.76), it is pertinent to note the following: these formulas were derived in the strictly regular model approximation which, as pointed out above (cf., for example, Fig. 43), assumes the complete dissociation of the compound. This is ultimately not self-consistent. On the other hand, the model assumes complete dissociation, and yet the same model evaluates the degree of dissociation, i.e., assumes partial dissociation. In order to resolve the inconsistency, we go back to the derivation of Eqs. (6.68)–(6.76). They were derived by using the six-parameter model, which contains the parameters ω_0, ω_{01}, ω_1, ω_{11}, ω_2, and ω_{21}, associated with the characteristics of the interatomic interaction, but which do not themselves have a direct physical meaning such as, for example, the exchange energy in the strictly regular model. If, therefore, one thinks of applying the model of strictly regular solutions to systems with partially dissociating compounds, then this can obviously be done only formally, since the physical meaning which pertains to the concept of the exchange energy, and which assumes the presence of only weak intermolecular forces (van der Waals type), is lost in the present case. Accordingly, ω_0 should be treated as a formal parameter which does not have the physical meaning of the energy of mixing in the strictly regular solution model applying to systems without intermediate phases (cf. also [252]).

To put it differently, the model of a strictly regular solution, in the sense of a physical model of the structure of the solution, in principle cannot be applied to systems with partially dissociating congruently melting compounds. Hence the application of this model to such a system is purely formal, and the parameter ω_0 represents all forms of interaction within the solution, whether weak or strong. Consequently, with regard to systems containing chemical compounds, one may speak only of mathematical representations of the phase equilibrium curves, rather than a physical description of the structure of solutions in a given approximation.

In accordance with the foregoing, the equation developed by Vieland, and its analog for the compound $A_m B_n$, enable one to determine some effective parameter ω_0 (excluding the case of complete dissociation), which may then be utilized in the appropriate calculations.

We have therefore resolved the inconsistency referred to above. It was important to mention this, because in certain studies [403, 404] the application of the formal apparatus of the theory of regular solutions to systems with partially dissociated compounds was taken in the literal sense, i.e., it was appraised as a rigorous physical model.

Furthermore, Dutchak et al. [403] and Kornev et al. [404], in considering the problem of dissociation in the limit of order–disorder structural theories, have actually identified the degree of disorder with the degree of dissociation. This is clearly wrong, since the melts of a partially dissociated compound may themselves be distinguished by various degrees of structural disorder, unrelated to the degree of dissociation.

We considered [402] the possibility of estimating the degree of dissociation of an $A_m B_n$ compound by adopting one of the most formal approaches to the description of the behavior of solutions, developed in the works of Krupkowski [140, 141]. According to Krupkowski, one may write, on the basis of expressions (5.3) and (4.172),

$$(\partial^2 g/\partial x_B^2)_{p,\,T} = RT/[x_B(1-x_B)] - k\omega(T)(1-x_B)^{k-2}. \qquad (6.77)$$

At the maximum, (6.77) may be transformed into the form

$$\left(\frac{\partial^2 g}{\partial x_B^2}\right)_{p,\,T} = \frac{(m+n)^k\,RT^\circ_{m,\,A_m B_n} - k\omega(T)\,m^{k-1}n}{(m+n)^{k-2}\,mn}. \qquad (6.78)$$

Substituting expression (6.78) into (6.65), we obtain

$$\alpha_0^L = \frac{(m+n)^k\,RT^\circ_{m,\,A_m B_n}}{(m+n)^k\,RT^\circ_{m,\,A_m B_n} - k\omega(T)\,m^{k-1}n}. \qquad (6.79)$$

Similarly, for the opposite form of asymmetry of the liquidus curve we have

$$\alpha_0^L = \frac{(m+n)^{k+1}\,RT^\circ_{m,\,A_m B_n}}{(m+n)^{k+1}\,RT^\circ_{m,\,A_m B_n} - k\omega(T)\,n^{k-1}m^2}. \qquad (6.80)$$

Equations (6.68)–(6.76), as well as (6.79) and (6.80), were derived for situations in which the maximum of the fusion curve corresponds to the stoichiometric composition. However, as pointed out by Kurnakov while considering the classification of compounds, if the degree of dissociation is significant, then the maximum of the fusion curve is shifted considerably (cf. Fig. 21e). We showed above that this shift could be taken into account in estimating the parameters of dissociation in a general way. Let us illustrate this possibility, using model representations. For simplicity we confine ourselves to the model of a strictly regular solution in a system with a congruently melting AB compound [405]. We have

$$(\partial G/\partial c)_{p,\, T} = (\mu_B^\circ - \mu_A^\circ) + RT \ln [c/(1-c)] + (1 - 2c)\, \omega_0, \qquad (6.81)$$

where

$$c \equiv x_B^L \equiv x_B^S. \qquad (6.82)$$

Noting that

$$c = 1/2 + \delta, \qquad (6.83)$$

we have from Eqs. (6.81) and (6.82)

$$\delta = \frac{\left(\mu_A^{\circ\,(L)} - \mu_A^{\circ\,(S)}\right) - \left(\mu_B^{\circ\,(L)} - \mu_B^{\circ\,(S)}\right)}{2\left(\omega_0^S - \omega_0^L\right)}. \qquad (6.84)$$

If the difference of the chemical potentials in the numerator of Eq. (6.84) is expressed in terms of the corresponding distribution coefficients of components A and B among the solid and liquid phases, we shall obtain

$$\delta = \frac{RT \ln (K_A/K_B)}{2\left(\omega_0^S - \omega_0^L\right)}. \qquad (6.85)$$

If the said difference is expressed in terms of the heat and temperature of melting of the components, then

$$\delta = \frac{\Delta h_{m,\,A}^\circ \left[\left(T_{\max}/T_{m,\,A}^\circ\right) - 1\right] - \Delta h_{m,\,B}^\circ \left[\left(T_{\max}/T_{m,\,B}^\circ\right) - 1\right]}{2\left(\omega_0^S - \omega_0^L\right)}. \qquad (6.86)$$

Equations (6.84)–(6.86) characterize the deviation from stoichiometry for the model of a strictly regular solution of the AB compound. They contain ω_0^S, which is difficult to determine. But if we are interested in estimating the degree of dissociation, this parameter may be eliminated.

We write the expression for the ratio of the radii of curvature of the liquidus and the solidus at the maximum in the form

$$\frac{\rho_{\max}^L}{\rho_{\max}^S} = \frac{(K^L + 4\delta^2)\,(K^S + 1)}{(K^S + 4\delta^2)\,(K^L + 1)}. \qquad (6.87)$$

From this equation we obtain

$$\left(\frac{K^S + 4\delta^2}{K^S + 1}\right)^{1/2} = \left(\frac{K^L + 4\delta^2}{K^L + 1} \cdot \frac{\rho_{\max}^S}{\rho_{\max}^L}\right)^{1/2}. \qquad (6.88)$$

Using Eq. (6.88), the expression for the radius of curvature of the liquidus assumes the form

$$\rho_{max}^L = -\frac{\bar{q}^{(S \to L)}}{4RT_{max}^2}(1 - 4\delta^2)\frac{[(K^L + 4\delta^2)/(K^L + 1)]^{1/2}}{1 - (\rho_{max}^S/\rho_{max}^L)^{1/2}}.$$ (6.89)

Let us express the radius of curvature of the liquidus in terms of the parameter ω_0 when there is a shift in the maximum with respect to the stoichiometric composition. To do this we must express the second derivative of the Gibbs free energy with respect to concentration in the assumed approximation.

From formula (5.3), we find for a strictly regular solution

$$(\partial^2 g/\partial x_B^2)_{p, T}^\circ = RT_{max}/[c(1 - c)] - 2\omega_0,$$ (6.90)

or, substituting the value of c from (6.83)

$$(\partial^2 g/\partial x_B^2)_{p, T}^\circ = RT_{max}/(1 - 4\delta^2) - 2\omega_0.$$ (6.91)

Substituting this formula into Eq. (6.29), we find

$$\rho_{max}^L = -\frac{\bar{q}^{(S \to L)}}{2T_{max}}\frac{2RT_{max} - \omega_0^S(1 - 4\delta^2)}{(\omega_0^S - \omega_0^L)[2RT_{max} - \omega_0^L(1 - 4\delta^2)]}.$$ (6.92)

Dividing expression (6.28) by (6.29), and using Eq. (6.91), we shall have

$$\left(\frac{\rho_{max}^S}{\rho_{max}^L}\right)^{1/2} = \frac{2RT_{max} - \omega_0^L(1 - 4\delta^2)}{2RT_{max} - \omega_0^S(1 - 4\delta^2)}$$ (6.93)

or, subtracting from unity both sides of Eq. (6.93),

$$1 - \left(\frac{\rho_{max}^S}{\rho_{max}^L}\right)^{1/2} = \frac{(\omega_0^L - \omega_0^S)(1 - 4\delta^2)}{2RT_{max} - \omega^S(1 - 4\delta^2)}.$$ (6.94)

Comparing Eqs. (6.89) and (6.92) and remembering (6.94), one may show that

$$[(K^L + 4\delta^2)/(K^L + 1)]^{1/2} = 2RT_{max}/[2RT_{max} - \omega_0^L(1 - 4\delta^2)]$$ (6.95)

or, putting

$$\{2RT_{max}/[2RT_{max} - \omega_0^L(1 - 4\delta^2)]\}^2 = M,$$ (6.96)

we find, on the basis of (6.95), an expression for calculating the degree of dissociation of the AB compound, having taken into account the shift in the maximum with respect to the stoichiometry of a strictly regular solution:

$$\alpha_0^L = \sqrt{(M - 4\delta^2)/(1 - 4\delta^2)}.$$ (6.97)

This expression also takes into account the behavior of the solid phase since, in the first place, the initial equation for the shift in the maximum introduced the formal exchange energy in the solid state (ω_0^S) and, second, the actual shift δ of the maximum presupposes dissociation in the solid phase.

We also remark that the above procedure for obtaining the expressions for the dissociation parameters illustrates the principle of the general approach to solving the problem of the dissociation of a compound of any complexity through the adoption of other models, ranging all the way to the six-parameter model. The technique of computation will be determined by the choice of the relevant model. The adoption of the quasiregular, subregular, and six-parameter models, while accounting for the shift of the maximum, leads only to a gradual increase in the complexity of the formulas, without adding anything new in principle. Hence we shall take into account the shift in the maximum only within the model of a strictly regular solution of an AB compound, and will not consider other approximations for more complex compounds.

We shall deal separately with the calculation of the degree of association of congruently melting A_mB_n compounds along the liquidus curve, within the regular associated solution model.

Scheider and Guillaume [406] proposed a method for estimating the degree of dissociation of a congruently melting binary AB compound possessing a narrow region of homogeneity along the liquidus curve. They used a regular associated solution [186], and incorporated thermodynamic e.m.f. data in their calculation. They also made an attempt to generalize this method to the situation where more complex A_mB_n compounds are formed, and demonstrated the possibility of using, in principle, other thermodynamic data for solving the stated problem.

We now consider the binary system A–B with congruently melting compounds of complex composition of the A_mB_n type having a narrow region of homogeneity [less than 1% (atomic)].

We suppose that in the liquid phase A_mB_n dissociates according to the reaction scheme in (6.39), whose equilibrium constant is determined by the equation

$$K = \left(x_{A_1}^m x_{B_1}^n \big/ x_{A_mB_n} \right) \left(\gamma_{A_1}^m \gamma_{B_1}^n \big/ \gamma_{A_mB_n} \right). \tag{6.98}$$

Therefore, the liquid solution may be represented as a molecular mixture of three sorts of particles: the monomers A_1 and B_1 and the aggregates A_mB_n. In order to describe the behavior of such a mixture, we shall adopt the model of a regular associated solution.

According to [24], for the case under consideration relation (5.223) is satisfied. Let us find expressions for the activity coefficients $a_A{}^L$ and $a_B{}^L$.

The condition of mass conservation of the components will read in our case

$$n_A = n_{A_1} + m n_{A_mB_n}; \tag{6.99}$$

$$n_B = n_{B_1} + n n_{A_mB_n}. \tag{6.100}$$

Using the definition of the mole fractions of the components and the molecular forms, it is simple to transform relations (6.99) and (6.100) into the form

$$x_{A_1} = x_A - [m - x_A (m + n - 1)] x_{A_m B_n},$$
$$x_{B_1} = x_B - [n - x_B (m + n - 1)] x_{A_m B_n}. \tag{6.101}$$

In the regular solution approximation the activity coefficients of the ternary mixture $A_1 + B_1 + A_m B_n$ may be computed, in principle, from an equation like (4.216). However, in order to use such an equation, one must have the parameters $\omega_{i(k)j(k)}$, which are not known.

To calculate the activity coefficients of the monomeric forms in the molecular mixture $A_1 + B_1 + A_m B_n$, one may use the approximations of Eqs. (5.233) and (5.234). To find the coefficients $\gamma_{A_m B_n}$, we use the Gibbs–Duhem relation

$$x_{A_1} d \ln \gamma_{A_1} + x_{B_1} d \ln \gamma_{B_1} + x_{A_m B_n} d \ln \gamma_{A_m B_n} = 0. \tag{6.102}$$

Substituting into this equation from (6.101), (5.233), and (5.234), we find, after simple algebra,

$$(2\omega^*/RT) [-x_B (m + n) + n] dx_B = -d \ln \gamma_{A_m B_n}. \tag{6.103}$$

Integrating this equation from $x_B = n/(m + n)$ up to x_B, and using the normalization condition $\gamma_{A_m B_n} = 1$ for $x_B = n/(m + n)$, we obtain the following expression for the activity coefficient $\gamma_{A_m B_n}$ in the regular associated model:

$$RT \ln \gamma_{A_m B_n} = [\omega^*/(m + n)] [n - (m + n) x_B]^2. \tag{6.104}$$

By means of Eqs. (5.233), (6.101), and (5.234) we obtain for the activity coefficients of A and B:

$$a_A^L = a_{A_1}^L = \left\{ x_A - [m - x_A (m + n - 1)] x_{A_m B_n} \right\} \exp \left(\omega^* x_B^2 / RT \right), \tag{6.105}$$

$$a_B^L = a_{B_1}^L = \left\{ x_B - [n - x_B (m + n - 1)] x_{A_m B_n} \right\} \exp \left(\omega^* x_A^2 / RT \right). \tag{6.106}$$

Assuming that ΔH_f^S and ΔS_f^S do not depend on temperature, which is valid for quite a wide temperature region, we obtain, on the basis of (5.203),

$$a_A^m a_B^n = a_A'^m a_B'^n, \tag{6.107}$$

where a_i and a_i' are the activities of component i in the solution pertaining to the liquidus, and lying to the left and right of the compound's composition $x_B = n/(m + n)$ on the same isotherm. The quantities characteristic of solutions lying to the right of the composition of the compound are denoted by a prime.

In accordance with the conclusions of Storonkin [13], applied to a situation in which the compound $A_m B_n$ has negligible dissociation in the liquid phase, the regular associated solutions lying to the left and right of the compound's composition should be characterized by different values of the parameter ω^*. Thus, combining Eqs. (6.105) and (6.107), we obtain

$$\{x_A - [m - x_A (m + n - 1)] x_{A_m B_n}\}^m$$
$$\times \{x_B - [n - x_B (m + n - 1)] x_{A_m B_n}\}^n \exp [\omega^* (m x_B^2 + n x_A^2)/RT]$$
$$= \{x_A' - [m - x_A' (m + n - 1)] x_{A_m B_n}'\}^m$$
$$\times \{x_B' - [n - x_B' (m + n - 1)] x_{A_m B_n}'\}^n$$
$$\times \exp [\omega^{**} (m x_B'^2 + n x_A'^2)/RT]. \tag{6.108}$$

This equation contains four unknown quantities: $x_{A_m B_n}$, $x_{A_m B_n}'$, ω^*, and $\omega^{*'}$. To determine them we need four equations. Since in the present case ($p = $ const) the equilibrium constant of the reaction (6.39) should be treated as a function of temperature only, we may write, on the basis of (6.98) and (5.223),

$$a_A^m a_B^n / a_{A_m B_n} = a_A'^m a_B'^n / a_{A_m B_n}. \tag{6.109}$$

Using this together with Eq. (6.107), we find

$$x_{A_m B_n} \gamma_{A_m B_n} = x_{A_m B_n}' \gamma_{A_m B_n}'. \tag{6.110}$$

Substituting from (6.104), we obtain a second equation

$$x_{A_m B_n} \exp \left\{ \frac{\omega^*}{(m+n) RT} [n - (m+n) x_B]^2 \right\}$$
$$= x_{A_m B_n}' \exp \left\{ \frac{\omega^*}{(m+n) RT} [n - (m+n) x_B']^2 \right\}. \tag{6.111}$$

Two more equations may be obtained by including experimental data. In particular, one may use data on the pressure of the saturated vapor along the line of three-phase equilibrium $GLS_{A_m B_n}$. For this case the additional equations have the following form:

$$p = p_A^\circ \{x_A - [m - x_A (m + n - 1)] x_{A_m B_n}\} \times \exp \left[\left(\frac{\omega^*}{RT} \right) x_B^2 \right]$$
$$+ p_B^\circ \{x_B - [n - x_B (m + n - 1)] x_{A_m B_n}\} \exp \left[\left(\frac{\omega^*}{RT} \right) x_A^2 \right], \tag{6.112}$$

$$p' = p_A^\circ \left\{ x_A' - [m - x_A' (m + n - 1)] x_{A_m B_n}' \right\} \exp \left[\left(\frac{\omega^{*'}}{RT} \right) x_B'^2 \right]$$
$$+ p_B^\circ \left\{ x_B' - [n - x_B' (m + n - 1)] x_{A_m B_n}' \right\} \exp \left[\left(\frac{\omega^{*'}}{RT} \right) x_A'^2 \right], \tag{6.113}$$

where p and p' are the total pressures of the saturated vapor on the saturated liquid solutions at a given temperature, lying to the left and right of the compound $A_m B_n$ along the composition axis, respectively.

If we utilize data on the heats of mixing of the components, then on the basis of formula (4.89), and also using (5.223), (5.233), (5.234), and (6.101), we shall have for a ternary regular associated solution:

TABLE 3. Dissociation Parameters of Semiconducting Compounds of Various Groups, Determined Experimentally and Estimated by Methods of the Thermodynamics of Solutions

Compound	Order of Chebyshev polynomial approximation	$\sqrt{\delta^2}$, deg	$-\rho_e \cdot 10^6$	α_0^e	α_0^{calc}	Solution model
Mg_2Pb	5	4,2	12,44	0,24	0,28	SR
Mg_2Sn	6	3,2	8,87	0,18	0,26	
Mg_2Ge	7	1,8	7,27	0,17	0,33	
Mg_2Si	8	1,5	10,93	0,26	0.30	
$CdSb$	3	2,0	29,04	0,32	0,47	R
$GaSb$	4	1,9	12,0	0,16	0,34	6P
$InSb$	5	2,7	12,97	0,14	0,52	
$InAs$	4	1,8	7,11	0,13	0,51	
$GaTe$	4	2,8	5,48	0,16	0,17	
$InTe$	6	4,0	2,089	0,04	0,11	
Ga_2Te_3	7	2,7	12,15	0,13	0,21	SR
In_2Te_3	7	2,2	16,89	0,14	0,18	6P
$GeTe$ *	—	—	—	—	0,16	QR
$SnTe$	5	3,9	3,960	0,07	—	
$PbTe$	10	3,7	4,908	0,08	—	
Sb_2Te_3	3	2,1	14,20	0,20	0,20	6P
Bi_2Te_3	6	1,0	40,42	0,43	0,57	

*The analysis of the liquidus in the Ge–Te system was carried out in a concentration interval ranging from 50 to 67% (atomic) of Te.

$$h^{E\,(L)} = \omega^* \left\{ x_A x_B - \frac{(m+n)\left(x_A^2 + x_B^2\right) x_{A_m B_n}}{(m+n-1)\, x_{A_m B_n} + 1} \right.$$
$$\left. + \frac{[n-(m+n)\,x_B]^2}{(m+n)\left[(m+n-1)\,x_{A_m B_n} + 1\right]} \right\} + \Delta H_f^L \frac{x_{A_m B_n}}{(m+n-1)\,x_{A_m B_n} + 1}, \qquad (6.114)$$

$$h'^{E(L)} = \omega^{*'} \left\{ x_A' x_B' - \frac{(m+n)\left(x_A'^2 + x_B'^2\right) x_{A_m B_n}'}{(m+n-1)\, x_{A_m B_n}' + 1} \right.$$
$$\left. + \frac{[n-(m+n)\,x_B']^2}{(m+n)\left[(m+n-1)\,x_{A_m B_n}' + 1\right]} \right\} + \Delta H_f^L \frac{x_{A_m B_n}'}{(m+n-1)\,x_{A_m B_n}' + 1}, \qquad (6.115)$$

where ΔH_f^L is the heat of formation of the liquid $A_m B_n$ from the liquid components. If one has e.m.f. data, one may write

$$-zFE = RT \ln a_i, \qquad (6.116)$$

$$-zFE' = RT \ln a_i, \qquad (6.117)$$

where a_i is the activity of the ith substance, determined by either (6.105) or (6.106).

If reliable data are available on the temperature dependence of ΔG_f^S, one may estimate the mole fraction of the aggregates and the parameter ω^* by using a system of only two equations, one of which has the form

$$\{x_A - [m - x_A(m+n-1)]x_{A_mB_n}\}^m$$
$$\times \{x_B - [n - x_B(m+n-1)]x_{A_mB_n}\}^n$$
$$\times \exp[\omega^*(mx_B^2 + nx_A^2)/RT] = \exp(\Delta G_f^S/RT), \qquad (6.118)$$

while the other could be any one of Eqs. (6.112)–(6.117).

An analogous system of equations holds for the other branch of the liquidus. In addition, the degree of association of the A_mB_n compound may be computed from the following formulas:

for the left branch of the liquidus

$$\alpha = \frac{-x_{A_mB_n} + (x_B/n)[(m+n-1)x_{A_mB_n}+1]}{(x_B/n)[(m+n-1)x_{A_mB_n}+1]}, \qquad (6.119)$$

for the right branch of the liquidus

$$\alpha' = \frac{-x'_{A_mB_n} + (x_A/m)[(m+n-1)x'_{A_mB_n}+1]}{(x_A/m)[(m+n-1)x'_{A_mB_n}+1]}. \qquad (6.120)$$

The above methods of estimating the dissociation parameters in various approximations of the thermodynamics of solutions may be employed in all cases for which, by a preliminary analysis, one has chosen the appropriate model of the solution for the system under study.

For systems containing semiconducting phases of various compositions we have already given a reliable foundation for choosing the appropriate model solutions (cf. Chapter 5, Section 9). Based on those results, and using the equations for the corresponding models, we have estimated the dissociation parameters of semiconducting compounds of various groups. The results of these estimates are presented in Table 3. They compare well with experimental data in those cases in which sufficiently detailed and reliable studies have been carried out on the liquidus curve.

The estimate of the degree of dissociation of a congruently melting compound along the liquidus curve requires an analytic representation of the experimental dependence $T_{liq} = f(x_B)$. This can usually be fitted by various polynomials. Random errors, whose effects appear especially in finding the second derivative $(\partial^2 T_{liq}/\partial x_B^2)_{max}$, will determine the feasibility of finding the $T_{liq} = f(x_B)$ dependence as described by the experimental data. Accordingly, a smoothing approximation has to be performed.

In approximating the experimental dependence by a Taylor series, one assumes an expansion in a series around some chosen experimental point, so that this is only useful if the series converges very fast. The application of interpolating polynomials like those of Lagrange or Newton might also be inappropriate, because of the random experimental errors.

If the approximating function used is a power series

$$T = a_0 + a_1x + a_2x^2 + \cdots + a_nx^n \qquad (6.121)$$

then finding the coefficients of the expansion requires solution of a system of linear equations [407]. However, in calculating higher-order coefficients (in practice $n >$

2), to match the accuracy of measurements, this system of equations turns out to be useless, on account of the loss of accuracy in its solutions, which grows with the order of the expansion.

The orthogonal Chebyshev polynomials $p_0(x)$, $p_1(x)$, $p_2(x)$, ... calculated at the set of points x_i, x_2, ..., x_N with weight $W_i > 0$ are free from these defects. In this case the approximating polynomial will have the form

$$T = s_0 p_0(x) + s_1 p_1(x) + \cdots + s_n p_n(x). \tag{6.122}$$

The parameters of this expression are calculated from the formula

$$s_j = \frac{\sum\limits_{i=1}^{N} T_i p_j(x_i) W_i}{\sum\limits_{i=1}^{N} p_j^2(x_i) W_i}, \quad (j = 0, 1, \ldots, N), \tag{6.123}$$

which are practically independent of the order of the required polynomial.

Using the Chebyshev polynomials as approximating functions does not enable one to obtain satisfactory data on the curvature of the liquidus at the melting point, because of the shift of the maximum of the approximating curve from the point $x = n/(m + n)$.

The algorithm developed by Peck [408, 409] enables one to approximate the experimental dependence by means of the Chebyshev polynomials, whereby the approximating curve passes through k predetermined points. By means of the Peck procedure one may transform the polynomial expansion of (6.122) into the form of a simple power series with coefficients c_0, c_1, ..., c_n, which allows one to find the radius of curvature of the liquidus in an analytic form, thereby increasing the accuracy of differentiation.

By using the coordinate transformation $x' = x - x_0$, where $x_0 = n/(m + n)$ is the composition of the compound, we may transform the polynomial into the series

$$T = \sum_{n=0}^{l} c_n (x - x_0)^n, \tag{6.124}$$

from which the derivative $(\partial^2 T/\partial x^2)_{x_B = n/(m+n)}$ will be determined by the expression

$$(\partial^2 T/\partial x^2)_{x=n/(m+n)} = 1/\rho = 2c_2. \tag{6.125}$$

We tested the preceding method on 16 compounds. The melting point of the compound, $x_B = n/(m + n)$, $T = T_{m,A_m B_n}°$, was fixed. It is usually a tabulated quantity. The maximum of the approximating curve coincided with the maximum of the liquidus in all 16 cases. The optimal degree of the polynomial was determined by stabilizing the dispersion according to the Fisher criterion.

Table 3 shows the computed values of the radius of curvature of the liquidus curves and the degree of dissociation of the compounds, compared with the values calculated by the methods of the thermodynamics of solutions, employing appropriate models indicated in the last column. As the table shows, there is entirely satisfactory agreement between the theoretical and experimental values.

We note, however, that the largest discrepancy relative to the other compounds occurs in the $A^{III}B^V$ group of compounds. The greatest disparity between theory and experiment occurs for gallium antimonide and gallium arsenide. Although the disagreement is not specific to the particular solution model chosen (if one allows for the consistent linking of all the thermodynamic properties of the solutions of In–Sb), it still merits additional consideration.

4. DISSOCIATION OF BINARY COMPOUNDS IN QUASIBINARY SOLUTIONS OF THREE-COMPONENT SYSTEMS

Above we have considered the behavior of compounds in binary solutions. It is of great interest to analyze the behavior of compounds when dissolved in other substances, i.e., to treat more complicated, so-called quasibinary systems, so that one can follow the change in the thermal stability of the compound not only with temperature, but also with the degree of dilution of the solution. This is a problem which is often met with in the practical alloying of semiconductors and metals, apropos of which our general discussion [396] proves to have great practical value.

Let us consider the simplest quasibinary system $AB–C$, in which a congruently melting compound dissociates into its constituent atoms. Let the solution consisting of the molecules A_1, B_1, AB, and C be an ideal mixture of the particles. This assumption is not too coarse, since practically all the nonideality is taken into account by the act of chemical interaction leading to the formation of the compound AB in the solution. Then, according to [24], one may write

$$\mu^{\circ\prime}_{AB} + RT \ln x'_{AB}\gamma'_{AB} = \mu^{\circ}_{AB} + RT \ln x_{AB}, \qquad (6.126)$$

where x_{AB}' is the analytic molar fraction of the compound in the solution; γ_{AB}' is the activity coefficient of AB; $\mu_{AB}^{\circ\prime}$ and μ_{AB}° are the standard chemical potentials of the substance AB and the molecular form AB, respectively.

From the preceding equation we have

$$x_{AB}/(x'_{AB}\gamma'_{AB}) = \exp\left[(\mu^{\circ\prime}_{AB} - \mu^{\circ}_{AB})/RT\right]. \qquad (6.127)$$

We select the normalization according to which $\gamma_{AB}' \to 1$ when $x_{AB}' \to 1$. One may then write

$$x^*_{AB} = \exp\left[(\mu^{\circ\prime}_{AB} - \mu^{\circ}_{AB})/RT\right]. \qquad (6.128)$$

Comparing Formula (6.127) with (6.128), we obtain

$$x_{AB} = x^*_{AB}x'_{AB}\gamma'_{AB}. \qquad (6.129)$$

Here and below, the asterisk refers to the pure substance AB.

We express $x_{AB}*$ in terms of the degree of dissociation of the compound AB. By definition,

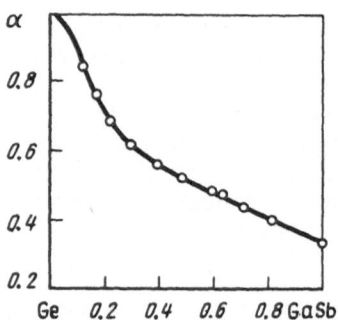

Fig.64. Change of the degree of dissociation of gallium antimonide along the liquidus curve in the Ge–GaSb system.

TABLE 4. Values of the Degree of Dissociation of GaSb along the Liquidus Curve of the State Diagram of GaSb–Ge, and Reference Data for Calculating Them [410, 411]

x'_{GaSb}	T_{liq}, K	a'_{GaSb}	a'_{Ge}	$\alpha°$	α	x'_{GaSb}	T_{liq}, K	a'_{GaSb}	a'_{Ge}	$\alpha°$	α
1.00	985	1.00	0.00	0.34	0.34	0.37	1014	0.28	0.49	0.35	0.57
0.80	952	0.72	0.14	0.33	0.40	0.28	1048	0.21	0.57	0.36	0.62
0.70	926	0.56	0.22	0.32	0.45	0.21	1066	0.16	0.61	0.37	0.68
0.62	905	0.46	0.29	0.31	0.48	0.16	1092	0.08	0.67	0.38	0.73
0.60	918	0.44	0.31	0.31	0.49	0.10	1115	0.06	0.73	0.38	0.82
0.48	968	0.36	0.40	0.34	0.52						

$$x^*_{AB} = n^*_{AB}/(n^*_{AB} + n^*_A + n^*_B). \qquad (6.130)$$

Using Eq. (6.130) and the equation of material balance for the number of particles in the solution while the compound AB dissociates to degree $\alpha°$, we find

$$x^*_{AB} = (1 - \alpha°)/(1 + \alpha°). \qquad (6.131)$$

Using the definition of the concept of the degree of dissociation of the compound AB in the solution, we may write

$$\alpha = n_A/n'_{AB} = x_{A\,(B)}/(x_{AB} + x_{A\,(B)}). \qquad (6.132)$$

In addition, from the relations $x_A + x_B + x_{AB} + x_C = 1$ and $x_C = x_C'\gamma_C'$, together with (6.129) and (6.131), we obtain

$$x_A = x_B = (1/2)\,[1 - x'_{AB}\gamma'_{AB}\,(1 - \alpha°)/(1 + \alpha°) - x'_C\gamma'_C]. \qquad (6.133)$$

Substituting the values of $x_{A(B)}$ from (6.133) and x_{AB} from (6.129) into (6.132) and using relation (6.131), we obtain the final expression for the degree of dissociation of a binary compound in a quasibinary solution:

$$\alpha^{(L)} = \frac{1 - x_C' \gamma_C' - x_{AB}' \gamma_{AB}' (1 - \alpha°)/(1 + \alpha°)}{1 - x_C' \gamma_C' + x_{AB}' \gamma_{AB}' (1 - \alpha°)/(1 + \alpha°)}. \tag{6.134}$$

This is a general thermodynamic result, and its application to concrete situations is contingent upon the choice of a definite solution model, which will then determine the explicit expressions for the activity coefficients γ_C' and γ_{AB}'.

Let us consider the possible use of Eq. (6.134) in the example of the GaSb–Ge system which is representative of the type we are discussing. The calculation requires information on the curves of monovariant equilibrium, in terms of data taken from a study of the quasibinary state diagram of GaSb–Ge [412]. This is presented in Table 4.

The degree of dissociation of pure gallium antimonide at temperature T was found from the equation

$$\ln K_T = -\Delta H°^{(L)}/RT + \Delta S°^{(L)}/R, \tag{6.135}$$

where K_T is the dissociation constant of pure GaSb, and $\Delta H°$ and $\Delta S°$ are the standard enthalpy and entropy of the dissociation reaction of GaSb, respectively.

The enthalpy $\Delta H°^{(L)}$ of the reaction is found from the equation

$$\Delta H°^{(L)} = 2h^{E\,(L)}/(1 - \alpha_0^{(L)}), \tag{6.136}$$

where, according to formula (5.212), $h^{E(L)}$ is equal to

$$h^{E\,(L)} = (\Delta H_f^S + \Delta h_{m,\,AB}°)/2.$$

The value of $\Delta H_f^{(S)}$ was calculated in the usual way from the standard heat of formation of GaSb [228]. The quantity α_0 was taken to be 0.34 (cf. Table 3). We also calculated $\Delta S°^{(L)}$, on the basis of Eq. (6.135). The values of $\Delta H°^{(L)}$ and $\Delta S°^{(L)}$ thus calculated for the reaction of dissociation of GaSb in the liquid phase turned out to be 19,590 J/mole and 3 J/mole·K, respectively.

The activity coefficient of germanium along the liquidus curve of this system was determined from the condition of phase equilibrium [cf. formula (5.142)], where we assumed that the heat capacities in the solid and supercooled liquid phase of germanium are the same, and the solubility of gallium antimonide in solid germanium is negligible. These are not crude assumptions, since $\Delta c_{p,Ge}°$ is usually utterly insignificant, while the solubility of GaSb in Ge does not exceed 1.5 mol. % according to the data of [413].

The activity of gallium antimonide was calculated by means of the Gibbs–Duhem equation, assuming that the small temperature dependence of the activity of germanium could be neglected. The resulting estimates of a_{Ge}' and a_{GaSb}' are presented in Table 4. Substituting these values and that of $a°$ into Eq. (6.134), we calculated the degree of dissociation of gallium antimonide along the liquidus curve in the GaSb–Ge system (Table 4). It follows from an analysis of the calculated data that the degree of dissociation of gallium antimonide grows with temperature and the degree of dilution. The behavior of the compound in its solutions with germanium is illustrated clearly in Fig. 64, which shows the temperature and concentra-

tion dependence (along the liquidus) of the degree of dissociation of gallium antimonide in the GaSb–Ge system. It is seen that in solutions containing up to 36–40
mol. % GaSb, the gallium antimonide dissociates quite substantially. This fully
elucidates the behavior of the germanium solutions alloyed with gallium and antimony during the drawing out of germanium monocrystals by the Czochralski
method, whereby the gallium and the antimony are trapped separately, corresponding to their individual distribution coefficients (cf. [66, 414]). In the present case,
the experimentally constructed state diagram does not reflect equilibrium between
the solid and liquid phases, since the lines connecting the compositions of the liquid
and solid phases at any given temperature are not conoids. The conclusions we
have drawn agree well with the data obtained by the viscosity method via a
physicochemical analysis for the present system [415] and for analogous systems
[416].

The substantial dissociation of gallium antimonide in liquid solutions based on
germanium was also confirmed directly in experimental x-ray studies of the
quenching (with a cooling rate of 10^6 K/sec) of alloys [417–419].

Our calculations on other systems of Ge–$A^{III}B^V$ showed the same general trend
as illustrated in Fig. 64.

These results, together with the experimental data [414–419], imply the absence of equilibrium between the solid and liquid phases in the same way as this is
observed in binary systems, since the process of dissociation of the pure substance,
proceeding always in solution, is enhanced by the dilution effect.

CONCLUSION

In conclusion, we would like to summarize some points and outline some avenues for further research in this interesting area of physical chemistry. First of all, it should be emphasized that definite progress has been achieved in the thermodynamic analysis of the phase diagrams of two-component systems. We have in mind primarily the use of phase diagrams to extract important information concerning the thermodynamic properties and structure of phases. The greatest progress in this respect has been achieved in studying liquid alloys for a wide range of compositions. This was facilitated by using fundamental ideas laid down in the works of Gibbs and van der Waals. The most fruitful among these for solving a series of problems has been the use of the differential equation of van der Waals, which describes the two-phase equilibrium of a two-component system. We would emphasize once again the fundamental character of this equation, especially with regard to the fact that, using it as the starting point, we have derived the differential equations of the curves of monovariant equilibrium (the liquidus and solidus curves), which correspond completely with the analogous equations derived by Prigogine on the basis of the chemical affinity introduced by De Donder.

The use of the rigorous thermodynamic differential equations of the curves of monovariant equilibrium for obtaining information on the thermodynamic properties and structure of phases of particular systems via the phase diagrams requires either the inclusion of additional thermodynamic information or the introduction of model representations. In the first case the problem may be maintained thermodynamically rigorous to the end.

In addition to the problems treated in the present monograph, in recent years numerous works have appeared in which problems have been successfully solved by this method [420, 421].*

---

*See also S. A. Degtyarev, "Development of methods for calculating thermodynamic properties of melts by using state diagrams," Dissertation, Moscow State University (1981).

In the second case one is dealing with the inverse problem, which belongs to the class of mathematically ill-posed problems [422]. As far as the problem of determining the thermodynamic properties solely on the basis of experimentally constructed phase diagrams is concerned, one meets with various sources of indeterminacy, the primary ones being the following: a) ignorance of the appropriate model for the solution; b) the presence of errors in determining the temperature and compositions of the figurative points in the curves of monovariant equilibrium; c) the ambiguity in estimating the parameters of the models of solutions, due to the presence of significant and, at times, insurmountable mathematical difficulties of a computational nature.

For this reason one sometimes encounters in the literature pessimistic statements concerning the chances of solving such problems successfully [423]. However, as pointed out by Tikhonov and Arsenin [422], we already have at our disposal some fairly reliable methods for solving ill-posed problems in terms of approximate data, which give stable results with respect to small variations of the latter.

Therefore, any attempt to develop methods for estimating thermodynamic properties from phase diagrams is of undoubted interest. The present authors are aware that their calculations and discussions of the various ways of estimating the thermodynamic properties of solutions in terms of phase diagrams do not exhaust all possible sources of improving the results, and that, therefore, they have limited predictive power. Nevertheless, the present means of successive system analysis, associated with the gradual increase in the complexity of the models, coupled with the choice of a suitable algorithm, already succeeds in providing stable estimates of the parameters and quite reliable values of the activity of the components, the integral Gibbs free energies, and the heat of mixing over a broad concentration regime. This cannot be said yet for the entropy of mixing, since this quantity is estimated quite inaccurately even if one uses experimental data. In light of the foregoing, the parameters of the model presented in Table 2 could, in principle, be improved, resulting in a corresponding improvement in the calculation of other quantities based on these parameters.

From among the problematic questions tackled and solved in detail in the present monograph, we would like to pick out the problem of the dissociation of congruently melting compounds, since the clarification of the problem of the thermal stability of compounds is of paramount importance in two situations: a) in developing the technology of producing semiconducting compounds with predetermined amounts of structural defects, and b) in developing the process of doping of donor or acceptor semiconducting elements, in the context of analyzing the reaction of the donor–acceptor interaction, which is in effect equivalent to the dissociation reaction of the corresponding donor–acceptor complex.

These questions need further study and could be solved on the basis of the methods and results discussed in the present book.

We are probably the first to give such a systematic treatment of the nature and methods of calculating the retrograde solidus. Nevertheless, the solution of this problem needs further refinement, in the first place on the basis of a more rational choice of the standard state of the solute.

Among the promising calculation methods used in estimating thermodynamic and structural properties by solving the direct problem on the basis of the phase diagram together with other thermodynamic properties, one should single out the method of spline functions [424, 425]. This has been applied with success to several metallic and semiconducting systems in the work of Voronin and Degtyarev [426]. We consider this to be especially favorable in estimating the radius of curvature of the liquidus at the melting point of a congruently melting compound.

Therefore, even a cursory list of the problems connected with the basic theme of this monograph, and the methods of solving them, demonstrates the need for further research aimed at a successful and complete solution of the major problems associated with the structure and properties of phases of varying physicochemical character, on the basis of phase equilibrium data.

REFERENCES

1. J. W. Gibbs, *Collected Works*, Yale Univ. Press, New Haven (1948).
2. M. Planck, *Treatise on Thermodynamics*, 3rd edn., Dover, New York (1927).
3. I. R. Krichevskii, *Concepts and Foundations of Thermodynamics* [in Russian], Khimiya, Moscow (1970).
4. L. K. Timiryazeva (ed.), *The Second Law of Thermodynamics* [in Russian] (collection of translated reprints of Sadi Carnot, W. Thomson Kelvin, R. Clausius, L. Boltzmann, and M. Smoluchowski), Gostekhizdat, Moscow (1934).
5. E. Fermi, *Thermodynamics*, Dover, New York (1937).
6. N. I. Karyakin, K. N. Bystrov, and P. S. Kireev, *Short Physics Handbook* [in Russian], Vysshaya Shkola, Moscow (1969).
7. V. M. Glazov, S. N. Chizhevskaya, and N. N. Glagoleva, *Liquid Semiconductors*, Plenum Press, New York (1969).
8. L. V. Radishkevich, *Course on Thermodynamics* [in Russian], Prosveshchenie, Moscow (1971).
9. V. A. Kireev, *Short Course on Physical Chemistry* [in Russian], Khimiya, Moscow (1970).
10. Ya. I. Gerasimov (ed.), *Course on Physical Chemistry* [in Russian], Vol. 2, Khimiya, Moscow (1972).
11. A. V. Storonkin, Vestn. Leningr. Univ., No. 16, 74–84 (1956).
12. J. D. van der Waals and P. Kohnstamm, *Lehrbuch der Thermodynamik*, Leipzig–Amsterdam (1908).
13. A. V. Storonkin, *Thermodynamics of Heterogeneous Systems* [in Russian], Izd. LGU, Leningrad (1967).
14. L. Boltzmann, *Lectures on Gas Theory*, Univ. of Calif. Press, Berkeley (1964).
15. A. V. Storonkin, V. T. Zharov, and A. N. Marinichev, *Zh. Fiz. Khim.*, 50, 3048–3957 (1976).

16. K. A. Putilov, *Thermodynamics* [in Russian], Nauka, Moscow (1971).
17. V. M. Glazov and I. I. Novikov, *Zh. Fiz. Khim.*, **48**, 1134–1137 (1974).
18. E. A. Guggenheim, *Thermodynamics*, Elsevier, New York (1967).
19. A. Münster, *Classical Thermodynamics*, Wiley, New York (1970).
20. W. Schottky, H. Ulich, and C. Wagner, *Thermodynamik*, Springer, Berlin (1973), p. 619.
21. A. V. Storonkin and V. P. Belousov, *Zh. Fiz. Khim.*, **39**, 174–177 (1965).
22. A. M. Toikka and M. P. Susarev, in: *Chemical Thermodynamics and Thermochemistry* [in Russian], Nauka, Moscow (1979), pp. 156–159.
23. A. M. Toikka and M. P. Susarev, *Zh. Fiz. Khim.*, **52**, No. 12, 2704–2709 (1978).
24. I. Prigogine and P. Defay, *Chemical Thermodynamics*, Longmans, London (1954).
25. T. De Donder and P. Van Rysselberghe, *Thermodynamic Theory of Affinity. A Book of Principles*, Stanford Univ. Press, Stanford, California (1936), p. 142).
26. P. S. Epstein, *Textbook on Thermodynamics*, Wiley, New York (1937).
27. P. Ehrenfest, *Zh. Russk. Fiz. Ova.*, **41**, 347 (1909).
28. A. I. Rusanov and M. M. Shul'tz, *Vestn. Leningr. Univ. Ser. Fiz. Khim.*, **15**, No. 4, 60–65 (1960).
29. U. I. Frankfurt and A. M. Frank, *Josiah Willard Gibbs* [in Russian], Nauka, Moscow (1964).
30. D. Konowalow, *Wied. Ann.*, **14**, 34–52 (1881).
31. M. S. Vrevskii, *Studies in the Theory of Solutions* [in Russian], Izd. Akad. Nauk SSSR, Moscow (1953).
32. A. V. Storonkin, *On the Conditions of Equilibrium of Multicomponent Systems* [in Russian], Izd. LGU, Leningrad (1948).
33. V. P. Zharov and S. V. Korobov, in: *Chemical Thermodynamics and Thermochemistry* [in Russian], Nauka, Moscow (1979), pp. 126–136.
34. V. T. Zharov and A. V. Storonkin, *Zh. Fiz. Khim.*, **49**, No. 12, 3048–3052 (1975).
35. V. A. Molochko, P. I. Fedorov, and G. M. Kurdyumov, *Tr. Khim. Khim. Tekhnol. Gor'kovsk. Gos. Univ.*, No. 3, 40–44 (1969).
36. A. V. Storonkin, *Zh. Fiz. Khim.*, **40**, No. 8, 1673–1679 (1966).
37. A. Hayes and J. Chipman, *Trans. AIME*, **135**, 85–132 (1939).
38. A. M. Marinichev and A. V. Storonkin, in: *Chemical Thermodynamics and Thermochemistry* [in Russian], Nauka, Moscow (1979), pp. 16–16.
39. V. K. Semenchenko, *Zh. Fiz. Khim.*, **49**, No. 1, 247–249 (1975).
40. V. M. Glazov and L. M. Pavlova, *Dokl. Akad. Nauk SSSR*, **240**, No. 1, 104–107 (1978).
41. A. B. Mlodzeevskii, *Thermodynamics and the Theory of Phases. Introduction to the Study of the States of Substances from the Point of View of Thermodynamics* [in Russian], Gosizdat, Moscow (1922).
42. A. B. Mlodzeevskii, *Theory of Phases with Applications to Solid and Liquid States* [in Russian], ONTI, Moscow (1937).
43. A. B. Mlodzeevskii, *Geometrical Thermodynamics* [in Russian], Izd. Mosk. Gos. Univ., Moscow (1956).
44. N. S. Kurnakov, *Introduction to Physicochemical Analysis* [in Russian], ONTI, Leningrad (1936).

45. D. A. Petrov, *Ternary Systems* [in Russian], Izd. Akad. Nauk SSSR, Moscow (1953).
46. V. P. R. Zlomanov, *T–x Diagrams of Two-Component Systems* [in Russian], Izd. Mosk. Gos. Univ., Moscow (1980).
47. I. I. Novikov, *Fiz. Metal. Metalloved.*, **6**, No. 4, 768 (1958).
48. A. A. Bochvar, *Metallography* [in Russian], Metallurgizdat, Moscow (1956).
49. A. N. Krestovnikov and V. N. Vigdorovich, *Chemical Thermodynamics* [in Russian], Metallurgiya, Moscow (1973).
50. V. V. Ogorodnikov and A. A. Ogorodnikova, *Zh. Fiz. Khim.*, **49**, No. 1, 30–34 (1975).
51. V. Ya. Anosov and S. A. Pogodin, *Fundamentals of Physicochemical Analysis* [in Russian], Izd. Akad. Nauk SSSR, Moscow (1947).
52. N. M. Wittorf, *Theory of Alloys Applied to Metallic Systems* (1909).
53. V. P. Dreving and Ya. A. Kalashnikov, *The Phase Rule and the Principles of Thermodynamics* [in Russian], Izd. Mosk. Gos. Univ., Moscow (1964).
54. M. V. Zakharov, M. V. Rumyantsev, and V. D. Turkin, *State Diagrams of Binary and Ternary Metallic Systems* [in Russian], Metallurgizdat, Moscow (1940).
55. F. Reines, *Phase Equilibrium Diagrams in Metallurgy* [Russian translation], Metallurgizdat, Moscow (1960).
56. G. Tamman, *Handbook on Heterogeneous Equilibrium* [Russian translation], ONTI, Moscow (1935)
57. A. Findlay, *The Phase Rule and Its Applciations*, Dover, New York (1951).
58. W. Hume-Rothery, G. Christian, and V. Pearson, *Equilibrium Diagrams of Metallic Systems* [Russian translation], Metallurgizdat, Moscow (1956).
59. A. A. Bochvar, *Investigation of the Mechanism and Kinetics of Crystallization of Eutectic Alloys* [in Russian], ONTI, Moscow (1935).
60. A. M. Zakharov, *State Diagrams of Binary and Ternary Systems* [in Russian], Metallurgiya, Moscow (1964).
61. J. J. Van Laar, *Z. Phys. Chem.*, **63**, No. 2, 216–253 (1908).
62. G. M. Kuznetsov and S. K. Kuznetsova, *Izv. Akad. Nauk SSSR, Neorg. Mater.*, **2**, No. 4, 643–648 (1966).
63. R. A. Swalin, *Thermodynamics of Solids*, Wiley, New York (1962).
64. J. L. Meejering, *Philips Res. Rep.*, **3**, No. 4, 281–302 (1948).
65. J. J. Van Laar, *Z. Phys. Chem.*, **B64**, 257–297 (1908).
66. V. M. Glazov and V. S. Zemskov, *Principles of Semiconductor Doping*, Israel Program for Scientific Translations, Jerusalem (1968).
67. W. Hume-Rothery, *Introduction to Physical Metallurgy* [Russian translation], Metallurgiya, Moscow (1965).
68. D. I. Mendeleev, *Collected Works* [in Russian], Vol. 3, ONTI, Leningrad (1934), p. 216.
69. I. I. Novikov, *Dokl. Akad. Nauk SSSR*, **100**, No. 6, 1119–1121 (1955).
70. A. N. Krestovnikov and V. N. Vigdorovich, *Zh. Fiz. Khim.*, **31**, No. 6, 1345–1351 (1957).
71. A. N. Krestovnikov and V. N. Vigdorovich, in: *Pure Metals and Semiconductors* [in Russian], Metallurgizdat, Moscow (1959), p. 93.
72. I. L. Aptekar' and L. G. Isaeva, *Dokl. Akad. Nauk SSSR*, **224**, No. 1, 120–123 (1975).
73. B. G. Lifshits, *Metallography* [in Russian], Metallurgiya, Moscow (1971).

74. T. A. Lebedev, *Some Problems of the General Theory of Alloys* [in Russian], Lenizdat, Leningrad (1951).
75. A. Prince, *Alloy Phase Equilibria*, Elsevier, Amsterdam (1966), p. 74.
76. V. M. Glazov and V. N. Vigdorovich, *Microhardness of Metals and Semiconductors*, Consultants Bureau, New York (1971).
77. V. M. Glazov, N. N. Glagoleva, and G. A. Korol'kov, *Izv. Akad. Nauk SSSR, Otd. Tekh. Nauk*, No. 8, 89–94 (1957).
78. D. A. Petrov and A. A. Bukhanova, *Zh. Fiz. Khim.*, **28**, No. 1, 161–172 (1954).
79. G. N. Lewis and M. Randall, *Chemical Thermodynamics*, McGraw-Hill, New York (1961).
80. R. Fowler and E. Guggenheim, *Statistical Thermodynamics*, Cambridge Univ. Press, Cambridge (1949).
81. W. Nernst, *Theoretical Chemistry from the Point of View of Avogadro's Law and Thermodynamics* (1904).
82. E. Hala, *Collect. Czech. Chem. Commun.*, **44**, No. 8, 2455–2459 (1979).
83. M. A. Evseev and G. F. Voronin, *Thermodynamics and the Structure of Liquid Metal Alloys* [in Russian], Izd. Mosk. Gos. Univ., Moscow (1966).
84. E. V. Biron, *Zh. Russk. Fiz. Khim. Ova.*, **41**, 569 (1909).
85. J. J. Van Laar, *Z. Phys. Chem.*, **72**, No. 6, 723–751 (1910).
86. J. J. Van Laar, *Z. Phys. Chem.*, **185**, 35 (1929).
87. J. H. Hildebrand, *Proc. Natl. Acad. Sci. USA*, **13**, 267 (1927).
88. J. H. Hildebrand, *J. Am. Chem. Soc.*, No. 51, 66–75 (1929).
89. J. H. Hildebrand, J. M. Prausnitz, and R. L. Scott, *Regular and Related Solutions*, Van Nostrand Reinhold, New York (1970), p. 226.
90. H. C. Carlson and A. P. Colburn, *Ind. Eng. Chem.*, **34**, No. 5, 581–589 (1942).
91. G. Scatchard, *Chem. Rev.*, **8**, No. 2, 321–333 (1931).
92. J. H. Hildebrand and R. L. Scott, *Regular Solutions*, Prentice-Hall, Englewood Cliffs, New Jersey (1962), p. 186.
93. J. H. Hildebrand and R. L. Scott, *Solubility of Nonelectrolytes*, Reinhold, New York (1950).
94. K. F. Herzfeld and W. Heitler, *Z. Electrochem.*, **31**, 536–539 (1925).
95. J. H. Hildebrand and S. E. Wood, *J. Chem. Phys.*, **1**, No. 12, 817–822 (1933).
96. V. A. Kireev, *Zh. Fiz. Khim.*, **14**, No. 11, 1456–1468 (1940).
97. L. S. Darken, *Trans. Metall. Soc. AIME*, **239**, No. 1, 80–89 (1967).
98. A. G. Lesnik, *Models of Interatomic Interactions in the Statistical Theory of Alloys* [in Russian], Fizmatgiz, Moscow (1962).
99. E. A. Guggenheim, *Mixtures*, Oxford Univ. Press, Oxford (1952), p. 276.
100. G. S. Rushbrooke, *Introduction to Statistical Mechanics*, Clarendon, Oxford (1949), p. 334.
101. N. A. Smirnova, *Methods of Statistical Thermodynamics in Physical Chemistry* [in Russian], Vysshaya Shkola, Moscow (1973).
102. M. I. Shakhparonov, *Introduction to the Molecular Theory of Solutions* [in Russian], Gostekhizdat, Moscow (1956).
103. I. Prigogine, *The Molecular Theory of Solutions*, North-Holland, Amsterdam (1957), p. 446.
104. C. Kittel, *Elementary Statistical Physics*, Wiley, New York (1958).
105. E. A. Guggenheim, *Trans. Faraday Soc.*, **44**, 1007–1012 (1948).

106. R. P. Rastogi, *J. Indian Chem. Soc.*, **39**, No. 3, 215–219 (1962).
107. N. A. Kolosovskii, *Chemical Thermodynamics* [in Russian], Goskhimizdat, Leningrad (1932).
108. R. P. Rastogi and R. K. Nigam, *Trans. Faraday Soc.*, **55**, 2005–2012 (1959).
109. R. P. Rastogi and R. K. Nigam, *Proc. Natl. Inst. Sci. India*, **26A**, 184–194 (1960).
110. H. C. Longuet-Higgins, *Proc. R. Soc., London*, **205A**, 247 (1951).
111. J. S. Rowlinson, *Proc. R. Soc., London*, **214A**, 192–206 (1952).
112. O. A. Esin, *Tr. Inst. Metall. Ural'sk. Fil. Akad. Nauk SSSR*, No. 25, 15–20 (1971).
113. I. T. Sryvalin, O. A. Esin, N. A. Vatolin, et al., *Zh. Fiz. Khim.*, **42**, No. 3, 717–722 (1968).
114. H. K. Hardy, *Acta Metall.*, **1**, No. 12, 202–209 (1953).
115. Tien Teh-chêng and Chu Hsi-hsiung, *Acta Met. Sin.*, **8**, No. 1, 38–50 (1965).
116. I. T. Sryvalin and O. A. Esin, *Nauchn. Dokl. Vyssh. Shkoly, Ser. Metall.*, No. 1, 5–10 (1959).
117. I. T. Sryvalin, O. A. Esin, and N. A. Vatolin, in: *Physical Chemistry of Metallurgical Alloys* [in Russian], No. 18, Nauka, Sverdlovsk (1969), pp. 3–44.
118. M. V. Rao, R. Hiskes, and W. A. Tiller, *Acta Metall.*, **21**, No. 6, 733–740 (1973).
119. M. V. Rao and W. A. Tiller, *Mater. Sci. Eng.*, **11**, No. 2, 61–68 (1973).
120. R. L. Sharkey, M. I. Pool, and M. Hoch, *Metall. Trans.*, **2**, No. 11, 3039–3049 (1971).
121. I. R. Krichevskii, *Zh. Fiz. Khim.*, **18**, Nos. 3–4, 187–193 (1944).
122. I. Prigogine and J. Garikian, *Physica*, **16**, 239–248 (1950).
123. I. Prigogine and V. Mathot, *J. Chem. Phys.*, **20**, 49–57 (1952).
124. I. Prigogine and A. Bellemans, *Discuss. Faraday Soc.*, **15**, 80–93 (1953).
125. J. E. Lennard-Jones and F. Devonshire, *Proc. R. Soc., London*, **A163**, 53–70 (1937).
126. J. E. Lennard-Jones and F. Devonshire, *Proc. R. Soc., London*, **A165**, 1–11 (1938).
127. H. Buchowski, Colloq. Centre Natl. Rech. Sci., 1965; Sept. (1966), pp. 161–177.
128. P. J. Flory, *J. Chem. Phys.*, **9**, 660–661 (1941).
129. P. J. Flory, *J. Chem. Phys.*, **10**, 51–61 (1942).
130. P. J. Flory, *Principles of Polymer Chemistry*, Cornell Univ. Press, Ithaca, New York (1953).
131. M. L. Huggins, *Ann. N. Y. Acad. Sci.*, **43**, 1–32 (1942).
132. M. L. Huggins, *J. Chem. Phys.*, **9**, 440 (1941).
133. S. Kemeny, *Kem. Kozl.*, **51**, Nos. 3–4, 311–344 (1979).
134. M. Margules, *Sitzber. Akad. Wiss. Wien, Math. Nat., Abt. 11*, **104**, 1243–1278 (1895).
135. J. Eliezer and R. A. Howald, *U.S. Dept. Commerce Natl. Bur. Stand. Spec. Publ.*, No. 496, 846–906 (1978).
136. C. W. Bale and A. D. Pelton, *Can. Met. Q.*, **14**, No. 3, 213–219 (1975).
137. C. W. Bale and A. D. Pelton, *Metall. Trans.*, **5**, 2323–2337 (1974).
138. O. Williams, *Trans. Metall. Soc. AIME*, **245**, 2565–2570 (1969).

139. N. Brouwer and J. A. J. Oonk, *Z. Phys. Chem. (Neue Folge)*, **105**, Nos. 3/4, 113–123 (1977).
140. A. Krupkowski, *Bull. Acad. Pol. Sci., Lett., Ser. A*, No. 1, 15 (1950).
141. A. Krupkowski, *Bull. Acad. Pol. Sci., Ser. Sci. Tech.*, **7**, No. 5, 333–340 (1959).
142. W. Ptak, *Arch. Hutnictwa*, **13**, No. 3, 252–272 (1968).
143. Z. Mozer and K. Fitzner, *Thermodyn. Nucl. Mater.*, **2**, 379–390 (1975).
144. A. Krupkowski and K. Fitzner, *Arch. Hutnictwa*, **20**, No. 4, 529–537 (1975).
145. G. Knobloch, *Krist. Tech.*, **10**, No. 6, 605–616 (1975).
146. R. F. Brebrick, *Metall. Trans.*, **2**, No. 6, 1657–1662 (1971).
147. R. Hiskes and W. A. Tiller, *Mater. Sci. Eng.*, **2**, No. 6, 320–330 (1968).
148. N. A. Smirnova, in: *Chemistry and Thermodynamics of Solutions* [in Russian], No. 2, Izd. Leningr. Gos. Univ., Leningrad (1968), pp. 3–42.
149. N. A. Smirnova, *Teor. Éksp. Khim.*, **10**, No. 6, 781–786 (1974).
150. F. Dolezalek, *Z. Phys. Chem.*, **64**, 727–747 (1908).
151. J. H. Hildebrand and R. L. Scott, *The Solubility of Nonelectrolytes*, 3rd edn., Dover, New York (1964).
152. H. V. Kehiaian and K. Sosnowska-Kehiaian, *Bull. Acad. Pol. Sci., Ser. Sci. Chem.*, **11**, No. 10, 591–596 (1964).
153. L. Sarolea-Mathot, *Trans. Faraday Soc.*, **49**, 8–20 (1953).
154. J. Stecki, *Roczn. Chem.*, **33**, 255–257 (1959).
155. A. S. Jordan, *J. Electrochem.*, **119**, No. 1, 123–127 (1972).
156. M. Malinovsky and A. Safarikova, *Zb. Pr. Chemickotechnol. Fak. SVST*, 225–230 (1978).
157. I. F. Schröder, *Z. Phys. Chem.*, **11**, No. 4, 449–465 (1893).
158. I. F. Schröder, *Gorn. Zh.*, No. 11, 272 (1890).
159. E. Tyrkiel, *Metaloznawstwo Obrob. Ciepl.*, No. 34, 7–11 (1978).
160. V. M. Glazov, L. M. Pavlova, and N. A. Moskvinova, *Dokl. Akad. Nauk SSSR*, **225**, No. 5, 1096–1099 (1975).
161. A. N. Gorbunov, *Teor. Éksp. Khim.*, **5**, No. 6, 805–812 (1969).
162. R. Becker, *Z. Metallk.*, **29**, No. 8, 245–249 (1937).
163. B. Ya. Pines, *Zh. Éksp. Teor. Fiz.*, **13**, Nos. 11–12, 411–417 (1943).
164. B. Ya. Pines, *Outline of Metal Physics* [in Russian], Izd. Khar'kovsk. Gos. Univ., Kharkov (1961).
165. V. I. Danilov and D. S. Kamenetskaya, *Zh. Fiz. Khim.*, **22**, No. 1, 69–79 (1948).
166. D. S. Kamenetskaya, *Zh. Fiz. Khim.*, **22**, No. 1, 81–89 (1948).
167. D. S. Kamenetskaya, *Zh. Fiz. Khim.*, **38**, No. 1, 73–79 (1964).
168. H. A. J. Oonk and A. Sprenkels, *Recl. Trav. Chim.*, **88**, No. 11, 1313–1331 (1969).
169. D. S. Kamenetskaya, in: *Problems of Metallography and the Physics of Metals* [in Russian], Metallurgizdat (1949), pp. 113–121.
170. D. S. Kementskaya, *Zh. Neorg. Khim.*, **3**, 607–610 (1958).
171. V. M. Glazov, A. N. Krestovnikov, and A. Yu. Mendelevich, *Zh. Fiz. Khim.*, **43**, No. 12, 3067–3069 (1969).
172. V. M. Glazov, A. N. Krestovnikov, and A. Yu. Mendelevich, in: *Theoretical and Experimental Methods of Studying State Diagrams of Metallic Systems* [in Russian], Nauka, Moscow (1969), pp. 116–119.
173. A. A. Kostarev, *Zh. Fiz. Khim.*, **51**, No. 4, 929–931 (1977).

174. B. W. Bennet, G. J. Shiflet, and H. J. Aaronson, *CALPHAD*, **2**, No. 3, 281–284 (1978).
175. H. W. B. Rozeboom, *Z. Phys. Chem.*, **12**, No. 3, 359–389 (1893).
176. H. W. B. Rozeboom, *Z. Phys. Chem.*, **30**, No. 3, 385–412 (1899).
177. H. A. J. Oonk, *Recl. Trav. Chim.*, **87**, No. 12, 1345–1358 (1968).
178. T. Mager, H. L. Lukas, and G. Petrow, *Z. Metallk.*, **63**, No. 10, 638–646 (1972).
179. T. Mager and G. Petrow, *Z. Metallk.*, **63**, No. 11, 702–709 (1972).
180. L. Kaufman and H. Bernstein, *Computer Calculation of Phase Diagrams*, Academic Press, New York (1970).
181. L. M. Foster and J. F. Woods, *J. Electrochem. Soc.*, **118**, No. 7, 1175–1183 (1971).
182. R. F. Brebrick and R. J. Panlener, *J. Electrochem. Soc.*, **121**, No. 7, 932–942 (1974).
183. V. N. Romanenko and V. I. Ivanov-Omskii, *Dokl. Akad. Nauk SSSR*, **129**, No. 3, 553–555 (1959).
184. G. B. Stringellow and P. E. Green, *J. Phys. Chem. Solids*, **30**, 1779–1791 (1969).
185. A. S. Jordan, *Metall. Trans.*, **78**, 191–202 (1976).
186. A. S. Jordan, *Metall. Trans.*, **1**, 239–249 (1970).
187. A. S. Jordan, *Metall. Trans.*, **2**, 1959–1963 (1971).
188. A. S. Jordan, *Metall. Trans.*, **2**, 1965–1970 (1971).
189. E. A. Guggenheim, *Trans. Faraday Soc.*, **33**, 151–159 (1937).
190. D. Huber, in: *Growth Processes and Synthesis of Semiconducting Crystals and Films* [in Russian], Part 2, Nauka, Novosibirsk (1975), pp. 212–218.
191. A. Laugier, *Rev. Phys. Appl.*, **8**, No. 3, 259–270 (1973).
192. C. H. M. Jenkins, *J. Inst. Met.*, **36**, No. 1, 63–97 (1926).
193. D. Stockdale, *J. Inst. Met.*, **44**, No. 1, 75–80 (1930).
194. E. Raub, *Z. Metallforsch.*, **2**, 119–120 (1947).
195. E. Raub and A. Engel, *Z. Metallforsch.*, **1**, 76–81 (1946).
196. E. Raub and A. Von Polaczek-Wittek, *Z. Metallk.*, **34**, 93–96 (1942).
197. F. Kröger, *Chemistry of Unsaturated Solutions* [Russian translation], Mir, Moscow (1959).
198. C. D. Thurmond and J. D. Struthers, *J. Phys. Chem.*, **57**, 831–835 (1953).
199. P. A. Trumbore, *Bell Syst. Tech. J.*, **39**, No. 1, 205–234 (1960).
200. V. M. Glazov, *Zh. Fiz. Khim.*, **46**, No. 3, 606–611 (1972).
201. A. N. Kirgintsev, *Izv. Akad. Nauk SSSR, Ser. Khim.*, No. 6, 1313–1314 (1971).
202. A. N. Kirgintsev, V. A. Isaenko, and V. I. Kosyakov, *Izv. Akad. Nauk SSSR, Ser. Khim.*, No. 2, 439–441 (1970).
203. G. M. Kuznetsov, "Studies in the metallurgy of metal–semiconductor contacts," Dissertation, Moscow Institute of Steel and Alloys, Moscow (1972).
204. C. D. Thurmond and M. Kowalchik, *Bell Syst. Tech. J.*, **39**, No. 1, 169–204 (1960).
205. V. M. Kozlovskaya and R. N. Rubinshtein, *Fiz. Tverd. Tela*, **3**, 3354–3362 (1961).
206. H. Reiss, *J. Chem. Phys.*, **21**, No. 7, 1209–1217 (1953).
207. H. Reiss and C. S. Fuller, *J. Met. AIME Trans.*, **8**, No. 206, 276–282 (1956).

208. H. Reiss, C. S. Fuller, and F. Morin, *Bell Syst. Tech. J.*, **35**, No. 3, 536–636 (1956).
209. V. M. Glazov and L. M. Pavlova, *Zh. Fiz. Khim.*, **52**, No. 4, 854–857 (1978).
210. J. S. Blakemore, *Proc. Phys. Soc.*, **71**, 692–697 (1958).
211. K. Lehovec, *J. Phys. Chem. Solids*, **23**, 695–709 (1962).
212. V. M. Glazov and L. M. Pavlova, *Dokl. Akad. Nauk SSSR*, **232**, No. 3, 607–610 (1977).
213. L. M. Pavlova and V. M. Glazov, *Zh. Fiz. Khim.*, **52**, No. 11, 2774–2778 (1978).
214. L. J. Vieland, *Acta Metall.*, **11**, 137–142 (1963).
215. P. Nougaret and A. Potier, *J. Chem. Phys. Phys. Chem. Biol.*, **66**, No. 4, 764–768 (1969).
216. G. M. Kuznetsov, M. P. Leonov, and A. S. Luk'yanov, *Dokl. Akad. Nauk SSSR*, **223**, No. 1, 124–126 (1975).
217. V. A. Rotenberg and G. M. Kuznetsov, *Zh. Fiz. Khim.*, **48**, No. 2, 464–466 (1974).
218. M. Hansen and K. Anderko, *Constitution of Binary Alloys*, McGraw-Hill, New York (1958).
219. R. P. Elliott, *Constitution of Binary Alloys*, McGraw-Hill, New York (1965).
220. J. M. Eldridge, E. Miller, and K. L. Komarek, *Trans. Metall. Soc. AIME*, **233**, No. 7, 1303–1308 (1965).
221. J. M. Eldridge, E. Miller, and K. L. Komarek, *Trans. Metall. Soc. AIME*, **236**, No. 1, 114–121 (1966).
222. R. Geffken and E. Miller, *Trans. Metall. Soc. AIME*, **242**, No. 11, 2320–2328 (1968).
223. R. F. Brebrick, *Metall. Trans.*, **7A**, No. 10, 1609–1610 (1976).
224. K. Osamura and Y. Murakami, *J. Phys. Chem. Solids*, **36**, No. 9, 931–937 (1975).
225. V. V. Kornev, Ya. I. Dumchak, and N. M. Korenchuk, *Zh. Fiz. Khim.*, **51**, No. 10, 2563–2567 (1977).
226. L. M. Pavlova and V. M. Glazov, *Dokl. Akad. Nauk SSSR*, **241**, No. 6, 1371–1374 (1978).
227. B. Legendre and C. Souleau, *Bull. Soc. Chim. Trans.*, Nos. 7–8, 2202–2206 (1973).
228. V. P. Glushko (ed.), *Thermal Constants of Materials* [in Russian], VINITI, No. 4, Moscow (1972), p. 184.
229. R. Hultgren, P. D. Desai, D. J. Hawkins, et al., *Selected Values of the Thermodynamic Properties of Binary Alloys*, Am. Soc. Met., Metals Park, Ohio (1973).
230. V. M. Glazov, K. Dovletov, and A. S. Malkova, *Élektron. Tekh., Ser. Mater.*, No. 9, 74–78 (1975).
231. V. M. Glazov, K. Dovletov, and A. S. Malkova, *Élektron. Tekh., Ser. Mater.*, No. 10, 64–68 (1975).
232. V. M. Glazov, L. V. Lebedeva, and L. M. Pavlova, *Élektron. Tekh., Ser. Mater.*, No. 7, 49–55 (1973).
233. V. M. Glazov and L. M. Pavlova, *Izv. Akad. Nauk SSSR, Neorg. Mater.*, **13**, No. 1, 15–17 (1977).

234. V. M. Glazov, L. M. Pavlova, and R. A. Akopyan, *Zh. Fiz. Khim.*, **52**, No. 7, 1811–1812 (1978).
235. V. M. Glazov, L. M. Pavlova, and N. L. Gryazeva, *Izv. Akad. Nauk SSSR, Neorg. Mater.*, **11**, No. 3, 418–423 (1975).
236. V. M. Glazov, L. M. Pavlova, and A. V. Evdokimov, *Zh. Fiz. Khim.*, **50**, No. 10, 2489–2493 (1976).
237. V. M. Glazov, L. M. Pavlova, and L. V. Lebedeva, in: *Thermodynamic Properties of Metallic Alloys* [in Russian], Izd. ÉLM, Baku (1975), pp. 372–375.
238. M. M. Popov, *Thermometry and Calorimetry* [in Russian], Izd. Mosk. Gos. Univ., Moscow (1954).
239. A. V. Novoselova and A. S. Pashinkin, *Vapor Pressure of Volatile Metal Chalcogenides* [in Russian], Nauka, Moscow (1978).
240. V. M. Glazov, A. S. Burkhanov, and N. M. Saleeva, *Zh. Fiz. Khim.*, **49**, No. 7, 1658–1661 (1975).
241. V. M. Glazov and N. M. Korenchuk, *Zh. Fiz. Khim.*, **45**, No. 8, 2107–2208 (1971).
242. V. M. Glazov, A. S. Pashinkin, and A. S. Bukhanov, *Zh. Fiz. Khim.*, **52**, No. 5, 1136–1138 (1978).
243. G. B. Stringfellow, *Mater. Res. Bull.*, **6**, 371–380 (1971).
244. L. C. Pauling, *The Nature of the Chemical Bond*, 3rd edn., Cornell Univ. Press, Ithaca, New York (1960).
245. Ya. I. Gerasimov, V. M. Glazov, V. M. Lazarev, et al., *Dokl. Akad. Nauk SSSR*, **235**, No. 4, 846–849 (1977).
246. Ya. I. Gerasimov, V. M. Glazov, V. B. Lazarev, and V. V. Zharov, *Dokl. Akad. Nauk SSSR*, **242**, No. 4, 868–871 (1978).
247. Ya. I. Gerasimov, V. M. Glazov, V. B. Lazarev, and V. V. Zharov, *Zh. Fiz. Khim.*, **53**, No. 6, 1361–1368 (1979).
248. V. M. Glazov, L. M. Pavlova, and L. I. Perederii, *Izv. Akad. Nauk SSSR, Neorg. Mater.*, **16**, No. 4, 595–598 (1980).
249. V. M. Glazov, L. M. Pavlova, and L. I. Perederii, *Izv. Akad. Nauk SSSR, Neorg. Mater.*, **13**, No. 2, 209–213 (1977).
250. A. A. Ugai, L. I. Sokolov, and E. G. Goncharov, in: *Semiconducting Materials and Their Applications* [in Russian], Izd. Voronezh. Gos. Univ., Voronezh (1974), pp. 71–76.
251. L. M. Pavlova and L. I. Perederii, in: *Problems of Microelectronics* [in Russian], Part 38, MIÉT, Moscow (1978), pp. 96–101.
252. C. D. Thurmund, *J. Phys. Chem. Solids*, **26**, No. 5, 785–802 (1965).
253. J. Boomgard and K. Schol, *Philips Res. Rep.*, **12**, 127–140 (1957).
254. L. N. Barmin, O. A. Esin, and I. E. Dobrovinskii, *Zh. Fiz. Khim.*, **44**, No. 10, 2560–2563 (1970).
255. I. Gertsriken and I. Ya. Dekhtyar, *Diffusion in the Solid Phase of Metals and Alloys* [in Russian], Fizmatgiz, Moscow (1960).
256. E. A. Moelwyn-Hughes, *Physical Chemistry*, Pergamon, New York (1961).
257. A. I. Bachinskii, *Selected Works* [in Russian], Izd. Akad. Nauk SSSR (1960), p. 51.
258. V. M. Glazov, V. U. Belousov, and O. D. Shchelikov, *Zh. Fiz. Khim.*, **51**, No. 9, 2288–2292 (1977).
259. V. M. Glazov, O. D. Shchelikov, and V. U. Belousov, *Fiz. Tekh. Poluprovodn.*, **10**, No. 10, 1998–2001 (1976).

260. O. A. Esin, *Zh. Fiz. Khim.*, **43**, No. 1, 228–231 (1969).
261. V. F. Ukhov, O. A. Esin, N. A. Vatolin, and É. L. Dubinin, *Tr. Inst. Metall. Ural'sk. Fil. Akad. Nauk SSSR*, No. 18, 87–103 (1969).
262. V. F. Ukhov, O. A. Esin, N. A. Vatolin, and É. L. Dubinin, in: *Thermodynamic and Thermochemical Constants* [in Russian], Nauka, Moscow (1970), pp. 114–119.
263. V. T. Bublik, S. S. Gorelik, and A. A. Zaitsev, *Solid Solutions of Elementary Semiconductors and Semiconducting Compounds* [in Russian], Scientific Transactions of the Moscow Institute of Steel and Alloys, No. 83, Metallurgiya, Moscow (1974), pp. 48–60.
264. M. A. Krivoglaz and A. A. Smirnov, *Theory of Ordered Alloys* [in Russian], Fizmatgiz, Moscow (1958).
265. M. A. Krivoglaz, *Theory of X-Ray and Thermal Neutron Scattering of Real Crystals* [in Russian], Nauka, Moscow (1967).
266. C. Wagner, *Thermodynamics of Alloys* [Russian translation], Metallurgizdat, Moscow (1957).
267. J. Lumsden, *Thermodynamics of Molten Salt Mixtures*, Academic Press, London (1966).
268. K. Hauffe and C. Wagner, *Z. Electrochem.*, **46**, No. 3, 160–170 (1940).
269. W. F. Schottky and M. B. Bever, *Acta Metall.*, **6**, No. 5, 320–326 (1958).
270. C. Wagner, *Acta Metall.*, **6**, No. 5, 309–319 (1958).
271. F. Sauerwald, P. Brand, and W. Menz, *Z. Metallk.*, **57**, No. 2, 103–108 (1966).
272. G. F. Voronin, in: *Current Problems in Physical Chemistry* [in Russian], Izd. Mosk. Gos. Univ., Vol. 9, Moscow (1976), pp. 29–31.
273. G. F. Voronin, *Dokl. Akad. Nauk SSSR*, **196**, No. 1, 133–135 (1971).
274. R. Hiskes, *Generation of Chemical Potential by Analysis of Phase Diagrams*, Stanford Univ. Press, Stanford, California (1968).
275. R. Hiskes and W. A. Tiller, *Mater. Sci. Eng.*, **4**, Nos. 2–3, 163–172 (1969).
276. R. Hiskes and W. A. Tiller, *Mater. Sci. Eng.*, **4**, Nos. 2–3, 173–184 (1969).
277. C. Lanczos, *Applied Analysis*, Prentice-Hall, Englewood Cliffs, New Jersey (1956).
278. J. Proks, *Chem. Zvesti*, **32**, No. 4, 433–440 (1978).
279. O. J. Kleppa and J. A. Weil, *J. Am. Chem. Soc.*, **73**, 4848–4850 (1951).
280. M. Malinowsky, *Chem. Zvesti*, **25**, No. 2, 92–96 (1971).
281. J. Ansara, *Int. Met. Rev.*, **24**, No. 1, 20–53 (1979).
282. L. H. Bennet, D. J. Kahan, and J. C. Carter, *Mater. Sci. Eng.*, **24**, No. 1, 1–17 (1976).
283. K. Hartmann, I. Letskii, and V. Schäffer, *Experimental Design of Technological Processes* [Russian translation], Mir, Moscow (1977).
284. Y. Bard, *Nonlinear Parameter Estimation*, Academic Press, New York (1974).
285. V. Bjorn, *Model Fitting and Discrimination for Vapor–Liquid Equilibrium Calculation. A Critical Evaluation of Liquid Mixture Models*, Göteborg (1978).
286. W. E. Deming, *Statistical Adjustment of Data*, Wiley, New York (1943).
287. Yu. V. Linnik, *Method of Least Squares and Elements of the Mathematical–Statistical Treatment of Error* [in Russian], Fizmatgiz, Moscow (1962).

288. N. P. Klepikov and S. N. Sokolov, *Analysis and Design of Experiments by the Method of Maximum Likelihood* [in Russian], Fizmatgiz, Moscow (1964).
289. C. R. Rao, *Linear Statistical Inference and Its Applications*, Wiley, New York (1965).
290. C. Gulglion, S. Domenech, and M. Enjalbert, *J. Phys. Chim. Phys. Chim. Biol.*, **74**, No. 9, 867–874 (1977).
291. A. A. Clifford, *Multivariate Error Analysis*, Wiley, New York (1973).
292. N. R. Draper and H. Smith, *Applied Regression Analysis*, Wiley, New York (1981).
293. V. V. Fedorov, *The Theory of Optimal Experiments* [in Russian], Nauka, Moscow (1971).
294. V. I. Belevantsev, V. I. Malkova, and B. I. Peshchevitskii, *Izv. Sib. Otd. Akad. Nauk SSSR, Ser. Khim.*, No. 6, 35–40 (1977).
295. R. Mezaki, N. R. Draper, and R. A. Johnson, *Ind. Eng. Chem. Fundam.*, **12**, No. 2, 251–254 (1973).
296. D. Himmelblau, *Applied Nonlinear Programming*, McGraw-Hill, New York (1972).
297. J. V. Reklaitis and D. T. Phillips, *AIIE Trans.*, No. 3, 235–256 (1976).
298. P. R. Powell and J. R. MacDonald, *Comput. J.*, **15**, 148–154 (1972).
299. T. F. Anderson, D. S. Abrams, and E. A. Grens, *AJChE J.*, **24**, No. 1, 20–28 (1978).
300. T. L. Sutton and J. F. MacGregor, *Can. J. Chem. Eng.*, **55**, No. 5, 602–608 (1977).
301. R. F. Brebrick, *Metall. Trans.*, **2**, No. 12, 3377–3383 (1971).
302. R. F. Brebrick, *Metall. Trans.*, **8A**, No. 3, 403–414 (1977).
303. M. V. Rao and W. A. Tiller, *J. Phys. Chem. Solids*, **31**, 191–198 (1970).
304. K. Damert, *Chem. Tech. (DDR)*, **28**, No. 2, 74–76 (1976).
305. M. Nowak and W. Flork, *Proceedings of the 5th Symposium on Computers in Chemical Engineering*, The High Tatras, Czechoslovakia (1977), pp. 111–114.
306. H. C. Van Ness, F. Pedersen, and P. Rasmussen, *AJChE J.*, **24**, No. 6, 1055–1063 (1978).
307. H. L. Lukas, E. Th. Henig, and B. Zimmermann, *CALPHAD*, **1**, NO. 3, 225–236 (1977).
308. M. B. Panish and M. Ilegems, in: *Progress in Solid State Chemistry*, H. Reiss and J. O. McCaldin (eds.), Pergamon, Oxford (1972), Vol. 7, Chapter 2, pp. 39–83.
309. M. V. Rao and W. A. Tiller, *J. Mater. Sci.*, **7**, 14–18 (1972).
310. M. V. Rao and W. A. Tiller, *J. Mater. Sci. Eng.*, **14**, 47–54 (1974).
311. H. Hoshino, Y. Nakamura, M. Shimoji, and K. Niwa, *Ber. Bunsenges. Phys. Chem.*, **69**, No. 2, 114–118 (1965).
312. V. M. Glazov and L. M. Pavlova, *Zh. Fiz. Khim.*, **50**, No. 11, 2764–2768 (1976).
313. F. A. Shunk, *The Constitution of Binary Alloys*, McGraw-Hill, New York (1969).
314. K. Itagaki and A. Yazawa, *Res. Inst. Miner. Dressing Metall., Tohoku Univ.*, **39**, No. 8, 880–887 (1975).
315. B. Predel and G. Oehme, *Z. Metallk.*, **67**, No. 12, 826–834 (1976).

316. V. M. Glazov, *Izv. Akad. Nauk SSSR, Otd. Tekh. Nauk*, No. 5, 190–194 (1960).
317. S. I. Gorbov, *Thermodynamics of $A^{III}B^V$ Semiconducting Compounds*. Scientific and Technical Summaries. Series on Chemical Thermodynamics and Equilibrium [in Russian], Ya. I. Gerasimov (ed.), Izd. VINITI Akad. Nauk SSSR, Moscow (1975).
318. V. V. Karataev, M. G. Mil'vidskii, and G. A. Nemtsova, *Izv. Akad.ʼNauk SSSR, Neorg. Mater.*, **11**, No. 5, 830–834 (1975).
319. B. Predel and D. W. Stein, *J. Less-Common Met.*, **24**, No. 4, 391–403 (1971).
320. N. Gambino and J.-P. Bros, *J. Chem. Thermodyn.*, **7**, 443–451 (1975).
321. Chao Pêng-nien and Mo Chin-chi, *Acta Metall. Sin.*, **8**, 32 (1965).
322. C. Bergman, M. Laffite, and Y. Maggianu, *High Temp. High Pressures*, **6**, No. 1, 53–60 (1974).
323. L. N. Gerasimenko, I. V. Kirichenko, L. N. Lozhkin, and A. G. Morachevskii, in: *Protective Metallic and Oxide Coatings, Metal Corrosion, and Studies in Electrochemistry* [in Russian], Nauka, Moscow (1965).
324. V. I. Danilin and S. P. Yatsenko, in: *Tr. Inst. Khim. Ural'sk. Fil. Akad. Nauk SSSR*, No. 20, 142–145 (1970).
325. A. A. Vecher, L. A. Mechkovskii, and A. S. Skoropanov, *Izv. Akad. Nauk SSSR, Neorg. Mater.*, **10**, No. 12, 2140–2143 (1974).
326. O. Kubashevskii and E. Evans, *Metallurgical Thermochemistry*, Pergamon, Oxford (1967).
327. N. Kh. Abrikosov, V. F. Bankina, L. V. Poretskaya, E. V. Skudnova, and S. N. Chizhevskaya, *Semiconducting Chalcogenides and Their Alloys* [in Russian], Nauka, Moscow (1975).
328. R. Castanet and C. Bergman, *J. Chem. Thermodyn.*, **9**, 1127–1132 (1977).
329. T. Maekawa, T. Yokokawa, and K. Niwa, *J. Chem. Thermodyn.*, **4**, 153–157 (1972).
330. B. Predel, J. Piehl, and M. J. Pool, *Z. Metallk.*, **66**, No. 5, 268–274 (1975).
331. V. M. Glazov and S. N. Chizhevskaya, *Dokl. Akad. Nauk SSSR*, **145**, No. 1, 115–118 (1962).
332. V. M. Glazov and S. N. Chizhevskaya, *Zh. Neorg. Khim.*, **7**, No. 8, 1933–1937 (1962).
333. A. S. Abbasov, A. V. Nikol'skaya, Ya. I. Gerasimov, and V. P. Vasil'ev, *Dokl. Akad. Nauk SSSR*, **156**, No. 6, 1399–1401 (1964).
334. H. Hahn and F. Burow, *Angew. Chem.*, **68**, No. 11, 382 (1956).
335. A. V. Rodionov and A. M. Evseev, in: *Thermodynamic and Thermochemical Constants* [in Russian], Nauka, Moscow (1970), pp. 67–73.
336. N. Kh. Abrikosov and L. E. Shelimova, *Semiconducting Materials Based on $A^{IV}B^{VI}$ Compounds* [in Russian], Nauka, Moscow (1975).
337. K. B. Sadykov and S. A. Semenkovich, in: *Chemical Bonds in Semiconductors and Thermodynamics* [in Russian], Nauka i Tekhnika, Minsk (1966), pp. 153–157.
338. C. Bergman and R. Castanet, *Ber. Bunsenges. Phys. Chem.*, **80**, No. 8, 774–775 (1976).
339. M. Kh. Karapet'yants and M. L. Karapet'yants, Fundamental *Thermodynamic Constants of Inorganic and Organic Substances* [in Russian], Khimiya, Moscow (1968).

340. T. Maekawa, T. Yokokawa, and K. Niwa, *J. Chem. Thermodyn.*, **3**, No. 1, 143–146 (1971).
341. R. Blachnik and E. Enninga, *Thermochim. Acta*, **9**, 83–86 (1974).
342. R. F. Brebrick, *J. Phys. Chem. Solids*, **30**, 719–731 (1969).
343. V. M. Glazov, A. N. Krestovnikov, and N. N. Glagoleva, *Izv. Akad. Nauk SSSR, Neorg. Mater.*, **2**, No. 3, 468–475 (1966).
344. L. V. Poretskaya, N. Kh. Abrikosov, and V. M. Glazov, *Zh. Neorg. Khim.*, **8**, No. 5, 1196–1198 (1963).
345. B. T. Melekh and S. A. Semenkovich, *Izv. Akad. Nauk SSSR, Neorg. Mater.*, **3**, No. 7, 1265–1266 (1967).
346. P. Beardmor, B. W. Howlett, and B. D. Lichter, *Trans. Metall. Soc. AIME*, **236**, No. 1, 102–108 (1966).
347. V. N. Eremenko and G. N. Lukashenko, *Izv. Akad. Nauk, Neorg. Mater.*, **1**, No. 8, 1296–1297 (1965).
348. V. A. Kireev, *Method of Practical Calculations in the Thermodynamics of Chemical Reactions* [in Russian], Khimiya, Moscow (1975).
349. G. V. Samsonov and V. M. Perminov, *Magnets* [in Russian], Naukova Dumka, Kiev (1971).
350. R. Hultgren, R. L. Orr, P. D. Anderson, and K. K. Kelly, *Selected Values of Thermodynamic Properties of Metals and Alloys*, Wiley, New York (1963).
351. V. N. Ermenko and G. M. Lukashenko, *Ukr. Khim. Zh.*, **29**, No. 9, 896–900 (1963).
352. J. M. Eldridge, E. Miller, and K. L. Komarek, *Trans. Metall. Soc. AIME*, **239**, No. 6, 775–781 (1967).
353. A. K. Nayak and W. Oelsen, *Trans. Indian Inst. Met.*, **24**, No. 6, 66–73 (1971).
354. R. A. Sharma, *J. Chem. Thermodyn.*, **2**, No. 3, 373–389 (1970).
355. M. F. Lantratov, *Zh. Neorg. Khim.*, **4**, 1415–1419 (1959).
356. I. T. Sryvalin, O. A. Esin, and B. M. Lepinskikh, *Zh. Fiz. Khim.*, **38**, No. 5, 1166–1172 (1964).
357. J. M. Eldridge, E. Miller, and K. L. Komarek, *Trans. Metall. Soc. AIME*, **236**, No. 8, 1094–1098 (1966).
358. A. Knappwost, *Z. Phys. Chem. (Frankfurt)*, **21**, 358–375 (1959).
359. V. B. Lazarev, V. Ya. Shevchenko, Ya. Kh. Grinberg, and V. V. Sobolev, *Semiconducting Compounds of the $A^{III}B^V$ Group* [in Russian], Nauka, Moscow (1978).
360. J. F. Elliott and J. Chipman, *Trans. Faraday Soc.*, **47**, 138–148 (1951).
361. R. Geffken, K. L. Komarek, and E. Miller, *Trans. Metall. Soc. AIME*, **239**, 1151–1160 (1967).
362. J. D. G. Masse, R. L. Orr, and R. Hultgren, *Trans. Metall. Soc., AIME*, **236**, No. 8, 1202–1204 (1966).
363. E. Scheil and H. L. Lukas, *Z. Metallk.*, **52**, 417–422 (1961).
364. H. Seltz and B. J. DeWitt, *J. Am. Chem. Soc.*, **60**, 1305–1308 (1938).
365. F. E. Wittig and E. Gehring, *Ber. Bunsenges. Phys. Chem.*, **70**, No. 7, 717–723 (1966).
366. E. H. Baker, *Trans. Inst. Min. Metall., Sect. C.*, **78**, 83–86 (1969).
367. J. R. Brayford and P. A. H. Wyatt, *Trans. Faraday Soc.*, **53**, 642–646 (1956).
368. E. Scheil, *Z. Metallk.*, **34**, 242–246 (1942).

369. N. S. Kurnakov and S. F. Zhemchuzhnyi, *Zh. Russk. Fiz. Khim. Ova*, **44**, 1964–1983 (1912).
370. O. A. Esin, *Izv. Sect. Fiz.-Khim. Anal.*, **17**, 38–63 (1949).
371. Le Chatelier, *Compt. Rend.*, **108**, 566–567, 801–803 (1889).
372. H. W. B. Rozeboom, *Z. Phys. Chem.*, **4**, No. 1, 31–65 (1889).
373. H. W. B. Rozeboom, *Z. Phys. Chem.*, **10**, No. 4, 477–593 (1892).
374. N. I. Patsukova, "Investigations in the thermodynamics of binary salts," Dissertation, N. S. Kurnakov Institute of General and Inorganic Chemistry, Moscow (1953).
375. R. Ruer, *Z. Phys. Chem.*, **59**, No. 1, 1–16 (1907).
376. J. J. Van Laar, *Z. Phys. Chem.*, **60**, No. 2, 197–220 (1909).
377. A. B. Mlodzeevskii, *Izv. Inst. Fiz.-Khim. Anal.*, **4**, No. 2, 247–271 (1928).
378. N. I. Shishkin, *Zh. Org. Khim.*, **10**, 12–13 (1940).
379. N. V. Ageev and E. S. Makarov, *Zh. Org. Khim.*, **13**, 242–248 (1943).
380. O. S. Ivanov, *Izv. Akad. Nauk SSSR, Otd. Khim. Nauk*, No. 4, 192–200 (1944).
381. N. N. Sirota, *Dokl. Akad. Nauk SSSR*, **44**, pp. 331–335 (1944).
382. A. B. Mlodzeevskii, *Izv. Inst. Fiz.-Khim. Anal.*, **6**, No. 1, 13–18 (1933).
383. A. B. Mlodzeevskii, *Izv. Sect. Fiz.-Khim. Anal.*, **16**, No. 1, 13–18 (1943).
384. O. A. Esin, *Izv. Sect. Fiz.-Khim. Anal.*, **19**, 151–154 (1949).
385. J. Zernike, *Chemical Phase Theory. A Comprehensive Treatise on the Deduction, the Applications, and the Limitation of the Phase Rule*, Kluver, Deventer, The Netherlands (1957).
386. R. J. Hodgkinson, *J. Electron.*, **1**, No. 6, 612–624 (1956).
387. R. J. Hodgkinson, *J. Electron.*, **2**, No. 2, 201–203 (1956).
388. N. V. Lipin, *Izv. Inst. Fiz.-Khim. Anal.*, **4**, 49–58 (1928).
389. N. V. Lipin, *Izv. Inst. Fiz.-Khim. Anal.*, **4**, 59–64 (1928).
390. N. I. Stepanov, *Izv. Inst. Fiz.-Khim. Anal.*, **4**, 327–330 (1928).
391. V. M. Glazov and L. M. Pavlova, *Dokl. Akad. Nauk SSSR*, **225**, No. 6, 1347–1350 (1975).
392. V. M. Glazov and L. M. Pavlova, *Izv. Akad. Nauk SSSR, Neorg. Mater.*, **13**, No. 2, 217–221 (1977).
393. G. A. Mauntford and P. A. H. Wyatt, *Trans. Faraday Soc.*, **62**, 3201–3216 (1966).
394. P. A. H. Wyatt, *Trans. Faraday Soc.*, **52**, 806–815 (1956).
395. P. A. H. Wyatt, *Trans. Faraday Soc.*, **56**, 490–497 (1960).
396. V. M. Glazov and L. M. Pavlova, *Zh. Fiz. Khim.*, **51**, No. 11, 2783–2787 (1977).
397. J. G. Dawber and P. A. H. Wyatt, *J. Chem. Soc.*, **52**, 3636–3638 (1958).
398. V. M. Glazov and L. M. Pavlova, *Dokl. Akad. Nauk SSSR*, **218**, No. 3, 600–603 (1974).
399. V. M. Glazov and L. M. Pavlova, *Phase Equilibrium in Heterogeneous Systems. Theory of Solutions Applied to the Analysis of Phase Equilibria* [in Russian], Izd. Mosk. Inst. Élektron. Tekh., Moscow (1975).
400. V. M. Glazov and L. M. Pavlova, *Izv. Akad. Nauk SSSR, Neorg. Mater.*, **12**, No. 12, 2114–2119.(1976).
401. V. M. Glazov and L. M. Pavlova, *Izv. Akad. Nauk SSSR, Neorg. Mater.*, **14**, No. 5, 824–826 (1978).
402. L. M. Pavlova, V. Ya. Shevchenko, S. F. Marenkin, et al., *Dokl. Akad. Nauk SSSR*, **238**, No. 1, 108–111 (1948).

403. Ya. I. Dutchak, V. V. Kornev, and N. M. Korenchuk, *Zh. Fiz. Khim.*, **51**, No. 10, 2568–2572 (1977).
404. V. V. Kornev, N. M. Korenchuk, and Ya. I. Dutchak, *Zh. Fiz. Khim.*, **51**, No. 1, 16–20 (1977).
405. V. M. Glazov and L. M. Pavlova, *Zh. Fiz. Khim.*, **51**, No. 4, 821–825 (1977).
406. M. Schneider and J. C. Guillaume, *J. Phys. Chem. Solids,* **35**, No. 4, 471–478 (1974).
407. L. Z. Rumshiskii, *Mathematical Analysis of Experimental Results* [in Russian], Nauka, Moscow (1971).
408. M. I. Ageev, V. P. Alik, and Yu. I. Markov, *Library of Algorithms* [in Russian], Sovetskoe Radio, Moscow (1976), p. 44.
409. T. E. L. Peck, *Soc. Ind. Appl. Math. Rev.*, **4**, No. 2, 135–141 (1962).
410. A. R. Regel' and V. M. Glazov, *The Periodic Law and the Physical Properties of Electronic Alloys* [in Russian], Nauka, Moscow (1978).
411. V. M. Glazov and L. M. Pavlova, *Thermal Dissociation of Semiconducting Compounds* [in Russian], Mosk. Inst. Élektron. Tekh., Moscow (1980).
412. V. M. Glazov and S. N. Chizhevskaya, *Dokl. Akad. Nauk SSSR*, **129**, No. 4, 869–872 (1959).
413. V. M. Glazov, D. A. Petrov, and S. N. Chizhevskaya, *Izv. Akad. Nauk SSSR, Otd. Tekh. Nauk,* No. 4, 153–155 (1959).
414. V. M. Glazov, V. S. Zemskov, B. G. Zhurkin, et al., *Tr. Inst. Metall. im. A. A. Baikova Akad. Nauk SSSR*, No. 14, 108–119 (1963).
415. V. M. Glazov, *Zh. Neorg. Khim.*, **6**, No. 4, 933–936 (1961).
416. V. M. Glazov, *Izv. Akad. Nauk SSSR, Otd. Tekh. Nauk*, No. 1, 89–93 (1962).
417. A. F. Belov, R. A. Akopyan, V. M. Glazov, and A. Ya. Potemkin, *Dokl. Akad. Nauk SSSR*, **238**, No. 5, 1128–1131 (1978).
418. V. M. Glazov, in: *Problems of the Metallurgy of Nonferrous Alloys* [in Russian], Nauka, Moscow (1978), pp. 181–192.
419. V. M. Glazov, L. M. Pavlova, and A. V. Evdokimov, *Dokl. Akad. Nauk SSSR*, **232**, No. 2, 371–374 (1977).
420. G. F. Voronin and S. A. Degtyarev, *Dokl. Akad. Nauk SSSR*, **254**, No. 5, 1146–1149 (1980).
421. A. D. Pelton, *Ber. Bunsenges. Phys. Chem.*, **84**, 212–218 (1980).
422. N. A. Tikhonov and V. Ya. Arsenin, *Methods for Solving Ill-Posed Problems*, Winston, Washington (1977).
423. V. V. Nalimov and T. I. Golikova, *Logical Foundations of Experimental Design* [in Russian], Metallurgiya, Moscow (1981).
424. E. Forsythe, M. M. Malcolm, and C. B. Moler, *Computer Methods for Mathematical Computations*, Prentice-Hall, Englewood Cliffs, New Jersey (1977).
425. Yu. S. Zav'yalov, B. I. Kvasov, and V. L. Miroshnichenko, *The Spline Function Method* [in Russian], Nauka, Moscow (1980).
426. S. A. Degtyarev and G. F. Voronin, in: *Thermodynamics of Metallic Systems* [in Russian], Part 1, Nauka, Alma-Ata (1979), pp. 38–43.